Edexcel GCSE

Mathematics B
Modular
Higher

Student Book
Unit 3

Series Director: Keith Pledger
Series Editor: Graham Cumming

Authors:
Chris Baston
Julie Bolter
Gareth Cole
Gill Dyer
Michael Flowers
Karen Hughes
Peter Jolly
Joan Knott
Jean Linsky
Graham Newman
Rob Pepper
Joe Petran
Keith Pledger
Rob Summerson
Kevin Tanner
Brian Western

143/12

A PEARSON COMPANY

Published by Pearson Education Limited, a company incorporated in England and Wales, having its registered office at Edinburgh Gate, Harlow, Essex, CM20 2JE. Registered company number: 872828

Edexcel is a registered trademark of Edexcel Limited

Text © Chris Baston, Julie Bolter, Gareth Cole, Gill Dyer, Michael Flowers, Karen Hughes, Peter Jolly, Joan Knott, Jean Linsky, Graham Newman, Rob Pepper, Joe Petran, Keith Pledger, Rob Summerson, Kevin Tanner, Brian Western and Pearson Education Limited 2010

The rights of Chris Baston, Julie Bolter, Gareth Cole, Gill Dyer, Michael Flowers, Karen Hughes, Peter Jolly, Joan Knott, Jean Linsky, Graham Newman, Rob Pepper, Joe Petran, Keith Pledger, Rob Summerson, Kevin Tanner and Brian Western to be identified as the authors of this Work have been asserted by them in accordance with the Copyright, Designs and Patent Act, 1988.

First published 2010

13 12 11 10
10 9 8 7 6 5 4 3 2 1

British Library Cataloguing in Publication Data
A catalogue record for this book is available from the British Library

ISBN 978 1 84690 808 8

Typeset by Tech-Set Ltd, Gateshead
Picture research by Rebecca Sodergren
Printed in Great Britain at Scotprint, Haddington

Acknowledgements
The publisher would like to thank the following for their kind permission to reproduce their photographs:
(Key: b-bottom; c-centre; l-left; r-right; t-top)

Alamy Images: Esa Hiltula 1, Jochen Tack 32, Robert Herrett 54, The Photolibrary Wales 290; **Getty Images:** AFP 13, Alan Copson 314-315, Alberto Arzoz 184, Bloomberg 19, Marcus Lyon 118, Robert Harding Picture Library 41; **iStockphoto:** 238, Jacom Stephens 147, Michael Svoboda 132, Natalia Lukiyanova 214; **NASA:** 207; **Photolibrary.com:** Aflo Foto Agency 75, Brian Lawrence 165, Carson Ganci 225, 225r; **Rex Features:** KPA / Zuma 194; **Science Photo Library Ltd:** 261, US Department of Energy 105; **Shutterstock:** 312bl, 312-313, 313r, 313br.

All other images © Pearson Education.

Every effort has been made to trace the copyright holders and we apologise in advance for any unintentional omissions. We would be pleased to insert the appropriate acknowledgement in any subsequent edition of this publication.

Disclaimer

This material has been published on behalf of Edexcel and offers high-quality support for the delivery of Edexcel qualifications.
This does not mean that the material is essential to achieve any Edexcel qualification, nor does it mean that it is the only suitable material available to support any Edexcel qualification. Edexcel material will not be used verbatim in setting any Edexcel examination or assessment. Any resource lists produced by Edexcel shall include this and other appropriate resources.

Copies of official specifications for all Edexcel qualifications may be found on the Edexcel website: www.edexcel.com

Contents

About this book

All set to make the grade!

Edexcel GCSE Mathematics is specially written to help you get your best grade in the exams.
Remember this is a calculator unit.

Section objectives show what you'll be learning.

Recap with a skills check at the start of a section – make sure you're up to speed.

Loads of practice to help you feel secure before you move on.

Crystal-clear worked examples – step-by-step guides to answering questions correctly, with helpful hints and reminders.

Full coverage of the new-style assessment objective questions – AO2 and AO3.

Graded questions – so you know what you're achieving.

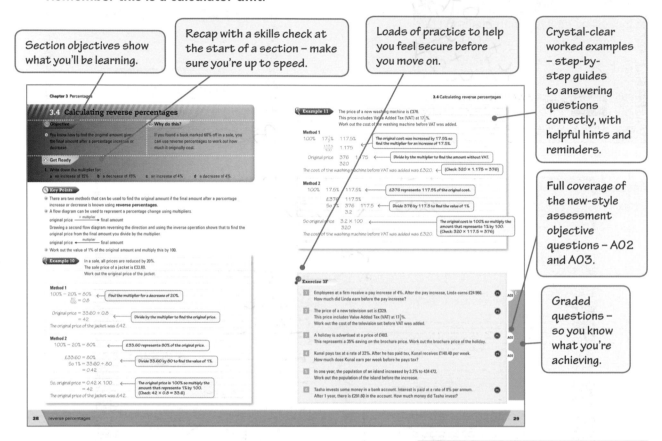

'Focus on AO2/3' pages demystify the new assessment objectives.

A fully worked example of an AO2/3 question...
...makes other AO2/3 questions on the same topic easy to tackle.

And:

- A pre-check at the start of each chapter helps you recall what you know.

- Functional elements highlighted – within ordinary exercises and on dedicated pages – so you can spend focused time polishing these skills.

- End-of-chapter graded review exercises consolidate your learning and include past exam paper questions indicated by the month and year.

About ActiveTeach

Use **ActiveTeach** to view and present the course on screen with exciting interactive content.

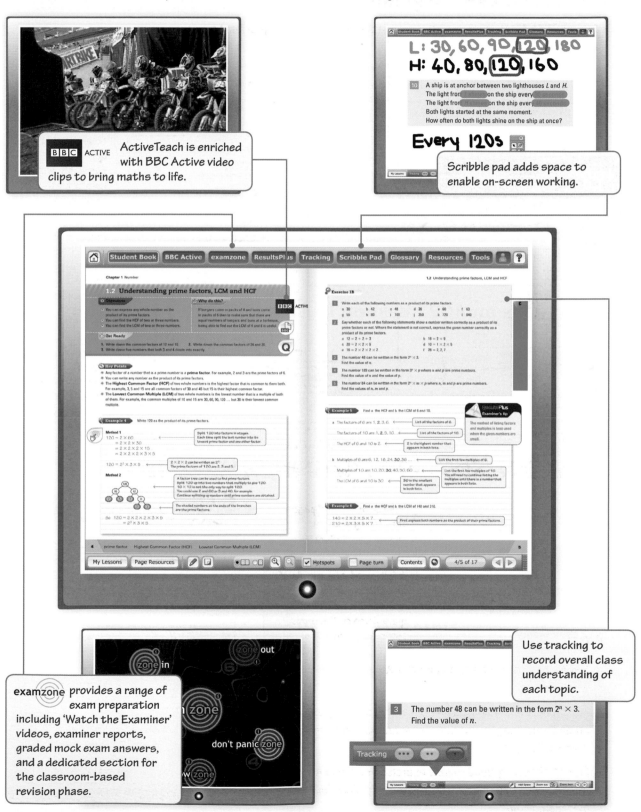

BBC ACTIVE — ActiveTeach is enriched with BBC Active video clips to bring maths to life.

Scribble pad adds space to enable on-screen working.

examzone provides a range of exam preparation including 'Watch the Examiner' videos, examiner reports, graded mock exam answers, and a dedicated section for the classroom-based revision phase.

Use tracking to record overall class understanding of each topic.

Assessment Objectives define the types of question that are set in the exam.

Assessment Objective	What it is	What this means	Range % of marks in the exam
A01	**Recall** and use knowledge of the prescribed content.	Standard questions testing your knowledge of each topic.	45-55
A02	**Select** and apply mathematical methods in a range of contexts.	Deciding what method you need to use to get to the correct solution to a contextualised problem.	25-35
A03	**Interpret** and analyse problems and generate strategies to solve them.	Solving problems by deciding how and explaining why.	15-25

The proportion of marks available in the exam varies with each Assessment Objective. Don't miss out, make sure you know how to do A02 and A03 questions!

What does an AO2 question look like?

D A02 **16** Katie wants to buy a car.
She decides to borrow £3500 from her father. She adds interest of 3.5% to the loan and this total is the amount she must repay her father. How much will Katie pay back to her father in total?

This just needs you to
(a) read and understand the question and
(b) decide how to get the correct answer.

What does an AO3 question look like?

D A03 **17** Rashida wishes to invest £2000 in a building society account for one year.
The Internet offers two suggestions. Which of these two investments gives Rashida the greatest return?

Here you need to read and analyse the question. Then use your mathematical knowledge to solve this problem.

CHESTMAN BUILDING SOCIETY
£3.50 per month
Plus **1% bonus** at the end of the year

DUNSTAN BUILDING SOCIETY
4% per annum. Paid yearly by cheque

Focus on
A02 A03

We give you extra help with AO2 and AO3 on pages 308–311.

About functional elements

What does a question with functional maths look like?

Functional maths is about being able to apply maths in everyday, real-life situations.

GCSE Tier	Range % of marks in the exam
Foundation	30-40
Higher	20-30

The proportion of functional maths marks in the GCSE exam depends on which tier you are taking. Don't miss out, make sure you know how to do functional maths questions!

In the exercises...

D AO3

20 The Wildlife Trust are doing a survey into the number of field mice on a farm of size 240 acres. They look at one field of size 6 acres. In this field they count 35 field mice.

a Estimate how many field mice there are on the whole farm.

b Why might this be an unreliable estimate?

> You need to read and understand the question. Follow your plan.
>
> Think what maths you need and plan the order in which you'll work.
>
> Check your calculations and make a comment if required.

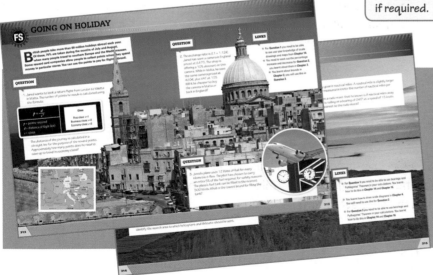

...and on our special functional maths pages: 312–315!

Quality of written communication

There will be marks in the exam for showing your working 'properly' and explaining clearly. In the exam paper, such questions will be marked with a star (*). You need to:

- use the correct mathematical notation and vocabulary, to show that you can communicate effectively
- organise the relevant information logically.

ResultsPlus

ResultsPlus features combine exam performance data with examiner insight to give you more information on how to succeed. ResultsPlus tips in the **student books** show students how to avoid errors in solutions to questions.

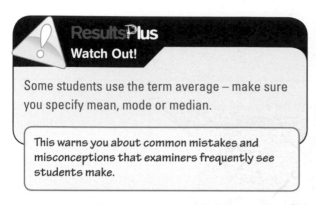

Watch Out!

Some students use the term average – make sure you specify mean, mode or median.

This warns you about common mistakes and misconceptions that examiners frequently see students make.

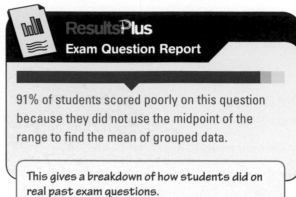

Exam Question Report

91% of students scored poorly on this question because they did not use the midpoint of the range to find the mean of grouped data.

This gives a breakdown of how students did on real past exam questions.

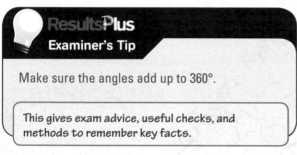

Examiner's Tip

Make sure the angles add up to 360°.

This gives exam advice, useful checks, and methods to remember key facts.

ResultsPlus in the **ActiveTeach** provides interactive practice for AO2 and AO3 questions…

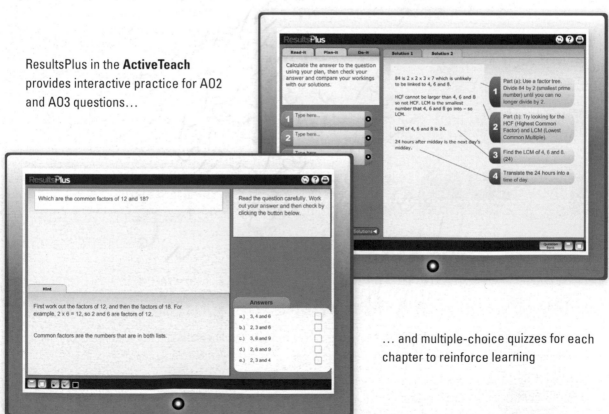

… and multiple-choice quizzes for each chapter to reinforce learning

1 NUMBER

The photo shows the Greek character, Pi, a mathematical constant used in calculations. Since Archimedes, people have used the fraction $\frac{22}{7}$ as an approximation for Pi. However, with a calculator, you can prove that the fraction is actually greater than the true value of Pi. Pi is an irrational number so it cannot be accurately converted into a fraction. Try $\frac{355}{133}$ – is this closer than $\frac{22}{7}$?

◉ Objectives

In this chapter you will:
- use a calculator
- solve problems involving fractions, decimals and proportion
- change recurring decimals into fractions
- make estimates to calculations using numbers in standard form
- use standard form on a calculator.

⬥ Before you start

You need to:
- understand decimals, fractions and easy proportion
- know how to convert between standard form and ordinary numbers
- be able to use the index laws.

1.1 Using a calculator

Objectives

- You can use a calculator.
- You can find and use reciprocals.

Why do this?

Many jobs require the accurate use of calculators, such as working in a bank or as an accountant.

Get Ready

1. Work out an estimate for   **a** 234×89 **b** $318.2 \div 2.98$ **c** $(7.2)^2$

Key Points

- A scientific calculator can be used to work out arithmetic calculations or to find the value of arithmetic expressions.
- Scientific calculators have special keys to work out squares and square roots. Some have special keys for cubes and cube roots.
- To work out other powers, your calculator will have a $\boxed{y^x}$ or $\boxed{x^y}$ or $\boxed{\wedge}$ key.
- The inverse of x^2 is $\sqrt{x}$ or $x^{\frac{1}{2}}$, and the inverse of x^3 is $\sqrt[3]{x}$ or $x^{\frac{1}{3}}$.
- The **inverse operation** of x^y is $x^{\frac{1}{y}}$.
- You can use the calculator's memory to help with more complicated numbers.

Example 1

Work out   **a** $4.6^2 + \sqrt{37}$ **b** $\dfrac{1.2^3 + 12.5}{(3.7 - 2.1)^2}$

Give your answers correct to 3 significant figures.

a $4.6^2 + \sqrt{37}$ ←   Key in $4.6\ x^2 + \sqrt{}\ 37 =$

$4.6^2 + \sqrt{37} = 27.242\,762\ldots$

$\qquad\qquad = 27.2$ ←   Round your answer to 3 significant figures.

b $\dfrac{1.2^3 + 12.5}{(3.7 - 2.1)^2}$ ←   Work out the sum on the top of the fraction.

$1.2^3 + 12.5 = 14.228$ ←   Key in $1.2\ x^3 + 12.5 =$

$(3.7 - 2.1)^2 = 2.56$ ←   Work out the sum on the bottom of the fraction. Key in $(3.7 - 2.1)\ x^2$

$14.228 \div 2.56 = 5.557\,8125$ ←   Divide your answers.

$\qquad\qquad = 5.56$ ←   Round the final answer to 3 significant figures.

ResultsPlus
Watch Out!

Do not round your numbers part way through a calculation; use all the figures shown on your calculator. Only round the final answer.

Exercise 1A

1 Work out:

a $\sqrt{961}$ b $\sqrt{40.96}$ c $\sqrt[3]{4913}$ d $\sqrt[3]{3.375}$ e $\sqrt{1024}$

2 Work out:

a $(3.7 + 5.9) \times 4.1$ b $3.1^2 + 4.8^2$ c $(-8.7 + 6.3)^2$ d $4.5^3 + 8^2$

3 Work out, giving your answers correct to one decimal place.

a $3.2^3 \times 6.7$ b $\sqrt{24} + 6.7^3$ c $9.2^2 \div \sqrt{14}$ d $7.5^3 - \sqrt{120}$

4 Work out, giving your answers correct to three significant figures.

a $\dfrac{5.63}{2.8 - 1.71}$ b $\dfrac{9.84 \times 2.6}{2.8 \times 1.71}$ c $\dfrac{6.78 + 9.2}{7.8 - 2.75}$ d $\dfrac{6.7^2}{5.6^2 - 2.1^2}$

5 Work out, giving your answers correct to three significant figures.

a $\sqrt{11.62} - \dfrac{6.3}{9.8}$ b $\dfrac{5.63}{2.8} + \dfrac{1.7}{0.3}$ c $\dfrac{\sqrt{342}}{1.8 - 1.71}$ d $\left(\sqrt{\dfrac{56}{0.18}} + 657\right)^2$

6 Work out, giving your answers correct to three significant figures.

a $\dfrac{\sqrt{45} + 6.3^2}{79.1 - 28.5}$ b $\sqrt{\dfrac{8.9 \times 2.3}{9.6 + 7.8}}$ c $\dfrac{4.2^3}{\sqrt{7.8^2 + 3.5^2}}$ d $\dfrac{(23.5 + 8.7)^2}{\sqrt{65^2 + 82}}$

D

C

Reciprocals

Key Points

- The **reciprocal** of the number n is $\dfrac{1}{n}$. It can also be written as n^{-1}.
- When a number is multiplied by its reciprocal the answer is always 1.
- All numbers, except 0, have a reciprocal.
- The reciprocal button on a calculator is usually $\boxed{1/x}$ or $\boxed{x^{-1}}$.

Example 2   Work out the reciprocal of   a 8   b 0.25   c $\dfrac{1}{4^3}$

a $\dfrac{1}{8} = 0.125$

b $1 \div 0.25 = 4$

c 4^3

Exercise 1B

1 Find the reciprocal of each of the following numbers.

a 4   b 0.625   c 6.4   d $\dfrac{2}{2^4}$

D

1.2 Number problems

◎ Objectives

● You can solve problems:
- involving fractions
- involving decimals
- involving proportions.

? Why do this?

You use decimals when adding up your mobile phone bill.

⬆ Get Ready

Use your calculator to work out

a $\frac{5}{9} + \frac{2}{7}$ b $\frac{5}{6}$ of £480   c 5.67×3.4

🔑 Key Points

● Problems can involve fractions or decimals or both.
● Two quantities are in **direct proportion** if their ratio stays the same as the quantities increase or decrease.
● Two quantities are said to be in **inverse proportion** if one quantity increases at the same rate as the other decreases.

A02 A03

🔍 Example 3

A school has 1800 pupils.
860 of these pupils are girls.
$\frac{3}{4}$ of the girls like swimming.
$\frac{2}{5}$ of the boys like swimming.
Work out the total number of pupils in the school who like swimming.

$\frac{3}{4} \times 860 = 645$ ← Work out the number of girls who like swimming.

$1800 - 860 = 940$ ← Work out the number of boys in the school.

$\frac{2}{5} \times 940 = 376$ ← Work out the number of boys who like swimming.

$645 + 376 = 1021$ ← Work out the total number of pupils who like swimming.

1021 pupils like swimming.

A02 A03

🔍 Example 4

Mr Bolton is working out his travel expenses.
He gets 42.5p for every mile he travels plus an additional allowance of £35 per month.
In June, Mr Bolton travels 280 miles.
Work out his total expenses.

ResultsPlus

Watch Out!

$42.5 \times 280 = 11\,900p$ ← Multiply the number of miles by 42.5p.
$= £119$ ← Change the answer into pounds.

Take care when working with different units. Think carefully about your answer to check it makes sense.

$119 + 35 = 154$
Mr Bolton gets £154.

Exercise 1C

1 A washing machine usually costs £420. In a sale all prices are reduced by $\frac{1}{3}$. Work out the sale price of the washing machine.

2 An MP3 player usually costs £119.95. In a sale all prices are reduced by $\frac{3}{5}$. Work out the sale price of the MP3 player.

3 The cost of an adult ticket for a museum is £6.50. A teacher buys tickets for 3 adults and 28 students. The total cost of the tickets is £137.10. Work out the cost of a student ticket for the museum.

4 Mr and Mrs Allan buy two adult tickets and some child tickets for the cinema. Adult tickets cost £7.95, child tickets cost £4.50. The total cost of all the tickets is £42.90. How many child tickets do they buy?

5 Mrs Jones is working out her gas bill. She has to pay 61p for every unit of gas she uses and a fixed charge of £7.50. In one month she uses 230 units of gas. Work out how much she has to pay.

6 Sarah has a mobile phone. Each month she pays 15.2p for each minute that she uses her phone and a fixed charge of £15.80. In July Sarah used her phone for 324 minutes. Work out the total amount that Sarah paid.

7 Barry's gas meter on 1 January reads 5602. It reads 7632 on 1 April.
Barry pays 10.8p per unit of gas and a standing charge of £5.50 per month.
Work out how much Barry must pay for the gas he used between 1 January and 1 April.

A02 D
A02
A02
A03
A02
A03
A03 C
A03
A03

Example 5 The weight of card is directly proportional to its area.
A piece of card has an area of 36 cm^2 and a weight of 15 grams.
A larger piece of the same card has an area of 48 cm^2.
Calculate the weight of the larger piece of card.

$\frac{15}{36} = \frac{5}{12}$ ← Work out the weight of 1 cm^2 of the card.
The answer is less than 1, so write this as a fraction in its simplest form.

$\frac{5}{12} \times 48^4 = 20$ ← Work out the weight of 48 cm^2 piece of card. Multiply the weight of 1 cm^2 by 48.

The weight of the larger piece of card is 20 grams.

Example 6 It takes 3 cleaners 6 hours to clean a school.
Work out how long it would take 9 cleaners to clean the school.

3 cleaners take 6 hours.

$9 \div 3 = 3$ ← Divide the new number of cleaners by the original number of cleaners. The original number of cleaners has been multiplied by 3 (3 × 3 = 9).

$6 \div 3 = 2$ ← Divide the number of hours by 3.

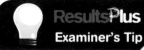

ResultsPlus
Examiner's Tip

Think about the problem: if there are more people then the work should take less time.

9 cleaners would take 2 hours. ← Check: 3 × 6 = 18 and 9 × 2 = 18.

Exercise 1D

B
A03

1. The cost of ribbon is directly proportional to its length. A 2.5 m piece of ribbon costs £1.35. Work out the cost of 6 m of this ribbon.

A03

2. It takes 10 men 2 days to cut a hedge.
   Work out how long it will take to cut the hedge if there are:
   a 5 men   **b** 4 men.

A02
A03

3. 5 computers process a certain amount of information in 10 hours.
   Work out how long it will take 25 computers to process the same amount of information.

A02
A03

4. A factory uses 3 machines to complete a job in 15 hours.
   If 2 extra machines are used, how long will the job take?

A02
A03

5. It takes 6 machines 3 days to harvest a crop. How long would it take 2 machines?

A02
A03

6. A document will fit onto exactly 32 pages if there are 500 words on a page.
   If the number of words on each page is reduced to 400, how many more pages will there be in the document?

A
A03

7. The length of the shadow of an object, at noon, is directly proportional to the height of the object.
   A lamp-post of height 5.4 m has a shadow of length 2.1 m at noon.
   Work out the length of the shadow, at noon, of a man of height 1.8 m.

1.3 Converting recurring decimals to fractions

Objectives

- You can convert recurring decimals to fractions.
- You understand and know how to use the recurring decimal proof.

Get Ready

1. Determine whether these fractions will be represented by a terminating or recurring decimal.

   a $\frac{14}{64}$ **b** $\frac{18}{35}$ **c** $\frac{20}{44}$

Key Points

- All recurring decimals can be converted to fractions.
- To convert a recurring decimal to a fraction:
  - introduce a **variable**, usually x
  - form an equation by putting x equal to the recurring decimal
  - multiply both sides of the equation by 10 if 1 digit recurs, by 100 if 2 digits recur, by 1000 if 3 digits recur, and so on
  - subtract the original equation from the new equation
  - rearrange to find x as a fraction.

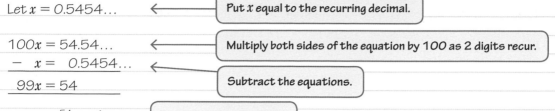

Example 7   Convert the recurring decimal $0.5\overset{\bullet\bullet}{4}$ to a fraction. Give your fraction in its simplest form.

Let $x = 0.5454...$ ←   Put x equal to the recurring decimal.

$100x = 54.54...$ ←   Multiply both sides of the equation by 100 as 2 digits recur.

$-\ \ x = \ \ 0.5454...$ ←   Subtract the equations.

$99x = 54$

$x = \dfrac{54}{99}$ ←   Divide both sides by 99.

$= \dfrac{6}{11}$ ←   Simplify the fraction.

$0.\overset{\bullet}{5}\overset{\bullet}{4} = \dfrac{6}{11}$

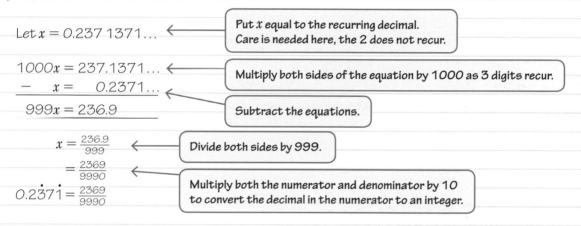

Example 8   Convert the recurring decimal $0.2\overset{\bullet}{3}7\overset{\bullet}{1}$ to a fraction.

Let $x = 0.237\,1371...$ ←   Put x equal to the recurring decimal. Care is needed here, the 2 does not recur.

$1000x = 237.1371...$ ←   Multiply both sides of the equation by 1000 as 3 digits recur.

$-\ \ \ \ x = \ \ \ \ 0.2371...$ ←   Subtract the equations.

$999x = 236.9$

$x = \dfrac{236.9}{999}$ ←   Divide both sides by 999.

$= \dfrac{2369}{9990}$ ←   Multiply both the numerator and denominator by 10 to convert the decimal in the numerator to an integer.

$0.2\overset{\bullet}{3}7\overset{\bullet}{1} = \dfrac{2369}{9990}$

Exercise 1E

Using the algebraic method above, convert each recurring decimal to a fraction. Give each fraction in its simplest form.

Use a calculator to check your answers.

1   0.777 77...   **2**   0.343 434...   **3**   0.915 915...

4 $0.\overset{\bullet}{1}\overset{\bullet}{8}$ **5** $0.3\overset{\bullet}{1}\overset{\bullet}{7}$ **6** $0.0\overset{\bullet}{5}$

7 $0.3\overset{\bullet}{2}\overset{\bullet}{6}$ **8** $0.7\overset{\bullet}{0}\overset{\bullet}{1}$ **9** $0.2\overset{\bullet}{3}$

10 $6.8\overset{\bullet}{3}$ **11** $2.1\overset{\bullet}{0}\overset{\bullet}{6}$ **12** $7.35\overset{\bullet}{2}$

1.4 Using standard form

Objectives

- You can convert to standard form to make sensible estimates for calculations.
- You can use standard form on your calculator.

Why do this?

Engineers sending spacecraft to other planets need to do calculations using very large numbers to work out how long the journey will take.

Get Ready

1. Write in standard form   **a** 40 000   **b** 0.000 7   **c** 56 700   **d** 0.503
2. Write as ordinary numbers   **a** 2×10^3 **b** 9×10^{-4} **c** 8.4×10^5 **d** 3.8×10^{-3}
3. Write as a single power of 10   **a** $10^6 \times 10^9$ **b** $10^4 \div 10^{-3}$

Key Points

- When working out an estimate using very large or very small numbers, it is often easier to convert the numbers into standard form first.
- To input numbers in standard form into your calculator, use the $\boxed{10^x}$ or $\boxed{\text{EXP}}$ key
  To enter 4.5×10^7 press the keys $\boxed{4}\,\boxed{\cdot}\,\boxed{5}\,\boxed{\times}\,\boxed{10^x}\,\boxed{7}$.

Example 9

By writing 56 900 and 0.0313 in standard form correct to 1 significant figure, estimate the value of $56\,900 \times 0.0313$.

$56\,900 \times 0.0313 = 6 \times 10^5 \times 3 \times 10^{-2}$ ⟵ Write both numbers correct to 1 significant figure.

$\qquad\qquad = 6 \times 3 \times 10^5 \times 10^{-2}$ ⟵ Rearrange the expression so the powers of 10 are together.

$\qquad\qquad = 18 \times 10^{5-2}$ ⟵ Multiply the numbers.

$\qquad\qquad = 18 \times 10^3$ ⟵ Use $a^m \times a^n = a^{m+n}$ to multiply the powers of 10.

$\qquad\qquad = 1.8 \times 10^4$ ⟵ Write your final answer in standard form.

Example 10

By writing 760 000 000 and 0.000 19 in standard form correct to one significant figure, work out an approximation for $760\,000\,000 \div 0.000\,19$.

$760\,000\,000 = 8 \times 10^8$ correct to one significant figure.
$0.000\,19 = 2 \times 10^{-4}$ correct to one significant figure.

$\dfrac{760\,000\,000}{0.000\,19} \approx \dfrac{8 \times 10^8}{2 \times 10^{-4}}$

$\qquad = \dfrac{8}{2} \times \dfrac{10^8}{10^{-4}} = 4 \times 10^{8--4}$ ⟵ Rearrange the expression so the powers of 10 are together.
Divide the numbers.
Use $a^m \div a^n = a^{m-n}$ to divide the powers of 10.

$\qquad = 4 \times 10^{12}$

B

A

A03

A02
A03

Exercise 1F

1   Write these numbers in standard form correct to 1 significant figure.
    a   5280   b   0.003 42   c   0.000 809   d   49 200 000
    e   0.5603   f   49.23   g   395   h   0.000 099

2   By writing these numbers in standard form correct to one significant figure, work out an estimate of the value of these expressions. Give your answer in standard form.
    a   600 008 × 598   b   78 018 × 4180   c   699 008 ÷ 198   d   8 104 660 000 ÷ 0.000 078

3   Work out an estimate for the value of each of these expressions.
    a $\dfrac{64\,000 \times 0.000\,492}{0.0342}$ b $\dfrac{0.002\,83 \times 0.006\,3}{209}$ c $\dfrac{193\,000}{820\,000 \times 0.000\,496}$

4   The distance of the Earth from the Sun is approximately 149 000 000 kilometres.
    Light travels at a speed of approximately 300 000 metres per second.
    Work out an estimate of the time it takes light to travel from the Sun to the Earth.

5   An atomic particle has a lifetime of 3.86×10^{-5} seconds. It travels at a speed of 4.2×10^6 metres per second. Calculate an approximation for the distance it travels in its lifetime.

Example 11   Use a calculator to work out
    a $(3.4 \times 10^6) \times (7.1 \times 10^4)$
    b $(4.56 \times 10^8) \div (3.2 \times 10^{-3})$

a $(3.4 \times 10^6) \times (7.1 \times 10^4)$ $= 2.414 \times 10^{11}$ ← Use the [EXP] or [10^x] button on your calculator.
b $(4.56 \times 10^8) \div (3.2 \times 10^{-3})$ $= 1.425 \times 10^{11}$

Example 12 $x = 3.1 \times 10^{12}, y = 4.7 \times 10^{11}$
    Use a calculator to work out the value of $\dfrac{x + y}{xy}$.
    Give your answer in standard form correct to 3 significant figures.

$\dfrac{(3.1 \times 10^{12} + 4.7 \times 10^{11})}{(3.1 \times 10^{12} \times 4.7 \times 10^{11})}$ ← Substitute the values into the expression.

$= \dfrac{3.57 \times 10^{12}}{1.457 \times 10^{24}}$ ← Write the number from your calculator correctly in standard form showing more than 3 significant figures.

$= 2.4502\ldots \times 10^{-12}$ ←

$= 2.45 \times 10^{-12}$ ← Give your answer correct to 3 significant figures.

 Results**Plus**
Examiner's Tip

Include brackets here to ensure that the answer from the calculation on the top of the fraction is divided by the answer to the calculation on the bottom of the fraction.

A

Exercise 1G

1. Find the value of these expressions, giving your answers in standard form.
   Give your answers to 4 significant figures where necessary.

   a $500 \times 600 \times 700$

   b 0.006×0.004

   c $\dfrac{65 \times 120}{1500}$

   d $\dfrac{8.82 \times 5.007}{10\,000}$

   e $(12.8)^4$

   f $(2.46 \times 10^{10}) \div (2.5 \times 10^6)$

   g $(3.6 \times 10^{20}) \div (3.75 \times 10^6)$

   h $(2.46 \times 10^{-10}) \div (2.5 \times 10^6)$

   i $(3.6 \times 10^{-20}) \div (3.75 \times 10^{-6})$

2. Evaluate these expressions. Give your answers in standard form correct to 3 significant figures.

   a $(3.5 \times 10^{11}) \div (6.5 \times 10^6)$

   b $(1.33 \times 10^{10}) \times (4.66 \times 10^4)$

   c $(3.5 \times 10^{11}) \div (6.5 \times 10^{-6})$

   d $(1.33 \times 10^{-10}) \times (4.66 \times 10^4)$

3. $x = 3.5 \times 10^9, y = 4.7 \times 10^5$

   Work out the following. Give your answer in standard form correct to 3 significant figures.

   a $\dfrac{x}{y}$

   b $x(x + 800y)$

   c $\dfrac{xy}{x + 800y}$

   d $\left(\dfrac{x}{2000}\right)^2 + y^2$

4. $x = 2.4 \times 10^{-5}, y = 9.6 \times 10^{-6}$

   Evaluate these expressions.

   Give your answer in standard form correct to 3 significant figures where necessary.

   a $\dfrac{x^2}{y}$

   b $\dfrac{x^2 + y^2}{x + y}$

   c $\dfrac{xy}{x - y}$

5. The distance of the Earth from the Sun is 1.5×10^8 km.
   The distance of the planet Neptune from the Sun is 4510 million km.
   Write in the form $1 : n$ the ratio
   distance of the Earth from the sun : distance of the planet Neptune from the Sun

6. The mass of a uranium atom is 3.98×10^{-22} grams.
   Work out the number of uranium atoms in 2.5 kilograms of uranium.

Chapter review

- A scientific calculator can be used to work out arithmetic calculations or to find the value of arithmetic expressions.
- Scientific calculators have special keys to work out squares and square roots.
  Some have special keys for cubes and cube roots.
- To work out other powers, your calculator will have a $\boxed{y^x}$ or $\boxed{x^y}$ or $\boxed{\wedge}$ key.
- The inverse of x^2 is $\sqrt{x}$ or $x^{\frac{1}{2}}$, and the inverse of x^3 is $\sqrt[3]{x}$ or $x^{\frac{1}{3}}$.
- The **inverse operation** of x^y is $x^{\frac{1}{y}}$.
- You can use the calculator's memory to help with more complicated numbers.
- The **reciprocal** of the number n is $\dfrac{1}{n}$. It can also be written as n^{-1}.
- When a number is multiplied by its reciprocal the answer is always 1.
- All numbers, except 0, have a reciprocal.

- The reciprocal button on a calculator is usually $^{1}/_{x}$ or x^{-1}.
- Problems can involve fractions or decimals or both.
- Two quantities are in **direct proportion** if their ratio stays the same as the quantities increase or decrease.
- Two quantities are said to be in **inverse proportion** if one quantity increases at the same rate as the other decreases.
- All recurring decimals can be converted to fractions.
- To convert a recurring decimal to a fraction:
  - introduce a **variable**, usually x
  - form an equation by putting x equal to the recurring decimal
  - multiply both sides of the equation by 10 if 1 digit recurs, by 100 if 2 digits recur, by 1000 if 3 digits recur, and so on
  - subtract the original equation from the new equation
  - rearrange to find x as a fraction.
- When working out an estimate using very large or very small numbers, it is often easier to convert the numbers into standard form first.
- To input numbers in standard form into your calculator, use the 10^x or EXP key.
  To enter 4.5×10^7 press the keys 4 · 5 × 10^x 7 .

Review exercise

1. Find the reciprocal of the following numbers.

   a 12   b $\frac{3}{8}$ c 2.5   d $\dfrac{5^2 - 7}{\sqrt[3]{216}}$

2. A factory has 1430 workers. 513 of the workers are female. $\frac{4}{9}$ of the female workers are under the age of 30, $\frac{3}{7}$ of the male workers are under the age of 30. How many workers in total are aged under 30?

3. Ally is charged for electricity using two different rates.
   She pays 13.5p per unit for the first 120 units and then the remainder of the units she uses are changed at a rate of 8.2p each.
   Ally pays a total of £24.81 for her electricity one month.
   How many units of electricity does she use?

4. Bob lays 200 bricks in one hour. He always works at the same speed.
   Bob takes 15 minutes morning break and 30 minutes lunch break.
   Bob has to lay 960 bricks. He starts work at 9 am.
   Work out the time at which he will finish laying bricks.

   June 2006, adapted

5. Use your calculator to work out $\dfrac{\sqrt{19.2 + 2.6^2}}{2.7 \times 1.5}$
   Write down all the figures on your calculator display.

6. A large ball of wool is used to knit a scarf.
   The scarf is 40 stitches wide and 120 cm long. If the same size ball of wool is used to knit a scarf 25 stitches wide, work out the length of the new scarf.

B

7 Work out $(3.2 \times 10^5) \times (4.5 \times 10^4)$.
Give your answer in standard form correct to 2 significant figures.

June 2005

A

8 Convert each recurring decimal to a fraction.
Give each fraction in its simplest form.
Use a calculator to check your answers.
a $0.\dot{4}$
b $0.1\dot{6}$
c $0.\dot{2}\dot{7}$
d $0.\dot{3}11\,68\dot{8}$

9 Estimate the value of each of the following.
a $672\,000 \times 0.003\,42$
b $(0.0543 \times 693)^2$
c $\dfrac{87\,000 \times 0.000\,198}{27\,850}$

10 Work out $\dfrac{(2 \times 2.2 \times 10^{12}) \times (1.5 \times 10^{12})}{(2.2 \times 10^{12}) - (1.5 \times 10^{12})}$
Give your answer in standard form correct to 3 significant figures.

Nov 2007

11 $x = \sqrt{\dfrac{p+q}{pq}}$ $\qquad p = 4 \times 10^8 \qquad q = 3 \times 10^6$
Find the value of x.
Give your answer in standard form correct to 2 significant figures.

Mar 2005

12 A floppy disk can store $1\,440\,000$ bytes of data.
a Write the number $1\,440\,000$ in standard form.
A hard disk can store 2.4×10^9 bytes of data.
b Calculate the number of floppy disks needed to store the 2.4×10^9 bytes of data.

Nov 2003

13 a Write $5\,720\,000$ in standard form.
$p = 5\,720\,000 \qquad q = 4.5 \times 10^5$
b Find the value of $\dfrac{p-q}{(p+q)^2}$
Give your answer in standard form, correct to 2 significant figures.

Jan 2005

2 UPPER AND LOWER BOUNDS

China has the largest population of any country in the world, but its population is constantly changing so it is difficult to give an accurate figure. If the estimated population of China in 2009 was 1,338,613,000 to the nearest thousand, what are the upper and lower bounds that it could be? To put this population into perspective, the population of the United Kingdom at the same time was estimated to be 61,113,000 to the nearest thousand.

◉ Objectives

In this chapter you will:
- write down upper bounds and lower bounds
- calculate the bounds of expressions.

◈ Before you start

You need to be able to:
- substitute numbers into an algebraic expression and work out its value.

2.1 Upper and lower bounds of accuracy

◉ Objective

● You know the upper bound and the lower bound of a number, given the accuracy to which it has been written.

⊘ Why do this?

If you knew how tall the Blackpool Tower was to 2 significant figures, then you could work out the maximum and minimum possible heights that it could be.

⬨ Get Ready

1. Write the following numbers correct to 1 decimal place (1 d.p.).
   a 6.05   b 6.99   c 6.49   d 6.51
2. Write the following numbers correct to 1 significant figure.
   a 0.33   b 0.339   c 0.26   d 0.349

🌑 Key Points

● The **upper bound** of a number written to 1 decimal place is the highest value which rounds down to that number.
● The **lower bound** of a number written to 1 decimal place is the lowest value which rounds up to that number.

Example 1

a $x = 6.4$ (correct to 1 decimal place)
   Write down the upper bound and the lower bound of x.
b $y = 248$ (correct to 3 significant figures)
   Write down the upper bound and the lower bound of y.

Results Plus
Examiner's Tip

Remember that the upper bound is the same distance above x as the lower bound is below x.

a Upper bound of $x = 6.45$ ← For 1 decimal place the upper bound is 0.05 above the stated value.

   Lower bound of $x = 6.35$ ← For 1 decimal place the lower bound is 0.05 below the stated value.

b Upper bound of $y = 248.5$ ← 248.5 is the largest value which will round down to 248 correct to 3 significant figures.

   Lower bound of $y = 247.5$

🛠 Exercise 2A

Questions in this chapter are targeted at the grades indicated.

A

1 Write down: i the upper bound and ii the lower bound of these numbers.
   a 84 (2 significant figures)   b 84.0 (3 significant figures)
   c 84.00 (4 significant figures)

2 Write down: i the upper bound and ii the lower bound of these numbers.
   a 0.9 (1 decimal place)   b 0.90 (2 decimal places)
   c 0.09 (2 decimal places)

3 The length of a line is 118 cm correct to the nearest cm. Write down:
  a the upper bound
  b the lower bound of the length of the line.
  Give your answers in cm.

4 The mass of a stone is 6.4 kg correct to the nearest one tenth of a kg. Write down:
  a the upper bound
  b the lower bound of the mass of the stone.
  Give your answers in grams.

5 The amount of fuel in a tank is 48.0 litres correct to the nearest tenth of a litre. Write down:
  a the upper bound
  b the lower bound of the amount of fuel in the tank.
  Give your answers in litres.

6 The length of a piece of wood is 1 metre correct to the nearest cm. Write down:
  a the upper bound
  b the lower bound of the length of the piece of wood.
  Give your answers in metres.

2.2 Calculating the bounds of an expression

Objective

● You can work out the upper bound and the lower bound of an expression when the numbers in it are approximate.

Why do this?

If you had rough estimates for its length and width, you could work out the maximum and minimum possible areas of a basketball court.

Get Ready

1. Work out $6.45 + 3.65$.
2. $a = 3.4$ (1 decimal place), $b = 5.6$ (1 decimal place)
   Work out the difference between the upper bound of a and the lower bound of b.
3. c, d, e and f are four positive numbers written in order of increasing size. Write down the pairs of numbers which will have the greatest:
   a sum   **b** difference   **c** product   **d** quotient.

Key Points

● Let $S = x + y$. Then upper bound of S = upper bound of x + upper bound of y.
● Let $D = x - y$. Then upper bound of D = upper bound of x − lower bound of y.
● Let $P = xy$. Then upper bound of P = upper bound of $x \times$ upper bound of y.
● Let $Q = \frac{x}{y}$. Then upper bound of Q = upper bound of $x \div$ lower bound of y.
● For the lower bounds of S and P, use the lower bounds of x and y.
● The lower bound of D = lower bound of x − upper bound of y.
● The lower bound of Q = lower bound of $x \div$ upper bound of y.

Example 2 $x = 3.4$ correct to 1 d.p. $y = 1.8$ correct to 1 d.p.

Find **i** the upper bound **ii** the lower bound of

a $x + y$ **b** $x - y$ **c** xy **d** $\dfrac{x}{y}$

a i $3.45 + 1.85 = 5.30$
  ii $3.35 + 1.75 = 5.10$ ⟵ Add the upper bounds.

b i $3.45 - 1.75 = 1.7$
  ii $3.35 - 1.85 = 1.50$ ⟵ Lower bound – upper bound.

c i $3.45 \times 1.85 = 6.3825$
  ii $3.35 \times 1.75 = 5.8625$

d i $3.45 \div 1.75 = 1.971$ (3 d.p.)
  ii $3.35 \div 1.85 = 1.811$ (3 d.p.)

Example 3 Here is a formula from science.

$$H = \frac{V^2(\sin x)^2}{2g}$$

$V = 250$ correct to 3 significant figures
$x = 72$ correct to 2 significant figures
$g = 9.81$ correct to 2 decimal places

a Find the upper bound and the lower bound of H.
b Write the value of H correct to an appropriate number of significant figures.

a Upper bound $\dfrac{250.5^2 \times (\sin 72.5)^2}{2 \times 9.805} = 2910.5621$

Lower bound $= \dfrac{249.5^2 \times (\sin 71.5)^2}{2 \times 9.815} = 2851.897\,864....$

b $H = 2900$ correct to 2 significant figures.

ResultsPlus
Examiner's Tip

The lower and upper bounds may agree when written correct to a certain number of significant figures. Make sure you choose an appropriate degree of accuracy.

Example 4 The length of a rectangle is measured as 8.3 cm to the nearest mm and its width as 3.6 cm to the nearest mm.
  a Write down the lower bound of the width.
  b Write down the lower bound for the perimeter.
  c Work out the lower bound of the area. Give your answer correct to 1 decimal place.

a 3.55 mm

b $2 \times 3.55 + 2 \times 8.25 = 23.6$ mm

c $3.55 \times 8.25 = 29.2875$
   $= 29.3$ mm^2 (1 d.p.)

Exercise 2B

1 $p = 480$ (3 s.f.), $q = 56$ (2 s.f.) Work out the lower bounds of

    a $p + q$ **b** pq **c** p^2

2 $c = 2.44$ (2 d.p.), $d = 4.45$ (2 d.p.) Work out the lower bounds of

    a $c + d$ **b** cd **c** $10c$

3 $r = 200$ (3 s.f.), $s = 250$ (3 s.f.), $t = 224$ (3 s.f.) Work out the upper bounds of

    a rst **b** $rs + st + tr$

4 $c = 42$ (2 s.f.), $d = 30$ (2 s.f.) Work out the lower bounds of

    a $c - d$ **b** $c \div d$

5 $e = 3.4$ (1 d.p.), $f = 2.5$ (1 d.p.) Work out the lower bounds of

    a $e - f$ **b** $\dfrac{e}{f}$

6 $p = q + rs$ $q = 18.7, r = -6.4, s = 7.7$ all correct to 1 d.p.

    a Find the lower bound of p.

    b Find the upper bound of p.

    c Write the value of p correct to an appropriate number of significant figures.

7 $E = mc^2$ $c = 3.0 \times 10^8$ (2 s.f.), $m = 2.4$ (2 s.f.)

    Calculate the lower bound of E. Calculate the upper bound of E.

8 $A = lw$

    a Suppose $l = 10, w - 5$, both exact. Work out the exact value of A.

    b Suppose $l = 10, w = 5$, both written correct to the nearest whole number.

    Work out the percentage difference between the upper bound of A and the exact value of A.

Chapter review

- The **upper bound** of a number written to 1 decimal place is the highest value which rounds down to that number.
- The **lower bound** of a number written to 1 decimal place is the lowest value which rounds up to that number.
- Let $S = x + y$. Then upper bound of S = the upper bound of x + upper bound of y.
- Let $D = x - y$. Then upper bound of D = upper bound of x − lower bound of y.
- Let $P = xy$. Then upper bound of P = upper bound of $x \times$ upper bound of y.
- Let $Q = \dfrac{x}{y}$. Then upper bound of Q = upper bound of $x \div$ lower bound of y.
- For the lower bounds of S and P, use the lower bounds of x and y.
- The lower bound of D = lower bound of x − upper bound of y.
- The lower bound of Q = lower bound of $x \div$ upper bound of y.

Review exercise

1 $x = 4.9$ (1 decimal place), $y = 12.1$ (1 decimal place). Work out the upper bounds of
  a $x + y$ **b** $4x + y$ **c** xy **d** x^2

2 $p = 450$ (2 significant figures), $q = 240$ (2 significant figures). Work out the lower bounds of
  a $p - q$ **b** $2p - q$ **c** $p \div q$ **d** $2 - \dfrac{p}{q}$

3 Write down the upper bounds of these numbers.
  a 7.4 (1 decimal place) **b** 8.0 (1 decimal place) **c** 6.43 (2 decimal places)
  d 460 (2 significant figures) **e** 3450 (3 significant figures)

4 Write down the lower bounds of these numbers.
  a 8.3 (1 decimal place) **b** 10.0 (1 decimal place) **c** 8.00 (2 decimal places)
  d 45 (2 significant figures) **e** 4000 (3 significant figures)

5 The length of a metal rod is 200 cm correct to the nearest cm.
Explain why the rod may be able to fit into a slot with a stated length of 199.8 cm.

6 Katy drove for 238 km, correct to the nearest mile.
She used 27.3 litres of petrol, to the nearest tenth of a litre.
Work out the upper bound for the petrol consumption in km per litre for Katy's journey.
Give your answer correct to 2 decimal places.

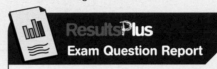

ResultsPlus
Exam Question Report

76% of students answered this question poorly because they selected the wrong bound for the numerator or the denominator.

June 2008, adapted

7 A cylinder has the label 'Contents 330 ml' printed on it. The diameter of the cylinder is 6.5 cm correct to the nearest mm and the height 10.0 cm to the nearest mm. Is the label correct?

8 The area of a square is 230 cm² correct to 2 significant figures. Work out the least possible perimeter of the square.

9 The length, L m, of a tyre skid when a car suddenly brakes from a speed s metres per second is given by $L = \dfrac{s^2}{2d}$ where d metres per second is the deceleration.
L was measured to be 15.6 correct to 1 decimal place and d was 7 correct to the nearest whole number.
Calculate the lower bound of s. Give your answer correct to 1 d.p.

3 PERCENTAGES

Have you heard of 10:10? It is a movement that is asking people, companies and organisations to sign up to say that they will reduce their carbon emissions by 10% by 2010. They think that the message is more powerful if it is a reduction by 2010 rather than 2020 or 2050, and 10% is about the amount that Britain needs to cut their emissions by 2010 if they are going to play their part in controlling climate change.

◎ Objectives

In this chapter you will:
- use a calculator to find quantities after a percentage increase or decrease
- express one quantity as a percentage of another and find the percentage profit or loss
- find an amount after repeated percentage changes
- find the original amount, given the final amount, after a percentage change.

◈ Before you start

You should know:
- how to find a percentage of a quantity
- find quantities after a percentage increase or decrease, without using a calculator.

3.1 Finding the new amount after a percentage increase or decrease

<table>
<tr><td>

◎ Objective

● You can use a calculator to find quantities after a percentage increase or decrease.

</td><td>

⊘ Why do this?

A pay rise can be given as a percentage of current earnings.

</td></tr>
</table>

⊕ Get Ready

1. Work out 30% of £400.   2. Work out 25% of £120.   3. Write 8 as a fraction of 20.

⊕ Key Points

◎ There are two methods that can be used to increase an amount by a percentage.
   ● You can find the percentage of that number and then add this to the starting number.
   ● You can use a multiplier.
◎ There are two methods that can be used to decrease an amount by a percentage.
   ● You can find the percentage of that number and then subtract this from the starting number.
   ● You can use a multiplier.

Example 1   Mark's salary is £38 000 a year. His salary is increased by 7%.
Work out his new salary.

Method 1

$$7\% \text{ of } £38\,000 = \frac{7}{100} \times 38\,000$$
$$= 2660$$

The increase in his salary is £2660.

$$38\,000 + 2660 = 40\,660$$

Add the increase to his original salary.

Mark's new salary is £40 660.

Method 2

$$100\% + 7\% = 107\%$$

His new salary is 107% of £38 000.

$$107\% = \frac{107}{100} = 1.07$$

1.07 is the multiplier.

$$1.07 \times 38\,000 = 40\,660$$

Multiply 38 000 by 1.07.
(This increases 38 000 by 7%.)

Hugh's new salary is £40 660.

Example 2   a Write down the single number that you can multiply by to increase an amount by 12.5%.
b Increase £56 by 12.5%.

a $100\% + 12.5\% = 112.5\%$
$112.5\% = \frac{112.5}{100} = 1.125$

b $56 \times 1.125 = £63$

Use the multiplier worked out in part a.

Exercise 3A

Questions in this chapter are targeted at the grades indicated.

1 Write down the single number you can multiply by to work out an increase of
    a 64%   **b** 3%   **c** 14%   **d** 40%
    e 13.4%   **f** $12\frac{1}{2}$%   **g** 15%   **h** 2.36%

2   **a** In order to increase an amount by 40%, what single number should you multiply by?
    b The cost of a theatre ticket is increased by 40% for a special concert.
       What is the new price if the normal price was £15.40?

3 The table shows the salaries of three workers.
Each worker receives a 4.2% salary increase.
Work out the new salary of each worker.

Helen	£12 000
Tom	£24 000
Sandeep	£32 000

4 Jenny puts £600 into a bank account. At the end of one year 3.5% interest is added.
How much is in her account at the end of 1 year?

5   **a** Increase £120 by 20%.   **b** Increase 56 kg by 25%.   **c** Increase 2.4 m by 16%.
    d Increase £1240 by 10.5%.   **e** Increase 126 cm by 2%.

D

Example 3   The value of a van depreciates by 18% each year.
                The value of a van when new is £19 000.
                Work out the value of the van after 1 year.

Depreciates means that the value of the van decreases.

Method 1

18% of £19 000 $= \frac{18}{100} \times 19\,000$ ←   The depreciation in 1 year is £3420.
               $= 3420$

$19\,000 - 3420 = 15\,580$ ←   Subtract to work out the new value.

Value after 1 year $=$ £15 580.

Method 2

   $100\% - 18\% = 82\%$ ←   The final value is 82% of the original value.
      $82\% = \frac{82}{100} = 0.82$ ←   0.82 is the multiplier.

$0.82 \times 19\,000 = 15\,580$ ←   Multiply the original amount by 0.82.

Value after 1 year $=$ £15 580.

Exercise 3B

1   Write down the single number you can multiply by to work out a decrease of
    a 7%   **b** 20%   **c** 16%   **d** 27%
    e 5.6%   **f** $2\frac{1}{2}$%   **g** $7\frac{1}{4}$%   **h** 0.8%

2   In a sale all prices are reduced by 15%. Work out the sale price of each of the following:
    a a television set that normally costs £300
    b a CD player that normally costs £40
    c a computer that normally costs £1200.

3   Alan weighs 82 kg before going on a diet. He sets himself a target of losing 5% of his original weight.
    What is his target weight?

4   A holiday normally costs £850. It is reduced by 12%. How much will the holiday now cost?

5   Ria buys a car for £7300. The value of the car depreciates by 20% each year.
    Work out the value of the car at the end of: **a** 1 year   **b** 2 years.

3.2 Working out a percentage increase or decrease

Objectives

- You can express one quantity as a percentage of another.
- You can find percentage loss or profit.

Why do this?

To work out how well a business is doing, you might want to work out what percentage profit or loss they have made.

Get Ready

1. Work out £24.80 − £12.05.   **2.** Work out £15 000 − £13 700.   **3.** Write 45 out of 200 as a fraction.

Key Points

- To write one quantity as a percentage of another quantity:
  - write down the first quantity as a fraction of the second quantity
  - convert the fraction to a percentage.

- Percentage problems sometimes involve percentage profit or percentage loss, where:
  - percentage profit (or increase) $= \dfrac{\text{profit (or increase)}}{\text{original amount}} \times 100\%$
  - percentage loss (or decrease) $= \dfrac{\text{loss (or decrease)}}{\text{original amount}} \times 100\%$

Example 4

a Convert 11 out of 20 to a percentage.

b Convert 23 cm out of 4 m to a percentage.

a $\frac{11}{20}$ ← Write the first number as a fraction of the second number.

$\frac{11}{20} \times 100 = 55\%$ ← To convert a fraction to a percentage, multiply by 100.

b $4\,m = 4 \times 100$ ← Multiply by 100 to convert 4 m into centimetres.
$= 400\,cm$

$\frac{23}{400} \times 100 = 5.75\%$ ← Convert the fraction to a percentage.

Results Plus

Watch Out!

When working with quantities in different units, first make sure that all the units are the same.

Exercise 3C

1 Write:

a £3 as a percentage of £6

b 2 kg as a percentage of 8 kg

c 4p as a percentage of 10p

d 8 cm as a percentage of 40 cm

e 60p as a percentage of £2.40

f 15 mm as a percentage of 6 cm

g 36 minutes as a percentage of 1 hour

h 50 cm as a percentage of 4 m.

2 Janet scored 36 out of 40 in a German test. Work out her score as a percentage.

3 Jerry took 60 bottles to a bottle bank. 27 of the bottles were green.
What percentage of the bottles were green?

4 A 40 g serving of cereal contains 8 g of protein, 24 g of carbohydrates, 4.5 g of fat and 3.5 g of fibre.
What percentage of the serving is:

a protein   b carbohydrates   c fat   d fibre?

D

Example 5

Karen bought a car for £1200.
One year later, she sold it for £840.
Work out her percentage loss.

$1200 - 840 = 360$ ← Subtract the selling price from the original price to find her loss.

$\frac{360}{1200}$ ← Write down the fraction $\frac{loss}{original\ price}$

$\frac{360}{1200} \times 100 = 30\%$ ← Multiply $\frac{360}{1200}$ by 100 to change it to a percentage.

Her percentage loss is 30%.

A02
A03

Example 6 ▶ Tony bought a box of 24 oranges for £4.

He sold all the oranges for 21p each.

Work out his percentage profit.

24 × 21 = 504p ⟵ | Work out the total amount, in pence, Tony received from selling all the oranges.

504 − 400 = 104p profit ⟵ | Subtract the original price from the selling price to find his profit in pence.

$\frac{104}{400}$ ⟵ | Write down the fraction $\frac{profit}{original\ price}$

$\frac{104}{400} \times 100 = 26\%$ ⟵ | Multiply $\frac{104}{400}$ by 100 to change it to a percentage.

Percentage profit = 26%.

Exercise 3D

C

1 Calculate the percentage increase or decrease to the nearest 1%:

    a £24 to £36   **b** 12.5 kg to 20 kg   **c** 45 cm to 39.5 cm

    d 2 minutes to 110 seconds.

2 In a sale, the price of a clock is reduced from £32 to £27.20. Work out the percentage reduction.

A02
A03

3 Rob bought a crate of 40 melons for £30. He sold all the melons for £1.05 each.

Work out his percentage profit.

A03

*****4** David owns three shops selling DVDs.

He tells the staff in each of the shops that some of them will receive a bonus.

He will give the bonus to the staff who work in the shop that has the biggest percentage increase in the number of DVDs sold from the first half to the second half of the year.

	DVDs sold Jan–Jun	DVDs sold Jul–Dec
Shop A	12 893	13 562
Shop B	9 875	10 346
Shop C	11 235	11 853

Which shop should receive the bonus? You must show how you decided on your answer.

A02
A03

5 Martin goes to a discount centre.

He buys 10 trays of drinks.

Each tray holds 24 cans of cola and costs £9.50.

Martin sells 150 cans of cola at a fair for 55p each.

He sells the rest of the cans for 35p each the next day at a car boot sale.

Work out the profit or loss percentage that Martin makes.

6 A badminton club has 44 members.

Each member pays £85 per year as a membership fee.

The club has to pay a total of £3700 to the sports centre to hire the badminton courts.

The sports centre decides to increase the cost of hiring courts by 8.5%.

The badminton club will have 46 members next year.

Work out the smallest possible percentage rise in club membership fees so that the club can afford to pay the sports centre. Give your answer correct to 1 decimal place.

A03 C

3.3 Working out compound interest

◎ Objective

● You can find an amount after repeated percentage changes.

② Why do this?

Percentage changes can happen over a period of time. You may want to work out how much money you will have in two years if you put £100 in your bank account now.

◆ Get Ready

1. Write

 a $2 \times 2 \times 2 \times 2 \times 2$ as a power of 2   **b** 30% as a decimal   **c** 125% as a decimal

🔑 Key Points

● Banks and building societies pay **compound interest**.

● At the end of the first year, interest is paid on the money in an account. This interest is then added to the account. At the end of the second year, interest is paid on the total amount in the account, that is, the original amount of money plus the interest earned in the first year.

● At the end of each year, interest is paid on the total amount in the account at the start of that year.
   For example, if £200 is invested in a bank account and interest is paid at a rate of 5% then

Year	Amount at start of year	Amount plus interest	Total amount at year end
1	£200	200×1.05	£210
2	£210	$210 \times 1.05 = 200 \times 1.05^2$	£220.50
3	£220.50	$220.50 \times 1.05 = 200 \times 1.05^3$	£231.52
4	£231.52	$231.52 \times 1.05 = 200 \times 1.05^4$	£243.10
5	£243.10	$243.10 \times 1.05 = 200 \times 1.05^5$	£255.26
6	£255.26	$255.26 \times 1.05 = 200 \times 1.05^6$	£268.02

● To calculate compound interest, find the multiplier:

   ● Amount after n years = original amount $\times$ multiplier^n

Example 7 £4000 is invested for 2 years at 5% per annum compound interest.
Work out the **total interest** earned over the 2 years.

Method 1

$$100\% + 5\% = 105\%$$
$$105\% = 1.05$$

Work out the multiplier for an increase of 5%.

$$4000 \times 1.05^2 = 4410$$

Multiply the original amount by 1.05^2 to find the amount in the account after 2 years.

$$4410 - 4000 = 410$$

The total interest earned over the 2 years is £410.

Subtract the original amount to find the interest.

Results Plus
Watch Out!

Read this type of question carefully to determine whether you need to work out just the interest or the final amount in the account.

Method 2

$$\frac{5}{100} \times 4000 = 200$$

Work out the interest in the first year.

$$4000 + 200 = 4200$$

Add the interest to the original amount.

$$\frac{5}{100} \times 4200 = 210$$

Work out the interest in the second year.

$$200 + 210 = 410$$

Find the total interest.

The total interest earned over the 2 years is £410.

Example 8

a Each year the value of a car depreciates by 30%. Find the single number, as a decimal, that the value of the car can be multiplied by to find its value at the end of 4 years.

b The value of a house increases by 16% of its value at the beginning of the year. The next year its value decreases by 3% of its value at the start of the second year. Find the single number, as a decimal, that the original value of the house can be multiplied by to find its value at the end of the 2 years.

a
$$100\% - 30\% = 70\%$$
$$70\% = \frac{70}{100} = 0.7$$

Find the multiplier that represents a decrease of 30%.

$$0.7 \times 0.7 \times 0.7 \times 0.7 = 0.7^4$$
$$0.7^4 = 0.2401$$

The depreciation is over 4 years so the single multiplier is 0.7 raised to the power of 4.

0.2401 is the single number.

b
$$100\% + 16\% = 116\%$$
$$116\% = \frac{116}{100} = 1.16$$

Find the multiplier for an increase of 16%.

$$100\% - 3\% = 97\%$$
$$97\% = \frac{97}{100} = 0.97$$

Find the multiplier for a decrease of 3%.

$$1.16 \times 0.97 = 1.1252$$
1.1252 is the single number.

The value increases and then decreases so find the product of the two multipliers.

Example 9 The value of a machine when new is £8000.
The value of the machine depreciates by 10% each year.
Work out its value after 3 years.

Method 1

$100\% - 10\% = 90\%$ ← Work out the multiplier for a decrease of 10%.
$90\% = 0.9$

$8000 \times 0.9^3 = £5832$ ← Multiply the value when new by 0.9^3 to find the value after 3 years.
The value of the machine after 3 years is £5832.

Method 2

$\frac{10}{100} \times 8000 = 800$

$8000 - 800 = 7200$

$\frac{10}{100} \times 7200 = 720$

$7200 - 720 = 6480$

$\frac{10}{100} \times 6480 = 648$

$6480 - 648 = 5832$

The value of the machine after 3 years is £5832.

Exercise 3E

1   Work out the multiplier as a single decimal number that represents:
   a   an increase of 20% for 3 years
   b   a decrease of 10% for 4 years
   c   an increase of 4% followed by an increase of 2%
   d   a decrease of 35% followed by a decrease of 20%.

2   £1000 is invested for 2 years at 5% per annum compound interest.
   Work out the total amount in the account after 2 years.

3   Mrs Bell buys a house for £60 000. In the first year, the value of the house increases by 16%.
   In the second year, the value of the house decreases by 4% of its value at the beginning of that year.
   a   Write down the single number, as a decimal, that the original value of the house can be multiplied by
       to find its value after 2 years.
   b   Work out the value of the house after the 2 years.

4   Ben says that an increase of 40% followed by an increase of 20% is the same as an increase of 60%.
   Is Ben correct? You must give a reason for your answer.

5   Jeremy deposits £3000 in a bank account. Compound interest is paid at a rate of 4% per annum.
   Jeremy wants to leave the money in the account until there is at least £4000 in the account.
   Calculate the least number of years Jeremy must leave his money in the bank account.

3.4 Calculating reverse percentages

⊙ Objective

● You know how to find the original amount given the final amount after a percentage increase or decrease.

⦿ Why do this?

If you found a book marked 60% off in a sale, you can use reverse percentages to work out how much it originally cost.

⬥ Get Ready

1. Write down the multiplier for:

   a an increase of 15%   **b** a decrease of 15%   **c** an increase of 4%   **d** a decrease of 4%.

⦿ Key Points

● There are two methods that can be used to find the original amount if the final amount after a percentage increase or decrease is known using **reverse percentages**.

● A flow diagram can be used to represent a percentage change using multipliers.

original price $\xrightarrow{\times \text{ multiplier}}$ final amount

Drawing a second flow diagram reversing the direction and using the inverse operation shows that to find the original price from the final amount you divide by the multiplier.

original price $\xleftarrow{\div \text{ multiplier}}$ final amount

● Work out the value of 1% of the original amount and multiply this by 100.

🔑 Example 10

In a sale, all prices are reduced by 20%.
The sale price of a jacket is £33.60.
Work out the original price of the jacket.

Method 1

$100\% - 20\% = 80\%$ ⟵ | Find the multiplier for a decrease of 20%.

$\frac{80}{100} = 0.8$

Original price $= 33.60 \div 0.8$
$= 42$ ⟵ | Divide by the multiplier to find the original price.

The original price of the jacket was £42.

Method 2

$100\% - 20\% = 80\%$ ⟵ | £33.60 represents 80% of the original price.

$£33.60 = 80\%$
So $1\% = 33.60 \div 80$ ⟵ | Divide 33.60 by 80 to find the value of 1%.
$= 0.42$

So, original price $= 0.42 \times 100$ ⟵ | The original price is 100% so multiply the amount that represents 1% by 100.
$= 42$ (Check: $42 \times 0.8 = 33.6$)

The original price of the jacket was £42.

Example 11 The price of a new washing machine is £376.
This price includes Value Added Tax (VAT) at $17\frac{1}{2}\%$.
Work out the cost of the washing machine before VAT was added.

Method 1

$100\% + 17\frac{1}{2}\% = 117.5\%$

> The original cost was increased by 17.5% so find the multiplier for an increase of 17.5%.

$\frac{117.5}{100} = 1.175$

Original price $= 376 \div 1.175$

> Divide by the multiplier to find the amount without VAT.

$= 320$

The cost of the washing machine before VAT was added was £320.

> (Check: 320 × 1.175 = 376)

Method 2

$100\% + 17.5\% = 117.5\%$

> £376 represents 117.5% of the original cost.

$£376 = 117.5\%$

So $1\% = 376 \div 117.5$

> Divide 376 by 117.5 to find the value of 1%.

$= 3.2$

So original price $= 3.2 \times 100$

> The original cost is 100% so multiply the amount that represents 1% by 100.
> (Check: 320 × 117.5 = 376)

$= 320$

The cost of the washing machine before VAT was added was £320.

Exercise 3F

1. Employees at a firm receive a pay increase of 4%. After the pay increase, Linda earns £24 960. How much did Linda earn before the pay increase?

2. The price of a new television set is £329.
This price includes Value Added Tax (VAT) at $17\frac{1}{2}\%$.
Work out the cost of the television set before VAT was added.

3. A holiday is advertised at a price of £403.
This represents a 35% saving on the brochure price. Work out the brochure price of the holiday.

4. Kunal pays tax at a rate of 22%. After he has paid tax, Kunal receives £140.40 per week. How much does Kunal earn per week before he pays tax?

5. In one year, the population of an island increased by 3.2% to 434 472. Work out the population of the island before the increase.

6. Tasha invests some money in a bank account. Interest is paid at a rate of 8% per annum. After 1 year, there is £291.60 in the account. How much money did Tasha invest?

Chapter review

- There are two methods that can be used to increase (or decrease) an amount by a percentage.
  - You can find the percentage of that number and then add this to (or subtract this from) the starting number.
  - You can use a **multiplier**.
- To write one quantity as a percentage of another quantity, write down the first quantity as a fraction of the second quantity, then change the fraction to a percentage.
- Percentage problems sometimes involve percentage profit or percentage loss, where:
  - percentage profit (or loss) $= \dfrac{\text{profit (or loss)}}{\text{original amount}} \times 100\%$
- To calculate **compound interest**, find the multiplier:
  - Amount after n years = original amount $\times$ multiplier^n
- There are two methods that can be used to find the original amount if the final amount after a percentage increase or decrease is known using **reverse percentages**.
- A flow diagram can be used to represent a percentage change using multipliers.

  original price $\xrightarrow{\times \text{ multiplier}}$ final amount

  Drawing a second flow diagram reversing the direction and using the inverse operation shows that to find the original price from the final amount you divide by the multiplier.

  original price $\xleftarrow{\div \text{ multiplier}}$ final amount

- Work out the value of 1% of the original amount and multiply this by 100.

Review exercise

1. A man invests £20 000 with a guaranteed compound interest rate of 4%.
   How long will it be before he has doubled his money?

*2. Barry has ben asked to compare the pay for four similar jobs advertised in a newspaper.

Able Computer Sales	Beta IT Support
Sales Assistant	Sales Consultant
You will spend time in the field, working both from our Manchester headquarters and from home in the North West region.	Full time: 30 hours per week
	Pay: £15 per hour
	Tele-sales based in our new offices.
Pay: £23 000 per annum	Daily hours variable.
Compu Systems	Digital Hardware
Sales Agent	Sales Adviser
As a sales agent your pay will be £1800 per month, plus commission of 1% of monthly sales. You can expect to make monthly sales to a minimum value of £22 000.	You will be part of a team with a salary of £20 000 per annum + team bonus. Team bonus last year was 20% of salary.

Which job pays the most?

3. Linda's mark in a Maths test was 36 out of 50.
   Find 36 out of 50 as a percentage.

D

4 A hotel has **56** guests.

35 of the guests are male.

a Work out 35 out of 56 as a percentage.

40% of the 35 male guests wear glasses.

b Write the number of male guests who wear glasses as a fraction of the 56 guests.

Give your answer in its simplest form.

Nov 2007

5 In April 2004, the population of the European Community was 376 million.

In April 2005, the population of the European Community was 451 million.

Work out the percentage increase in population.

Give your answer correct to 1 decimal place.

Nov 2007

6 Bill buys a new lawn mower.

The value of the lawn mower depreciates by 20% each year.

a Bill says 'after 5 years the lawn mower will have no value'.

Bill is wrong.

Explain why.

Bill wants to work out the value of the lawn mower after 2 years.

b By what single decimal number should Bill multiply the value of the lawn mower when new?

Nov 2005

A02 A03 B

7 In a sale normal prices are reduced by 20%.

Andrew bought a saddle for his horse in the sale.

The sale price of the saddle was £220.

Calculate the normal price of the saddle.

8 The value of a car depreciates by 35% each year.

At the end of 2007, the value of the car was £5460.

Work out the value of the car at the end of 2006.

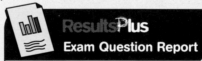

ResultsPlus

Exam Question Report

86% of students answered this question poorly because they did not find the value of the reduction first.

Nov 2008

9 Nimer got a pay rise of 5%. His new pay was £1680 per month.

Work out his pay per month before he got this pay rise.

10 Jim is a plumber. He has to work out the VAT on some equipment. VAT is charged at $17\frac{1}{2}\%$.

The total cost of the equipment including VAT is £4465. Calculate how much the VAT was.

11 Sophie is offered the following pay deals:

A '5% increase this year, followed by a 4% increase next year'

B '$4\frac{1}{2}\%$ increase this year, followed by $4\frac{1}{2}\%$ increase next year'

Which offer should Sophie accept?

A02 A03

4 LINEAR EQUATIONS

Old-fashioned scales like these are still used in many markets around the world. Just like linear equations, you must "do" the same to both sides to keep them balanced.

◎ Objectives

In this chapter you will:
- solve simple equations
- solve linear equations containing brackets and fractions
- solve linear equations in which the unknown appears on both sides of the equation
- set up and solve simple linear equations.

◈ Before you start

You need to be able to:
- collect like terms in an algebraic expression
- expand or multiply out brackets
- calculate using directed numbers
- apply the rules of BIDMAS.

4.1 Solving simple equations

**Objective**

⦿ You can solve simple equations.

Why do this?

If you have a number of albums on your mp3 player, as well as some individual tracks, you can solve a simple equation to find out how many songs you have in total.

Get Ready

1. Simplify

   a $4 + 3p - 2q - p + 7q$ **b** $-3(1 - 2z)$ **c** $4m(3m + 9)$

Key Points

⦿ In any equation, the value of the left-hand side is always equal to the value of the right-hand side. So whatever operation is applied to the left-hand side must also be applied to the right-hand side.

⦿ Two children sit on a see-saw, equally distant from the middle. The see-saw is level which means that the weight of each child is the same.

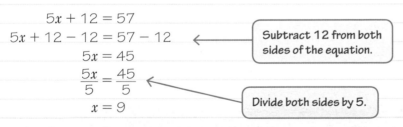

A   B

Weight of child A on the left-hand side $=$ weight of child B on the right-hand side.
The weight of child A is $5x + 12$.
The weight of child B is 57.
So, $5x + 12 = 57$.   This is a **linear equation**.

Example 1   Solve $5x + 12 = 57$

$$5x + 12 = 57$$
$$5x + 12 - 12 = 57 - 12$$
$$5x = 45$$
$$\frac{5x}{5} = \frac{45}{5}$$
$$x = 9$$

Subtract 12 from both sides of the equation.

Divide both sides by 5.

**Results****Plus**
Watch Out!

Always apply the same operation to **both** sides of the equation.

Exercise 4A

Questions in this chapter are targeted at the grades indicated.

Solve

 1 $2a + 5 = 13$ **2** $3b - 4 = 17$ **3** $2c + 6 = 11$

 4 $5d - 1 = 8$ **5** $13 + 4e = 7$ **6** $1 - 2f = 9$

 7 $13 = 3g - 5$ **8** $5 = 1 + 8h$ **9** $2 - 5k = 15$

 10 $17 = 3 - 4m$

4.2 Solving linear equations containing brackets

◉ Objective

○ You can solve linear equations that require prior simplification of brackets.

⬥ Get Ready

1. Solve

a $2x + 7 = 19$ **b** $3b + 8 = 20$ **c** $16 = 40 - 3q$

Key Points

◉ Follow the rules of BIDMAS when solving a linear equation. (See Section 1.3 in Unit 2 for BIDMAS).

◉ **Solutions** can be written as mixed fractions, improper fractions or as decimals.

Example 2   Solve $4(x + 1) = 11$

$4(x + 1) = 11$ ← Expand the left-hand side by multiplying out the brackets.
$4x + 4 = 11$

$4x + 4 - 4 = 11 - 4$ ← Solve the equation as in Section 13.1.
$4x = 7$
$\dfrac{4x}{4} = \dfrac{7}{4}$
$x = \dfrac{7}{4}$
$x = 1\dfrac{3}{4}$ or 1.75

ResultsPlus
Examiner's Tip

Your answer can be written as a fraction or as a decimal.

⚙ Exercise 4B

Solve

1 $2(x + 3) = 12$ **2** $5(y + 4) = 35$ **3** $4(x - 1) = 5$ **4** $3(2y - 1) = 9$

5 $13 = 4(x + 3)$ **6** $2(1 - w) = 10$ **7** $5(3 - 4z) = 20$ **8** $2 = 3(1 + 3x)$

9 $2(2x + 3) + 1 = 11$ **10** $17 = 6 - 3(5 - 2y)$

D

4.3 Solving linear equations with the unknown on both sides

⊙ Objective

⊙ You can solve linear equations, with integer coefficients, in which the unknown appears on both sides of the equation.

❓ Why do this?

Knowing how to solve equations helps you solve other problems such as finding one weight when given another.

⬆ Get Ready

1. Solve

   a $4 = 2(x - 6)$ **b** $3(3a + 4) + 3 = 33$ **c** $4 = 8 - 2(8 - 3b)$

🔑 Key Point

⊙ Collect terms so that the ones involving the unknown are on one side of the equation.

🔍 **Example 3**   Solve $5x + 5 = 3 - 3x$

$$5x + 5 = 3 - 3x$$
$$5x + 5 + 3x = 3 - 3x + 3x$$

> Add $3x$ to both sides of the equation.
> Remember $-3x + 3x = 0$.

$$8x + 5 = 3$$
$$8x + 5 - 5 = 3 - 5$$
$$8x = -2$$
$$\frac{8x}{8} = \frac{-2}{8}$$
$$x = -\frac{1}{4} \text{ or } -0.25$$

> Solve the equation as in Section 4.1.

ResultsPlus
Watch Out!

Always show each stage of your working.

🔍 **Example 4**   Solve $3 - 6x = 7 - 3x$

> Here both terms in x have a negative coefficient.

ResultsPlus
Examiner's Tip

Collect the terms in x on the side of the equation that gives them a positive coefficient.

$$3 - 6x = 7 - 3x$$
$$3 - 6x + 6x = 7 - 3x + 6x$$

> Add $6x$ to both sides of the equation.

$$3 = 7 + 3x$$
$$3 - 7 = 3x$$
$$-4 = 3x$$
$$x = -\frac{4}{3}$$

Exercise 4C

Solve

1 $4a + 3 = 8 + 2a$ **2** $5b + 3 = b - 7$ **3** $3c - 2 = 5c - 8$ **4** $d + 7 = 5d + 15$

5 $3 - 2e = 4 - 3e$ **6** $1 - 7f = 3f + 10$ **7** $2(x - 4) = x + 7$

8 $2x + 5 = 1 + 3(2 + x)$ **9** $3(4x + 1) + 2(1 - 5x) = 2 + x$

10 $6 + 2(x - 3) = x - 3(1 - 2x)$

4.4 Solving linear equations containing fractions

◎ Objective

● You can solve linear equations containing fractions.

◆ Get Ready

1. Solve

  a $4x + 6 = 2(4 + x) + 2$ **b** $2 - 8x = 20 + 3(4 + 9x)$ **c** $5 + 4(x - 1) = 2x - 4(2 - 3x)$

Key Point

◉ To solve an algebraic equation involving fractions, eliminate all fractions by multiplying each term by the LCM of the denominators.

Example 5

Solve $\dfrac{12}{p + 2} = 3$

$$\frac{12}{p + 2} = 3$$

$$\frac{12}{p + 2} \times (p + 2) = 3 \times (p + 2)$$

Multiply both sides of the equation by $(p + 2)$.
The terms in $(p + 2)$ on the left-hand side cancel out.

$$12 = 3(p + 2)$$
$$12 = 3p + 6$$
$$12 - 6 = 3p$$
$$3p = 6$$
$$p = 2$$

Results **Plus**
Examiner's Tip

Always try to remove the fraction first.

Example 6 Solve $\dfrac{x+1}{2} - \dfrac{4x-1}{3} = \dfrac{5}{12}$

$$\frac{x+1}{2} - \frac{4x-1}{3} = \frac{5}{12}$$

$$12 \times \frac{x+1}{2} - 12 \times \frac{4x-1}{3} = 12 \times \frac{5}{12}$$

Multiply each of the three terms by 12.

$${}^6\cancel{12} \times \frac{x+1}{\cancel{2}_1} - {}^4\cancel{12} \times \frac{4x-1}{\cancel{3}_1} = {}^1\cancel{12} \times \frac{5}{\cancel{12}_1}$$

$$6(x+1) - 4(4x-1) = 5$$
$$6x + 6 - 16x + 4 = 5$$
$$-10x + 10 = 5$$
$$-10x = -5$$
$$x = \frac{1}{2}$$

Remember: $-4 \times -1 = +4$

Exercise 4D

Solve

1. $\dfrac{p}{5} + 3 = 7$

2. $\dfrac{q+2}{3} = 4$

3. $\dfrac{m}{2} + \dfrac{m}{5} = 21$

4. $\dfrac{x}{6} + 1 = \dfrac{x-4}{4}$

5. $3(2y - 10) = \dfrac{4y-7}{2}$

6. $2\left(\dfrac{x}{3} - 3\right) = 16$

7. $\dfrac{1}{2n} + \dfrac{1}{3n} = 7$

8. $\dfrac{3t+6}{10} + \dfrac{5-2t}{5} = 6$

9. $2 - \dfrac{1-x}{3} = \dfrac{5x+2}{9}$

10. $\dfrac{3y-4}{2} - \dfrac{2y+1}{5} = \dfrac{1-y}{3}$

B

A

4.5 Setting up and solving simple linear equations

Objective

- You can set up and solve simple linear equations.

Why do this?

Businesses use linear equations to help them work out how much of a product needs to be produced to make a given profit.

Get Ready

1. Solve
   a $6a + 1 = 2a - 7$ b $5 - 6b = 10 + 4b$ c $\left(\dfrac{x}{5}\right) + 2 = \dfrac{(x-3)}{2}$

Key Points

- When setting up an equation, define all the unknowns used that have not already been defined.
- Make sure that units are consistent on both sides of the equation.

Example 7

Daniel makes some drinks to sell at the Summer Fair.

From one bottle costing £2.80 he can make 40 drinks.

Daniel wants to make a profit of £2 on each bottle.

a If c is the price of each drink, write an equation in terms of c.

b Solve your equation in **a** to find what the price of each drink should be.

a 40 drinks will cost $40 \times c = 40c$ pence ⟵ | Change all amounts to pence.

One bottle costs £2.80 = 280 pence

Daniel's profit = £2 = 200 pence

Profit = total sales − total costs

$200 = 40c − 280$

b $200 = 40c − 280$

$200 + 280 = 40c$

$40c = 480$

$c = \dfrac{480}{40}$

$c = 12$

So the cost of each drink is 12 pence.

Example 8

By setting up and solving an equation in terms of x, work out the size of the largest angle of this quadrilateral.

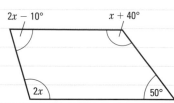

Since the sum of the interior angles of a quadrilateral is 360°, adding the angles gives:

$(2x − 10) + (x + 40) + (2x) + (50) = 360$

$2x − 10 + x + 40 + 2x + 50 = 360$ ⟵ | Collect like terms.

$5x + 80 = 360$

$5x = 360 − 80$

$5x = 280$

$x = 56$

ResultsPlus

Watch Out!

Read the question carefully; is the size of the largest angle or the smallest angle required?

Angles are: $2x − 10 = 2 \times 56 − 10 = 112 − 10 = 102°$

$x + 40 = 56 + 40 = 96°$

$2x = 2 \times 56 = 112°$ is the largest angle.

Exercise 4E

D

1 Viv thinks of a number. She multiplies the number by 5 and then subtracts 2.

Her answer is 23.

If x is the number that Viv was thinking of, work out the value of x.

A02

2 Michelle is 2 years younger than Angela.

If the sum of their ages is 64, work out Michelle's age.

3 In an exam, Jessica scored $p\%$.

Mason scored three-quarters of Jessica's score.

Zach scored 10% less than Jessica.

The total of their scores was 210%.

How much did each student score?

C

4 The length of a rectangle is $(2x + 3)$ cm.

The width of the rectangle is $\frac{1}{2}(x + 3)$ cm.

If the perimeter of the rectangle is 49 cm, find the value of x.

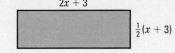

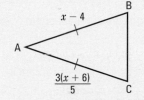

5 Joanna works for n hours each week for 5 weeks. In the sixth week she works an extra $4\frac{1}{2}$ hours.

In these six weeks she works a total of 117 hours.

Work out the number of hours Joanna works in the sixth week.

6 The diagram shows an isosceles triangle ABC.

The lengths of the sides are given in centimetres and AB = BC.

a Write down an equation in terms of x.

b Work out the lengths of AB and BC.

7 In one quarter Stuart uses 2512 units of electricity. Part of his electricity bill is shown below.

Units used = 2512

Cost of the first x units at 12.86p per unit = £_____

Cost of remaining units at 11.8p per unit = £_____

Total cost of electricity used = £ 298.

Work out the number of units, at 12.86p per unit, that Stuart used. Give your answer to the nearest unit.

Chapter review

- In any equation, the value of the left-hand side is always equal to the value of the right-hand side.
  So whatever operation is applied to the left-hand side must also be applied to the right-hand side.

- Follow the rules of BIDMAS when solving a **linear equation**.

- **Solutions** can be written as mixed fractions, improper fractions or as decimals.

- Collect terms so that the ones involving the unknown are on one side of the equation.

- To solve an algebraic fraction equation, eliminate all fractions by multiplying each term by the LCM of the denominators.

- When setting up an equation, define all unknowns used that have not already been defined.

- Make sure that units are consistent on both sides of an equation.

Review exercise

1 Solve $4t + 1 = 19$

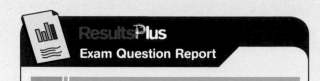

ResultsPlus

Exam Question Report

86% of students answered this question well because they remembered to do the same to both sides of the question.

Nov 2008

D

2 Solve $4(x - 2) = 10$

3 Solve   **a** $3x + 1 = x + 6$ **b** $4y - 3 = 2y - 8$

C

4 The sizes of the angles, in degrees, of the quadrilaterals are

$x + 10, 2x, x + 90$ and $x + 20$.

Work out the smallest angle of the quadrilateral.

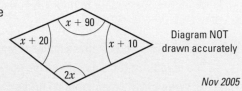

Diagram NOT drawn accurately

Nov 2005

A02

5 The diagram shows a rectangle.

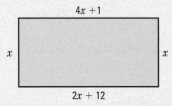

ResultsPlus

Exam Question Report

64% of students answered this question well.

All the measurements are in centimetres.
Work out the perimeter of the rectangle.

June 2009, adapted

A03

6 Uzma has £x. Hajra has £20 more than Uzma. Mabintou has twice as much as Hajra.
The total amount of money they have is £132.
Find how much money they each have.

A03

7

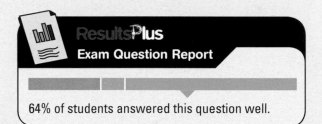

Here are 3 boxes. Box A has £x. Box B has £4 more than box A. Box C has one third of the money in box B.
Altogether there is £24 in the 3 boxes.
Find the amount of money in each box.

B

8 Solve $\dfrac{3x}{4} + 1 = 12$

9 Solve $\dfrac{x - 3}{5} = x - 5$

June 2005

A

10 Solve $\dfrac{y}{2} - \dfrac{y - 1}{3} = 2$

5 INEQUALITIES AND FORMULAE

The coldest ever recorded temperature in Antarctica was −89°C on 21 July 1983. The warmest ever temperature was 15°C on 5 January 1974. So using inequalities, you could say that the temperature in Antarctica is ⩾ −89°C and ⩽ 15°C. On top of these extremes of temperature, Antarctica has winds of up to 320 km/hour, and average precipitation of less than 5 cm per year.

◎ Objectives

In this chapter you will:
- represent inequalities on a number line using the correct notation
- solve simple linear inequalities in one variable
- solve graphically several inequalities in two variables and find the solution set
- change the subject of simple and more complex formulae.

◇ Before you start

You should already know how to:
- calculate using directed numbers
- set up and solve linear equations
- plot and read points on coordinate axes
- draw lines with the equation $y = mx + c$
- collect like terms in an algebraic expression
- know how to substitute into algebraic expressions
- use and derive algebraic formulae.

5.1 Representing inequalities on a number line

◎ Objective

● You can represent inequalities on a number line using the correct notation.

？ Why do this?

One way to check minimum and maximum temperatures in your greenhouse would be to represent them as an inequality on a number line.

⬆ Get Ready

1. What are the values of a, b, c, and d on the number line?

d   c   b   a

$$-2 \quad -1 \quad 0 \quad 1 \quad 2 \quad 3$$

🕐 Key Points

● In this chapter the symbols $<$, $\leqslant$, $>$ and $\geqslant$ are used.

  ◎ $<$ means 'less than'.

  ◎ $\leqslant$ means 'less than or equal to'.

  ◎ $>$ means 'greater than'.

  ◎ $\geqslant$ means 'greater than or equal to'.

Note: If you think of the sign '$>$' as an arrow, it always 'points' to the smaller value.

$2 < 7$ $(2 \leftarrow 7)$ and $7 > 2$ $(7 \rightarrow 2)$

● **Inequalities** have more than one value in their solution set. The solution set can be shown on a **number line**.

● When showing inequalities on a number line, an open circle shows the number is not included and a closed circle shows the number is included.

🔍 Example 1

Show the inequalities $x \leqslant 4$ and $x > -2$ on a number line.

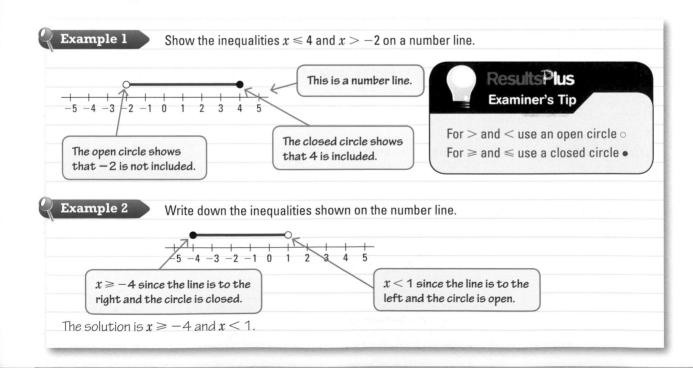

This is a number line.

The open circle shows that -2 is not included.

The closed circle shows that 4 is included.

Results Plus
Examiner's Tip

For $>$ and $<$ use an open circle ○
For $\geqslant$ and $\leqslant$ use a closed circle ●

🔍 Example 2

Write down the inequalities shown on the number line.

$x \geqslant -4$ since the line is to the right and the circle is closed.

$x < 1$ since the line is to the left and the circle is open.

The solution is $x \geqslant -4$ and $x < 1$.

C

⚙ **Exercise 5A**

Questions in this chapter are targeted at the grades indicated.

1 Show these inequalities on a number line.
  a $x \leqslant 3$ **b** $x > 0$ **c** $x < 3$ and $x > -4$
  d $x \leqslant 1$ and $x \geqslant -3$ **e** $x < 0$ and $x \geqslant -5$

2 Write down the inequalities shown on these number lines.

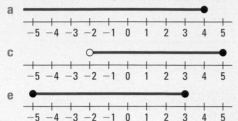

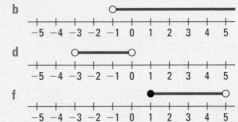

5.2 Solving simple linear inequalities in one variable

◉ **Objective**

◦ You can solve simple linear inequalities in one variable, and represent the solution set on a number line.

❓ **Why do this?**

All lifts have a maximum weight capacity. If there are 8 people in a lift, you can solve a simple inequality to find the maximum average weight that each person can be.

◈ **Get Ready**

Solve these equations.
1. $x + 10 = 4(x - 5)$ **2.** $2x + 1 = 6x - 1$

Show these on a number line.
3. $x \leqslant 7$ **4.** $x > 10$

🔍 **Key Points**

◉ You solve a linear inequality using a similar method to the one you use for solving a linear equation (see Section 4.5).

◉ If you multiply both sides of an inequality by a negative number, then you must reverse the inequality sign.

🔍 **Example 3**   Solve $3(x + 2) > 5 - x$ and show your answer on a number line.

$3(x + 2) > 5 - x$ ⟵ [Expand the brackets.]
$3x + 6 > 5 - x$

$3x + 6 + x > 5$ ⟵ [Add x to both sides.]
$4x + 6 > 5$

$4x > 5 - 6$ ⟵ [Subtract 6 from both sides.]
$4x > -1$

$x > -0.25$ ⟵ [Divide both sides by 4.]

Example 4   Solve $-3x \leqslant 12$.

$-3x \leqslant 12$

$-12 \leqslant 3x$ ← | Add $3x$ to both sides and subtract 12 from both sides.
$3x \geqslant -12$

$x \geqslant -4$ ← | Divide both sides by 3.

ResultsPlus
Examiner's Tip

Check your answer by putting a value that satisfies your answer into the inequality in the question.

Exercise 5B

C

1   Solve these inequalities and show each answer on a number line.

    a $x + 1 > 5$ **b** $x - 3 \leqslant -2$ **c** $2x + 5 \leqslant 1$ **d** $10x - 7 > 9$

B

2   Solve these inequalities.

    a $3x < x + 9$ **b** $5x - 3 > 2x + 9$ **c** $2(x + 3) \leqslant 11$ **d** $5x - 7 > 3(x + 2)$

A*

3   Solve these inequalities.

    a $x + 3 \geqslant 5(x - 2)$ **b** $3(x + 1) < 4(x - 5)$

    c $\dfrac{2 - 3x}{5} \leqslant 1 + \dfrac{x}{2}$ **d** $\dfrac{5x - 3}{4} + 1 \geqslant \dfrac{1 - 2x}{6}$

5.3   Finding integer solutions to inequalities in one variable

◉ Objective

● You can find integer solutions to inequalities in one variable.

⦾ Why do this?

You could find out the grade boundary to get a grade A in a test, and then find each individual exam mark that you could get that would give you an A.

⬆ Get Ready

Solve these inequalities.

1. $5x - 1 \geqslant 19$ **2.** $3x + 4 > 16$ **3.** $2x + 5 \leqslant 9$ **4.** $10x - 7 > 23$

◔ Key Point

◉ When a value has a **lower limit** and an **upper limit**, you need to seperate the two inequalities, and solve them separately.

Example 5 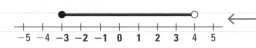 $-3 \leqslant n < 4$

n is an integer. Find all the possible values of n.

> Show the inequality $-3 \leqslant n < 4$ on a number line.

$n = -3, -2, -1, 0, 1, 2, 3.$

> Write down the integers from the number line.

ResultsPlus
Watch Out!

Remember that 0 is an integer.

Example 6 $-3 \leqslant 2p - 1 < 8$

p is an integer. Find all the possible values of p.

$-3 \leqslant 2p - 1$ and $2p - 1 < 8$

> Write the two inequalities seperately.

$\begin{aligned} 1 - 3 &\leqslant 2p & 2p &< 8 + 1 \\ -2 &\leqslant 2p & 2p &< 9 \\ -1 &\leqslant p & p &< 4.5 \end{aligned}$

> Solve each inequality.

So $-1 \leqslant p < 4.5$
$p = -1, 0, 1, 2, 3, 4.$

> Write down the integer values satisfying the inequality.

Exercise 5C

Find the possible integer values of x in these inequalities.

1.   **a** $-2 < x \leqslant 5$ **b** $-5 < x < 2$ **c** $0 < x \leqslant 3$ **d** $-5 \leqslant x \leqslant 4$

2.   **a** $-8 < 2x \leqslant 6$ **b** $-21 < 5x < 36$ **c** $-5 < 10x \leqslant 42$ **d** $-11 \leqslant 3x \leqslant 28$

3.   **a** $-5 < 2x + 1 < 9$ **b** $-7 < 3x - 2 \leqslant 11$ **c** $-12 < 4x - 7 \leqslant 10$ **d** $-9 \leqslant 2x + 5 \leqslant 13$

4.   **a** $-1 < \frac{x}{3} \leqslant 2$ **b** $-2 < \frac{2x}{5} \leqslant 3$ **c** $-1 < \frac{3x - 2}{4} \leqslant 2$ **d** $-3 < \frac{2 - x}{3} \leqslant 2$

C

B

A*

5.4 Solving graphically several linear inequalities in two variables

Objective

- You can solve graphically several linear inequalities in two variables and find the solution set.

Why do this?

If you're planning a party you could solve inequalities to work out who to invite. For example, if you wanted at least twice as many girls as boys and 60 people maximum you can express this as $G \geqslant 2B$ and $G + B \leqslant 60$.

Get Ready

Draw these lines on a coordinate grid.

1. $y = 4x + 1$ **2.** $y = 3x - 4$ **3.** $2y = 3x + 4$

Key Points

- You can show the points that satisfy an inequality on a graph.
- Lines which are boundaries for regions that do include values on the line are shown as solid lines. Lines which are boundaries for regions that do not include values on the line are shown as dotted lines.

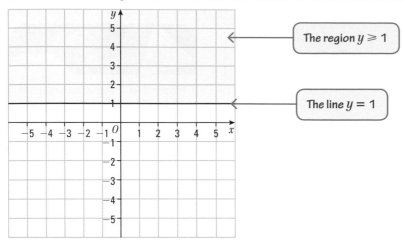

The region $y \geqslant 1$

The line $y = 1$

All points in the region $y \geqslant 1$ satisfy the inequality $y \geqslant 1$.
This includes points on the solid line.

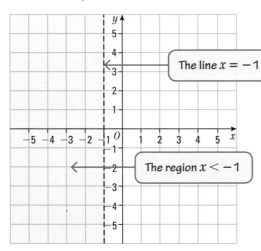

The line $x = -1$

The region $x < -1$

All points in the region $x < -1$ satisfy the inequality $x < -1$.
This does not include points on the dotted line.

Example 7   Write down the three inequalities satisfied by
the coordinates of all points in the shaded
region.
Write down the equations of each of the lines.

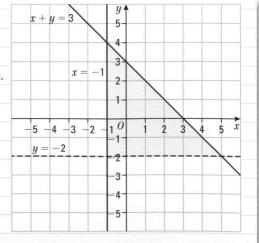

$x \geqslant -1$, since the shaded region is to the **right** of
the solid line $x = -1$.
$x + y \leqslant 3$, since the shaded region is **below** the
solid line $x + y = 3$.
$y > -2$, since the shaded region is **above** the
dotted line $y = -2$.

Example 8

a On the grid, shade the region of points whose coordinates satisfy these inequalities.
$y < 3$, $x < 2$, $y \geqslant 2x - 3$, and $x \geqslant -1$

b x and y are integers. Write down the coordinates of all points (x, y) which satisfy the inequalities in **a**.

a $y < 3$, $x < 2$, $y \geqslant 2x - 3$, and $x \geqslant -1$
Draw the dotted lines $y = 3$ and $x = 2$.
Draw the solid lines $y = 2x - 3$ and $x = -1$.
Shade the region:
$y < 3$ (points below the **dotted** line)
$x < 2$ (points to the left of the **dotted** line)
$y \geqslant 2x - 3$ (points above the **solid** line)
$x \geqslant -1$ (points to the right of the **solid** line).

b The points are:
$(-1, 2), (0, 2), (1, 2), (-1, 1), (0, 1),$
$(1, 1), (-1, 0), (0, 0), (1, 0), (-1, -1),$
$(0, -1), (1, -1), (-1, -2), (0, -2),$
$(-1, -3), (0, -3), (-1, -4), (-1, -5)$

Mark the points with a cross (x).

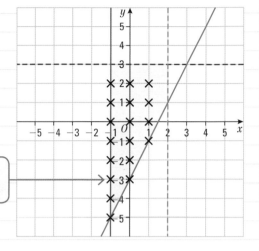

Exercise 5D

1 x and y are integers.
$-2 < x \leqslant 1$ $\qquad$ $y > -2$ $\qquad$ $y < x + 1$
On a copy of the grid, mark with a cross (x), each of the six points which satisfy all of these three inequalities.

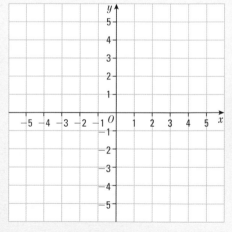

2 On a grid scaled from -6 to 6 on each axis, shade the region of points whose coordinates satisfy the following inequalities.

a $x \leqslant 4$ $\qquad$ b $y < 1$ $\qquad$ c $-2 < x \leqslant 5$ $\qquad$ d $-4 \leqslant y < 2$

3 On a grid scaled from -6 to 6 on each axis, shade the region of points whose coordinates satisfy the following inequalities.

a $-1 < x \leqslant 3$ and $-5 \leqslant y < 1$ $\qquad$ b $-3 \leqslant x \leqslant 2$ and $-1 \leqslant y < 4$

c $-2 \leqslant x \leqslant 3$, $y > -4$ and $y < x$ $\qquad$ d $x \geqslant 0$, $y + x < 4$ and $y > 3x - 2$

D

B

A

4 The diagram shows a shaded region bounded by three lines.

    a Write down the equation of each of these lines.

    b Write down the three inequalities satisfied by the coordinates of the points in this shaded region.

    c If x and y are integers, write down the coordinates of the points in this shaded region.

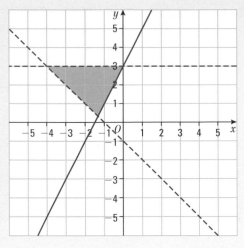

5 The diagram shows a shaded region bounded by four lines.

    a Write down the equation of each of these four lines.

    b Write down the four inequalities satisfied by the coordinates of the points in this shaded region.

    c If x and y are integers, write down the least value of y in this shaded region.

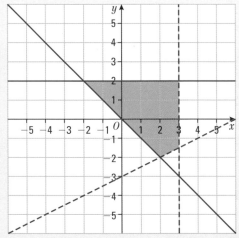

5.5 Changing the subject of a formula

◎ Objective

- You can change the subject of a simple formula.

◈ Why do this?

You can change the subject of the formula for density, which is useful if you knew the density and volume of something like a ball-bearing in a pinball machine, but still needed to know the mass.

◈ Get Ready

1. Use the formula distance = speed × time to find how far a car travelling at an average speed of 40 mph travels in 1.5 hours.
2. Use the formula $E = mc^2$ to find E when $m = 2$ and $c = 3 \times 10^8$.

◉ Key Point

- You can use the techniques you learnt in Chapter 4 to change the **subject** of a formula by isolating the terms involving the new subject.

Example 9 Make t the subject of the following formula

$$v = u + at$$ → The aim is to isolate the term in t.

$v = u + at$
$v - u = u + at - u$ → Just like you did when solving linear equations in Chapter 13, subtract u from both sides.

$v - u = at$ → Divide both sides by a.

$\dfrac{v - u}{a} = \dfrac{at}{a}$ → t is now the subject of the formula.

$t = \dfrac{v - u}{a}$

Example 10 Make p the subject of the following formula

$$T = \frac{2(p - 1)}{q}$$

ResultsPlus
Examiner's Tip

Only apply one operation in each stage of your working.

$T = \dfrac{2(p - 1)}{q}$

$T = \dfrac{2p - 2}{q}$ → Following the rules of BIDMAS (see section 1.3), first expand the brackets.

$T \times q = \dfrac{2p - 2}{q} \times q$ → Multiply both sides by q to remove the fraction.

$Tq = 2P - 2$

$Tq + 2 = 2p - 2 + 2$ → Add 2 to both sides.

$Tq + 2 = 2p$

$p = \dfrac{Tq + 2}{2}$ → Dividing both sides by 2 gives p as the subject of the formula.

Exercise 5E

In each of the following formulae, change the subject to the letter given in brackets.

1 $y = 5x + 3$ (x) 2 $c = 5d - 2$ (d) 3 $v = u + at$ (a)

4 $P = 5xy + y$ (x) 5 $E = 4 - 3m$ (m) 6 $f = 3(g - 10)$ (g)

7 $T = \dfrac{x + 2}{7}$ (x) 8 $Y = \dfrac{3(n - m)}{2}$ (n)

9 $W = \dfrac{2y}{3}(1 + p)$ (p) 10 $A = 1 + \dfrac{1 - w}{3}$ (w)

C

B

5.6 Changing the subject in complex formulae

◎ Objective

● You can change the subject of a formula where the subject appears twice, or where a power of the subject appears.

⑦ Why do this?

Physicists and engineers rearrange many complex formulae in order to find important measures.

⬆ Get Ready

1. Make u the subject of the formula $v = u + at$.
2. Make R the subject of the formula $V = IR$.
3. Make m the subject of the formula $E = mc^2$.

Example 11   Make a the subject of the following formula.

$$T = 2\pi\sqrt{a^2 - 4}$$

$T = 2\pi\sqrt{a^2 - 4}$ ←   Divide both sides by 2π.

$$\frac{T}{2\pi} = \frac{2\pi\sqrt{a^2 - 4}}{2\pi}$$

$$\frac{T}{2\pi} = \sqrt{a^2 - 4}$$

$\dfrac{T^2}{2^2 \times \pi^2} = a^2 - 4$ ←   Square both sides using the result that $\sqrt{x} \times \sqrt{x} = x$.

$\dfrac{T^2}{4\pi^2} + 4 = a^2$ ←   Add 4 to both sides.

$a = \sqrt{\left(\dfrac{T^2}{4\pi^2} + 4\right)}$ ←   Finally take the square root to give a as the subject of the formula.

Example 12   Make x the subject of the formula.

$$P = qx + 2x + 2a$$

Notice that x appears twice in this formula.

ResultsPlus

Examiner's Tip

Collect all terms in the 'new' subject together and then factorise.

$P = qx + 2x + 2a$

$P - 2a = qx + 2x$ ←   Subtract $2a$ from both sides.

$P - 2a = x(q + 2)$ ←   Factorising the right-hand side leaves x appearing once only.

$x = \dfrac{P - 2a}{q + 2}$ ←   Divide both sides by $(q + 2)$.

Exercise 5F

In each of the following formulae, change the subject to the letter given in brackets.

1 $A = \pi R^2$ (R)

2 $P = \sqrt{x - y}$ (x)

3 $v^2 = u^2 + 2as$ (u)

4 $T = \sqrt{\dfrac{2s}{g}}$ (g)

5 $y = 5\sqrt{3 - 2x^2}$ (x)

6 $f = 3g - 4 + g$ (g)

7 $T = \dfrac{1 + 3m}{m}$ (m)

8 Rearrange $4(p + 3) = q(1 - p)$ to make p the subject.

9 Make c the subject of $a - bc = 3 + 7c$.

10 Make T the subject of the formula $W = \sqrt{\dfrac{3T + 7}{2T}}$

Chapter review

- $<$ means 'less than'.
- $\leqslant$ means 'less than or equal to'.
- $>$ means 'greater than'.
- $\geqslant$ means 'greater than or equal to'.
- **Inequalities** have more than one value in their solution set. The solution set can be shown on **a number line**.
- When showing inequalities on a number line, an open circle shows that the number is not included and a closed circle shows that the number is included.
- You solve a linear inequality using a similar method to the one you use for solving a linear equation.
- If you multiply both sides of an inequality by a negative number, then you must reverse the inequality sign.
- When a value has a **lower limit** and an **upper limit**, you need to split the two inequalities and solve them separately.
- You can show the points that satisfy an inequality on a graph.
- Lines which are boundaries for regions that do include values on the line are shown as solid lines. Lines which are boundaries for regions that do not include values on the line are shown as dotted lines.
- You can change the **subject** of a formula by isolating the terms involving the new subject.

Review exercise

1 $S = 4p + 3q$

  a $p = 5, q = -4$

    Work out the value of S.

  b Make p the subject of the formula $S = 4p + 3q$.

2 Write down the inequalities represented on the number lines.

  a
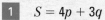

  b
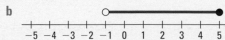

  c
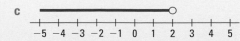

C

3 Draw a separate number line from −5 to 5 for each part. Show the inequality given in each case.

   a $-3 < x < 2$ **b** $x \geqslant 1$ **c** $x < -2$ **d** $-1 \leqslant x \leqslant 3$

4 $-3 \leqslant n < 2$

   n is an integer.

   Write down all the possible values of n.   *Nov 2006*

5 **a** Solve the inequality

      $3t + 1 < t + 12$

   b t is a whole number.

      Write down the largest value of t that

      satisfies

      $3t + 1 < t + 12$

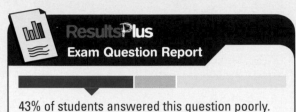

ResultsPlus
Exam Question Report

43% of students answered this question poorly. Remember to keep the inequality sign in throughout your working.

May 2009

6 $-6 \leqslant 2y < 5$ y is an integer. Write down all the possible values of y.   *Nov 2005*

B

7 Solve the inequality $4p - 8 < 7 - p$ *June 2006*

A02

8

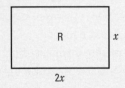

The perimeter of the rectangle R is less than the perimeter of the square S.

Write down the range of values of x.

A02

9

Diagram NOT accurately drawn

Here are 3 rods.

The length of rod A is x cm.

Rod B is 4 cm longer than rod A.

The length of rod C is twice the length of rod B.

The total length of all 3 rods is L cm

   a Show that $L = 4x + 12$

The total length of all 3 rods must be less than 50 cm.

   b Write down the inequality that must be satisfied.

   c Work out the range of possible values of x.

10 The region **R** satisfies the inequalities

$x \geqslant 2, y \geqslant -1, \ x + y \leqslant 6$

Draw a suitable graph and use shading to show the region **R**.

11 $\dfrac{1}{u} + \dfrac{1}{v} = \dfrac{1}{f}$ $\qquad u = 2\frac{1}{2}, v = 3\frac{1}{3}$

   a Find the value of f.

   b Rearrange $\dfrac{1}{u} + \dfrac{1}{v} = \dfrac{1}{f}$ to make u the subject of the formula.

     Give your answer in its simplest form.

ResultsPlus
Exam Question Report

93% of students answered this question poorly because they were not accurate in their calculations.

May 2009

12 Make x the subject of $5(x - 3) = y(4 - 3x)$.   *Nov 2005*

13 $P = \pi r + 2r + 2a$ $\qquad P = 84, r = 6.7$

   a Work out the value of a. Give your answer correct to 3 significant figures.

   b Make r the subject of the formula $P = \pi r + 2r + 2a$.   *June 2005*

14 **a** $4x + 3y < 12$

     x and y are both integers.
     Write down two possible pairs of values
     that satisfy this inequality.

       $4x + 3y < 12, \ y < 3x, \ y > 0, \ x > 0$

   b On the grid mark with a cross ($\times$) each of
     the three points which satisfy all these four inequalities.

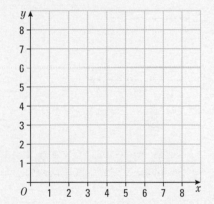

15 The cost of sweets is £2 per kg. The cost of chocolate is £5 per kg.

Jim buys x kg of sweets and y kg of chocolate.

He buys at least 2 kg of sweets.

He buys at least 3 kg of chocolate.

He spends at most £20.

   a Write down 3 inequalities in x and/or y.

   b Draw a suitable graph and show, by shading, the region that satisfies all 3 inequalities.

6 MORE GRAPHS AND EQUATIONS

Different-shaped curves are seen in many areas of mathematics, science and engineering. Galileo showed that if an object is thrown it traces out a type of curve called a parabola. This bridge is the Butterfly Bridge in Bedford and it has two parabolic arches.

⊙ Objectives

In this chapter you will:
- recognise and draw graphs of quadratic, cubic, reciprocal and exponential functions
- use graphs to solve quadratic and cubic equations
- find approximate solutions of equations by using a trial and improvement method.

◁ Before you start

You should be able to:
- draw straight-line graphs
- work out the value of a given expression by substituting values.

6.1 Graphs of quadratic functions

◎ Objectives

- You can recognise and draw graphs of quadratic functions.
- You can use graphs to solve quadratic equations.

◈ Why do this?

You can use quadratic functions to represent the path of projectiles, such as the trajectory of a cannonball or of a drop-goal in rugby.

◈ Get Ready

1. Draw the graph of $y = 2x + 3$ for values of x from -3 to $+3$.
2. Work out the value of $x^2 - 5$ when   **a** $x = 1$ **b** $x = -3$
3. Work out the value of $2x^2$ when   **a** $x = 1$ **b** $x = -3$

◉ Key Points

- A **quadratic function** (or expression) is one in which the highest power of x is x^2.
- All quadratic functions can be written in the form $ax^2 + bx + c$ where a, b and c represent numbers. Examples of quadratic functions include $x^2 + 1$, $x^2 - 2x + 3$, $3x^2 + x - 2$, and $3 - x^2$.
- The graph of a quadratic function is called a **parabola**. It has a smooth $\smile$ or $\frown$ shape according to whether $a > 0$ or $a < 0$.
- The lowest point of a quadratic graph is where the graph turns, and is called the **minimum point**.
- The highest point of a quadratic graph is where the graph turns, and is called the **maximum point**.
- All quadratic graphs have a line of symmetry.
- You can solve **quadratic equations** of the form $ax^2 + bx + c = 0$ by reading off the x-coordinate where the graph $y = ax^2 + bx + c$ crosses the x-axis.
- You can solve quadratic equations of the form $ax^2 + bx + c = mx + k$ by reading off the x-coordinate at the point of intersection of the graph $y = ax^2 + bx + c$ with the straight-line graph $y = mx + k$.

◎ Example 1

Draw the graph of $y = x^2 + 1$ taking values of x from -3 to $+3$.

When $x = 3$, $y = 3 \times 3 + 1 = 10$ ←   *Work out the value of y for each value of x.*
When $x = 2$, $y = 2 \times 2 + 1 = 5$
When $x = 1$, $y = 1 \times 1 + 1 = 2$
When $x = 0$, $y = 0 \times 0 + 1 = 1$
When $x = -1$, $y = (-1) \times (-1) + 1 = 2$ ←   *Put negative values of x in brackets when substituting them.*
When $x = -2$, $y = (-2) \times (-2) + 1 = 5$
When $x = -3$, $y = (-3) \times (-3) + 1 = 10$

x	-3	-2	-1	0	1	2	3
y	10	5	2	1	2	5	10

These results can be shown in a table of values.

quadratic function   parabola   minimum point   maximum point   quadratic equations   **55**

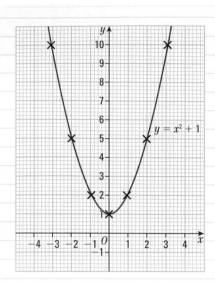

Be careful to plot the points accurately.
Use a sharp pencil cross for each point.
Join your points with a smooth curve.

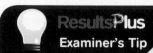

ResultsPlus
Examiner's Tip

Make sure your curve passes through all the points.

Example 2

a Complete the **table of values** for $y = x^2 - 2x - 4$.

x	-2	-1	0	1	2	3	4
y		-1	-4			-1	

b Draw the graph of $y = x^2 - 2x - 4$ for $x = -2$ to $x = 4$.

c Write down the equation of the line of symmetry of this curve.

d Write down the values of x where the graph crosses the x-axis.

a When $x = 4, y = 4^2 - 2 \times 4 - 4$ $\qquad = 16 - 8 - 4 = 4$
When $x = 2, y = 2^2 - 2 \times 2 - 4$ $\qquad = 4 - 4 - 4 = -4$
When $x = 1, y = 1^2 - 2 \times 1 - 4$ $\qquad = 1 - 2 - 4 = -5$
When $x = -2, y = (-2)^2 - 2 \times (-2) - 4 = 4 + 4 - 4 = 4$

Work out the value of y for each value of x in turn.

x	-2	-1	0	1	2	3	4
y	4	-1	-4	-5	-4	-1	4

b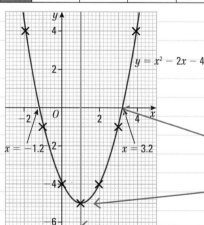

Look at the values of y in the table to determine the extent of the y-axis. Use values of -6 to $+5$.

Draw a smooth symmetrical curve through all the plotted points.

Read off the values where the curve crosses the x-axis.

This is the minimum point.

The curve has a line of symmetry.

c The line of symmetry has equation $x = 1$.

d $x = -1.2$ and $x = 3.2$

✲ Exercise 6A

Questions in this chapter are targeted at the grades indicated.

B

1 Here is the table of values for $y = x^2 - 3$.

x	-3	-2	-1	0	1	2	3
y	6		-2	-3			6

  a Copy and complete the table of values.
  b Draw the graph of $y = x^2 - 3$ for $x = -3$ to $x = 3$.
  c Write down the equation of the line of symmetry of your graph.
  d Write down the coordinates of the minimum point.

2 **a** Copy and complete the table of values for $y = 4 - x^2$.

x	-3	-2	-1	0	1	2	3
y		0	3	4			-5

  b Draw the graph of $y = 4 - x^2$ for $x = -3$ to $x = 3$.
  c Write down the coordinates of the maximum point.
  d Write down the values of x where the graph crosses the x-axis.

3 **a** Copy and complete the table of values for $y = 2x^2 + 2$.

x	-3	-2	-1	0	1	2	3
y	20		4	2			20

  b Draw the graph of $y = 2x^2 + 2$ for $x = -3$ to $x = 3$.
  c Use your graph to find:
    i the value of y when $x = 1.5$
    ii the two values of x when $y = 11$.

4 Draw the graph for each of the following equations:
  a $y = x^2 - 4x - 1$ for values of x from -2 to 6
  b $y = 2x^2 - 4x - 3$ for values of x from -2 to 4
  c $y = (x + 2)^2$ for values of x from -6 to 2
  d $y = 5 + 3x - 2x^2$ for values of x from -2 to 4.

For each case use your graphs to:
    i write down the values of x when the graph crosses the x-axis
    ii draw in and write down the equation of the line of symmetry.

Example 3

Here is the graph of $y = x^2 - 2x - 2$.

a Use the graph to solve the equation
$x^2 - 2x - 2 = 0$.
Give your answers correct to 1 decimal place.

b Use the graph to solve the equation
$x^2 - 2x - 5 = 0$.
Give your answers correct to 1 decimal place.

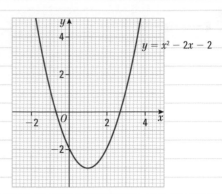

a

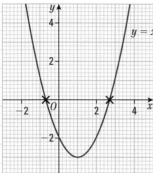

$x^2 - 2x - 2 = 0$ ← Find where the graph crosses the x-axis – that is where $y = 0$.

$x = -0.7$ and ← Read off the values.
$x = 2.7$

b

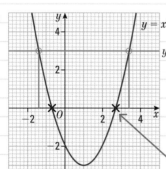

$x^2 - 2x - 5 = 0$ Rearrange the equation so that one side is $x^2 - 2x - 2$.

$x^2 - 2x - 2 = 3$ ← Add 3 to each side of the equation. Find where $x^2 - 2x - 2$ intersects $y = 3$.

$x = -1.4$ and
$x = 3.4$

Read off the x values.

Example 4

The diagram shows the graph of $y = x^2$.

a By drawing a suitable straight line
on this graph, solve the equation
$x^2 = 2x + 3$.

b By drawing another suitable straight
line on this graph, solve the equation
$x^2 + x - 3 = 0$.

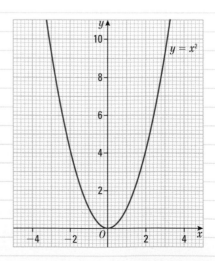

a

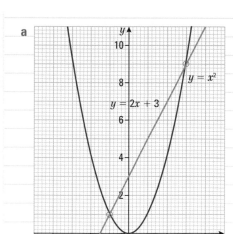

$x^2 = 2x + 3$ ← Draw the graph of $y = 2x + 3$.

$x = -1$ and $x = 3$ ← Read off the x values at the points of intersection.

Results Plus
Examiner's Tip

If the equation in the question is in terms of x remember to give only the value of the x-coordinate.

b

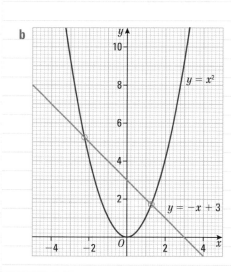

$x^2 + x - 3 = 0$ ← Rearrange the equation so that one side is x^2.

$x^2 = -x + 3$ ← Draw the line $y = -x + 3$.

$x = -2.3$ and $x = 1.3$ ← Read off the x values at the points of intersection.

Exercise 6B

1 Here are four graphs.

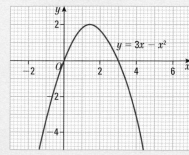

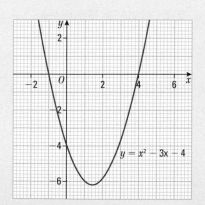

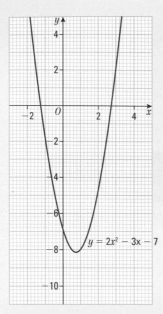

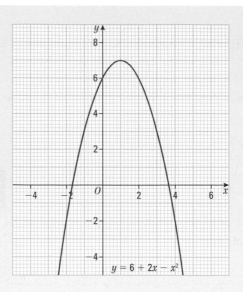

Use these graphs to solve the equations

a $3x - x^2 = 0$

b $x^2 - 3x - 4 = 0$

c $2x^2 - 3x - 7 = 0$

d $6 + 2x - x^2 = 0$

2 Use the graphs in question 1 to solve the equations

a $3x - x^2 = 1$

b $x^2 - 3x - 4 = -5$

c $2x^2 - 3x - 10 = 0$

d $4 + 2x - x^2 = 0$

3 Here is a table of values for $y = 1 + 2x - x^2$.

x	-2	-1	0	1	2	3	4
y		-2		2	1		-7

a Copy and complete the table.

b Draw the graph of $y = 1 + 2x - x^2$.

c By drawing a suitable line on your graph, solve the equation $2 + 4x - 2x^2 = 2 - 2x$.

4 a Make a table of values for $y = 3x^2 - x + 2$, taking values of x from -3 to $+3$.

b Draw the graph of $y = 3x^2 - x + 2$.

c By drawing a suitable line on your graph, solve the equation $3x^2 - 3x - 2 = 0$.

6.2 Graphs of cubic functions

⊙ Objectives

● You can recognise and draw graphs of cubic functions.

● You can use graphs to solve cubic equations.

⊗ Why do this?

Engineers use cubic models, for example, when testing the strength of rubber in car tyres.

⊕ Get Ready

1. Write down the first five cube numbers.

2. Work out the value of x^3 when

a $x = 100$ b $x = -10$

Key Points

- A **cubic function** (or expression) is one in which the highest power of x is x^3.
- All cubic functions can be written in the form $ax^3 + bx^2 + cx + d$ where a, b, c and d represent numbers. Examples of cubic functions include $4 - x^3$ and $x^3 - 2x^2 + 3$.
- The graph of a cubic function has one of the following shapes.

for $a > 0$ for $a < 0$

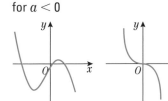

- To draw the graph of a cubic function, make a table of values, then plot the points from your table and join them with a smooth curve.

 Example 5

a Draw the graph of $y = x^3 - 4x^2 + 5$ for $-2 \leqslant x \leqslant 4$.

b Use your graph to solve the equation $x^3 - 4x^2 - x + 4 = 0$.

a When $x = 4$, $y = 4^3 - 4 \times 4^2 + 5$ $= 5$ ← | Work out y for each value of x. |

  When $x = 3$, $y = 3^3 - 4 \times 3^2 + 5$ $= -4$

  When $x = 2$, $y = 2^3 - 4 \times 2^2 + 5$ $= -3$

  When $x = 1$, $y = 1^3 - 4 \times 1^2 + 5$ $= 2$

  When $x = 0$, $y = 0^3 - 4 \times 0^2 + 5$ $= 5$

  When $x = -1$, $y = (-1)^3 - 4 \times (-1)^2 + 5 = 0$

  When $x = -2$, $y = (-2)^3 - 4 \times (-2)^2 + 5 = -19$ | Make a table of values. |

x	-2	-1	0	1	2	3	4
y	-19	0	5	2	-3	-4	5

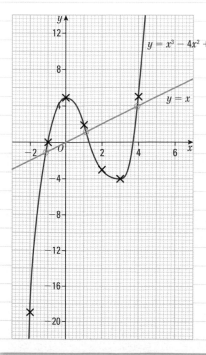

$y = x^3 - 4x^2 + 5$

| Plot the points and draw a smooth curve through all the points. |

b $x^3 - 4x^2 - x + 5 = 0$ ← | Compare the expression to $y = x^3 - 4x^2 + 5$. Add x to each side of the equation. |

   $x^3 - 4x^2 + 5 = x$

ResultsPlus
Examiner's Tip

Take care drawing the graph of $y = x$ if there are different scales on the axes.

The solutions are $x = -1.1$,   ← | Draw $y = x$ on the graph and find where it intersects with $y = x^3 - 4x^2 + 5$. |

$x = 1.1$ and $x = 3.8$.

Exercise 6C

1 **a** Copy and complete the table of values for $y = x^3 + 2$.

x	-3	-2	-1	0	1	2	3
y							

b Draw the graph of $y = x^3 + 2$ for $-3 \leqslant x \leqslant 3$.

c Use your graph to find the value of y when $x = 2.5$.

2 Here is a table of values for $y = x^3 - 9x$.

x	-4	-3	-2	-1	0	1	2	3	4
y	-28	0		8	0		-10		28

a Copy and complete the table.

b Draw the graph of $y = x^3 - 9x$ for $-4 \leqslant x \leqslant 4$.

c Use your graph to find the solutions to the equation $x^3 - 9x = 0$.

3 **a** Copy and complete the table of values for $y = 12x + 3x^2 - 2x^3$.

x	-3	-2	-1	0	1	2	3	4
y	$+45$		-7		$+13$	$+20$		-32

b Draw the graph of $y = 12x + 3x^2 - 2x^3$ for $-3 \leqslant x \leqslant 4$.

c By drawing a suitable line on your diagram, solve the equation $12x + 3x^2 - 2x^3 = 2x - 1$.

4 Here are four graphs.

A   **B**

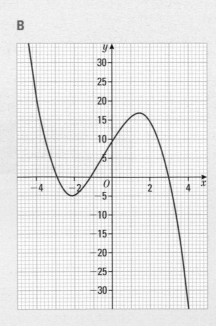

A

C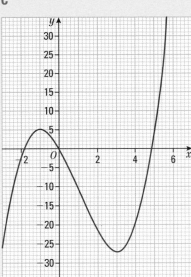

D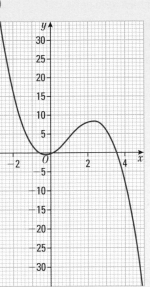

Here are four equations

  i $y = x^3 - 3x^2 - 9x$

  iii $y = 2x + 3x^2 - x^3$

  ii $y = x^3 - x^2 - 8x + 12$

  iv $y = 9 + 9x - x^2 - x^3$

Match each equation to one of the graphs.

Give reasons for your answers.

6.3 Graphs of reciprocal functions

⊙ Objective

⊙ You can recognise and draw graphs of reciprocal functions.

⑦ Why do this?

You might see a reciprocal graph if you're doing an experiment on volume and pressure. If you compress gas in a container, the volume will decrease but the pressure will increase, and vice-versa.

⊕ Get Ready

1. Work out the value of $\frac{1}{x}$ when

   a $x = 4$ **b** $x = \frac{1}{4}$ **c** $x = 2.5$ **d** $x = 0.4$

2. Explain what happens to the value of $\frac{1}{x}$ as x gets bigger.

Key Points

◉ The reciprocal of x is $\frac{1}{x} = 1 \div x$.

◉ Expressions of the form $\frac{k}{x}$, where k is a number are called reciprocal functions.

◉ The reciprocal of 0 is not defined since division by 0 is not possible.

◉ This means that the graph of $y = \frac{1}{x}$ does not have a point on the y-axis where $x = 0$.

⦿ The graphs of reciprocal functions have similar shapes.

⦿ They are **discontinuous** and have two parts.

⦿ They do not cross or touch the x-axis or the y-axis, but get nearer and nearer to them. We say that the axes are asymptotes to the graphs.

⦿ Here are the general shapes of reciprocal functions of the form $y = \dfrac{k}{x}$:

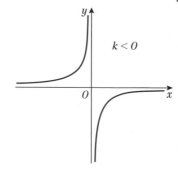

Example 6

a Draw the graph of $y = \dfrac{1}{x}$ where $x \neq 0$.

b Write down the equations of any lines of symmetry of the graph.

a

x	-4	-3	-2	-1	$-\frac{1}{2}$	$-\frac{1}{4}$	$\frac{1}{4}$	$\frac{1}{2}$	1	2	3	4
y	$-\frac{1}{4}$	$-\frac{1}{3}$	$-\frac{1}{2}$	-1	-2	-4	4	2	1	$\frac{1}{2}$	$\frac{1}{3}$	$\frac{1}{4}$

Complete a table of values.

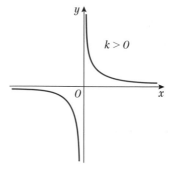

Plot the points and join the two parts with smooth curves.

b

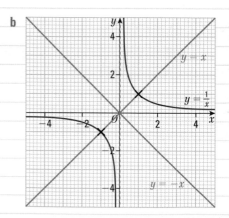

The equations of the lines of symmetry of the graph are $y = x$ and $y = -x$.

discontinuous

Exercise 6D

B A03

1 **a** Copy and complete the table of values for $y = \dfrac{5}{x}$ for $0 < x \leqslant 20$.

x	0.2	0.4	0.5	1	2	4	5	10	20
y	25	12.5		5	2.5			0.5	

b Using your answer to part **a**, copy and complete the following table of values for $y = \dfrac{5}{x}$ for $-20 \leqslant x < 0$.

x	-20	-10	-5	-4	-2	-1	-0.5	-0.4	-0.2
y									

c Draw the graph of $y = \dfrac{5}{x}$ for $-20 \leqslant x \leqslant 20$.

2 Draw the graph of $y = -\dfrac{2}{x}$ for $-10 \leqslant x \leqslant 10$.

3 **a** Draw the graph of $y = \dfrac{12}{x + 1}$ for $-5 \leqslant x \leqslant 3$.

A A03 A*

b Write down the value of x for which $y = \dfrac{12}{x + 1}$ is not defined.

6.4 Graphs of exponential functions

◎ Objective

● You can recognise and draw graphs of exponential functions.

◈ Why do this?

Scientists work out how quickly radioactive materials will break down using a graph of their radioactive half-life, which is an exponential graph.

◈ Get Ready

1. Work out the values of
   a 3^4 **b** 3^0 **c** 3^{-2}
2. Find the value of x in each of these equations.
   a $2^x = 16$ **b** $5^x = 25$ **c** $10^x = 1000$

🔍 Key Points

● Expressions of the form a^x, where a is a positive number are called **exponential functions**. Examples are 2^x, 10^x, $\left(\frac{1}{2}\right)^x$ and $(1.05)^x$.
● The graphs of exponential functions have similar shapes.
● They are continuous and always lie above the x-axis.
● They increase very quickly at one end and get nearer and nearer to the x-axis at the other end.
● They cross the y-axis at (0, 1) since $a^0 = 1$ for all values of a.
● Here are the general shapes of exponential functions of the form $y = a^x$ and $y = a^{-x}$.

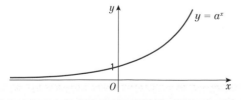
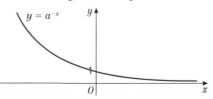

Example 7

a Draw the graph of $y = 2^x$, for values of x from -3 to $+3$.

b Use your graph to find an estimate for the solution of the equation $2^x = 6$.

a When $x = 0$, $y = 2^0 \;= 1$ ← | Substitute $x = 0$ into $y = 2^x$ and work out the value. |

When $x = 1$, $y = 2^1 \;= 2$

When $x = 2$, $y = 2^2 \;= 4$ ← | Repeat the process for other integer values of x. |

When $x = 3$, $y = 2^3 \;= 8$

When $x = -1$, $y = 2^{-1} = \frac{1}{2^1} = 0.5$

When $x = -2$, $y = 2^{-2} = \frac{1}{2^2} = 0.25$ ← | Use the result $a^{-n} = \frac{1}{a^n}$ to work out the values for negative values of x (see Section 25.1). |

When $x = -3$, $y = 2^{-3} = \frac{1}{2^3} = 0.125$

| Complete a table of values. |

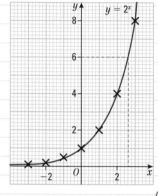

x	-3	-2	-1	0	1	2	3
y	0.125	0.25	0.5	1	2	4	8

| Draw the graph from the table of values. |

ResultsPlus
Examiner's Tip

Make sure your curve gets nearer and nearer to the x-axis without touching it.

b $x = 2.6$ ← | Use your graph to find the value of x when $y = 6$. |

Example 8

The sketch shows part of the graph of $y = pq^x$.

The points with coordinates $(0, 5)$ and $(2, 45)$ lie on the graph.

a Work out the value of p and of q.

b Find the value of y when $x = 3$.

a $y = pq^x$

The point $(0, 5)$ lies on the graph so

$5 = p \times q^0$ ← | Substitute $x = 0, y = 5$ into $y = pq^x$. Use the result $q^0 = 1$. |

$5 = p \times 1$

$p = 5$

So the equation of the curve is $y = 5q^x$.

The point $(2, 45)$ also lies on the graph so ← | Substitute $x = 2, y = 45$. Solve the equation. Work out the positive value of q. |

$45 = 5 \times q^2$

$9 = q^2$

$q = 3$

$p = 5, q = 3$

b $y = pq^x$ so $y = 5 \times 3^x$ ← | Substitute $p = 5, q = 3$ into $y = pq^x$. |

When $x = 3$ $y = 5 \times 3^3$

         $= 5 \times 27$ ← | Put $x = 3$ into $y = 5 \times 3^x$. |

         $= 135$

     $y = 135$

Exercise 6E

1 **a** Copy and complete the table of values for $y = 3^x$. Give the values correct to 2 decimal places.

x	-3	-2	-1	0	1	2	3
y	0.04		0.33		3		27

b Draw the graph of $y = 3^x$ for $-3 \leqslant x \leqslant 3$.

c Use your graph to find an estimate for:
   i the value of y when $x = 1.5$
   ii the value of x when $y = 15$.

2 The diagram shows the graphs of $y = 2^x$, $y = 5^x$, $y = \left(\frac{1}{2}\right)^x$ and $y = 3^{-x}$.

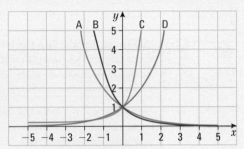

Match each graph to its equation.

3 The number of bacteria, n, after time t minutes is modelled by the equation $n = 10 \times 2^t$.
   a Work out the number of bacteria initially (when $t = 0$).
   b Work out the number of bacteria after 5 minutes.
   c Find the time taken for the number of bacteria to increase to one million. Give your answer to the nearest minute.

4 The points with coordinates (1, 10) and (3, 2560) lie on the graph with equation $y = pq^x$ where p and q are constants. Work out the values of p and q.

6.5 Solving equations by the trial and improvement method

◎ Objective

◉ You can use a systematic method to solve an equation to any degree of accuracy.

❓ Why do this?

Computers can be programmed to solve complex equations using the trial and improvement method.

◈ Get Ready

1. Write these numbers correct to 1 decimal place (1 d.p.).
   a 4.613   **b** 2.157   **c** 1.498
2. Show that $x = 1$ is a solution of the equation $x + \frac{1}{x} = 2$.
3. Work out the value of $x^3 - 3x$ when   **a** $x = 1$ **b** $x = 2$

Key Points

- A **trial and improvement** method is a systematic way of finding solutions of equations to any degree of accuracy.
- A first **approximation** is found for the solution then the **method of trial and improvement** is used to obtain a more accurate answer. The process can be repeated in order to get closer to the correct value.
- The root(s) of an equation are found between the two points where the value of the equation changes style.

Example 9

a Show that the equation $x^3 - 2x = 15$ has a solution between 2 and 3.

b Use a trial and improvement method to find this solution correct to 1 decimal place.

a

x	$x^3 - 2x$	Too high or too low	Comment
2	$2^3 - 2 \times 2 = 4$	too low since 4 is less than 15	x is greater than 2
3	$3^3 - 2 \times 3 = 21$	too high since 21 is more than 15	x is between 2 and 3

> Substitute $x = 2$ into the left-hand side of the equation and compare your answer with 15.

> Substitute $x = 3$ into the left-hand side of the equation and compare your answer with 15.

So the solution is between $x = 2$ and $x = 3$ since when $x = 2$, $x^3 - 2x$ is too low and when $x = 3$, $x^3 - 2x$ is too high.

b

x	$x^3 - 2x$	Too high or too low	Comment
2.5	$2.5^3 - 2.5 \times 2$ $= 10.625$	too low	x is between 2.5 and 3
2.6	$2.6^3 - 2 \times 2.6$ $= 12.376$	too low	x is between 2.6 and 3
2.7	$2.7^3 - 2 \times 2.7$ $= 14.283$	too low	x is between 2.7 and 3
2.8	$2.8^3 - 2 \times 2.8$ $= 16.352$	too high	x is between 2.7 and 2.8
2.75	$2.75^3 - 2 \times 2.75$ $= 15.297$	too high	x is between 2.7 and 2.75

> Substitute $x = 2.5$ into the left-hand side of the equation and decide whether your answer is too high or too low. Record the interval in which it lies.

> Choose a value between 2.5 and 3 and decide on a new interval.

x lies between 2.7 and 2.75.

So the solution is $x = 2.7$ correct to 1 decimal place.

We write $x = 2.7$ (1 d.p.).

trial and improvement   approximation   method of trial and improvement

Example 10 Use a trial and improvement method to find a solution of the equation $x^2 - \dfrac{3}{x} = 1$ correct to 2 decimal places.

x	$x^2 - \dfrac{3}{x} = 1$	H or L	Comment
1	$1^2 - \dfrac{3}{1} = -2$	L	$x > 1$
2	$2^2 - \dfrac{3}{2} = 2.5$	H	$1 < x < 2$
1.5	$1.5^2 - \dfrac{3}{1.5} = 0.25$	L	$1.5 < x < 2$
1.7	$1.7^2 - \dfrac{3}{1.7} = 1.125\ldots$	H	$1.5 < x < 1.7$
1.6	$1.6^2 - \dfrac{3}{1.6} = 0.685$	L	$1.6 < x < 1.7$
1.65	$1.65^2 - \dfrac{3}{1.65} = 0.904\ldots$	L	$1.65 < x < 1.7$
1.67	$1.67^2 - \dfrac{3}{1.67} = 0.992\ldots$	L	$1.67 < x < 1.7$
1.68	$1.68^2 - \dfrac{3}{1.68} = 1.036\ldots$	H	$1.67 < x < 1.68$
1.675	$1.675^2 - \dfrac{3}{1.675} = 1.014\ldots$	H	$1.67 < x < 1.675$

Try substituting whole-number values until you find two consecutive integers between which the solution lies: in this case $x = 1$ and $x = 2$.

Substitute values until you find two consecutive numbers with a difference of 0.1 between which the solution lies: in this case $x = 1.6$ and $x = 1.7$.

Substitute values until you find two consecutive numbers with a difference of 0.01 between which the solution lies: in this case $x = 1.67$ and $x = 1.68$.

Substitute the value halfway between 1.67 and 1.68 to find out whether the solution is nearer 1.67 or nearer to 1.68.

x lies between 1.670 and 1.675
so $x = 1.67$ (2 d.p.).

Exercise 6F

1. For each of the following equations find two consecutive whole numbers between which a solution lies.

   a $x^3 + 2x = 4$ b $x^3 + x^2 = 1$ c $x + \dfrac{1}{x^2} = 5$ d $x^2 - \dfrac{10}{x} = 0$

2. Use a trial and improvement method to find one solution of these equations correct to 1 decimal place.

   a $x^3 + x = 7$ b $x^3 - x^2 + 4 = 0$ c $x + \dfrac{1}{x} = 5$

3. Use a trial and improvement method to find a positive solution of these equations correct to 2 decimal places.

   a $x^3 - x = 25$ b $2x^2 - \dfrac{1}{x} = 9$ c $x^2(x + 2) = 150$

4. A cuboid has height x cm.
   The length of the cuboid is 2 cm more than its height.
   The width of the cuboid is 2 cm less than its height.
   The volume of the cuboid is 600 cm³.
   a Show that x satisfies the equation $x^3 - 4x = 600$.
   b Use a trial and improvement method to solve the equation $x^3 - 4x = 600$ correct to 1 decimal place.
   c Write down the length, width and height of the cuboid.

Chapter review

- A **quadratic function** is an expression of the form $ax^2 + bx + c$ where the highest power of x is x^2.
- The graph of a quadratic function is called a **parabola**. It has one of the following shapes.

$y = ax^2 + bx + c, a > 0$ $\qquad$ $y = ax^2 + bx + c, a < 0$

- The graph of a quadratic function has one line of symmetry.
- The lowest point of a quadratic graph is where the graph turns, and is called the **minimum point**.
- The highest point of a quadratic graph is where the graph turns, and is called the **maximum point**.
- You can solve **quadratic equations** of the form $ax^2 + bx + c = 0$ by reading off the x-coordinate where the graph $y = ax^2 + bx + c$ crosses the x-axis.
- You can solve quadratic equations of the form $ax^2 + bx + c = mx + k$ by reading off the x-coordinate at the point of intersection of the graph $y = ax^2 + bx + c$ with the straight-line graph $y = mx + k$.
- A **cubic function** is an expression of the form $ax^3 + bx^2 + cx + d$ where the highest power of x is x^3.
- The graph of a cubic function has one of the following shapes.

$y = ax^3 + bx^2 + cx + d, a > 0$ $\qquad$ $y = ax^3 + bx^2 + cx + d, a < 0$

- A reciprocal function is an expression of the form $\dfrac{k}{x}$.
- The graph of a reciprocal function has one of the following shapes.

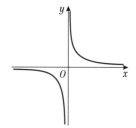 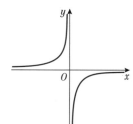

$y = \dfrac{k}{x}, k > 0$ $\qquad$ $y = \dfrac{k}{x}, k < 0$

- An **exponential function** is an expression of the form a^x or a^{-x}, where $a > 0$.
- The graph of an exponential function has one of the following shapes.

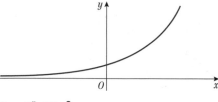 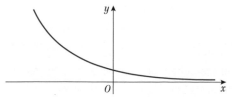

$y = a^x, a > 0$ $\qquad\qquad$ $y = a^{-x}, a > 0$

- The graphs cross the y-axis at $(0, 1)$ since $a^0 = 1$ for all values of a.
- You can find approximate solutions to equations which cannot be solved exactly by using a **trial and improvement** method.

Review exercise

1 A load, fitted with a parachute, is dropped from an aeroplane.

The parachute opens when the load is 400 m above the ground.

The distance fallen, in metres, t seconds after the parachute has opened, is given by the equation $s = 20t + 2.25t^2$.

a Copy and complete the table of values for $s = 20t + 2.25t^2$.

t	0	2	3	4	5	10
s						

b Draw the graph of $s = 20t + 2.25t^2$ for $t = 0$ to $t = 10$.

c Use your graph to find out

   i how far the load falls in the first three seconds immediately after the parachute opens

   ii when the load hits the ground.

2 **a** Copy and complete the table of values for $y = x^2 + 2x$.

x	−4	−3	−2	−1	0	1	2
y	8		0	−1			8

b Draw the graph of $y = x^2 + 2x$ for $x = -4$ to $x = 2$.

c Write down the equation for the line of symmetry of this curve.

d Use your graph to find:

   i the value of y when $x = 0.5$

   ii the values of x when $y = 6$.

3 **a** Make a table of values for $y = 2 + x - x^2$ for $-3 \leqslant x \leqslant 3$.

b Draw the graph of $y = 2 + x - x^2$ for $-3 \leqslant x \leqslant 3$.

c Solve the equations

   i $2 + x - x^2 = 0$

   ii $5 + x - x^2 = 0$

d Write down the coordinates of the maximum point of the graph of $y = 2 + x - x^2$

4 **a** Show that the equation $x^2 - 3x - 2 = x - 2$ can be rewritten as $x^2 - 4x = 0$.

b Solve the equation $x^2 - 4x = 0$.

c The equation $x^2 - 2x - 4 = 0$ can be solved by finding the intersection of the graph of $y = x^2 - 3x - 2$ with the graph of a suitable straight line. Find the equation of this straight line.

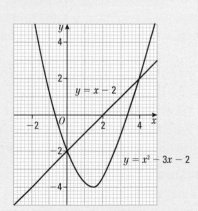

5 **a** Show that there is a solution of the equation $x^2 + \dfrac{2}{x} + 3 = 0$ between $x = -0.5$ and $x = -1$.

b Use a trial and improvement method to find this solution correct to 2 decimal places.

6 a Copy and complete the table of values for $y = x^3 - 2x^2 - 4x$.

x	−2	−1	0	1	2	3	4
y		1	0	−5		−3	

b Draw the graph of $y = x^3 - 2x^2 - 4x$ for $x = -2$ to $x = 4$.

c Solve the equations:

  i $x^3 - 2x^2 - 4x = 0$

  ii $x^3 - 2x^2 - 4x + 5 = 0$

7 a Copy and complete the table of values for $y = 3 - \dfrac{2}{x}$ $x \neq 0$

x	−3	−2	−1	−0.5	−0.1	0.1	1	2	3
y	3.7		5	7		−17		2	

b Draw the graph of $y = 3 - \dfrac{2}{x}$ for $-3 \leqslant x \leqslant 3$.

c This graph approaches two lines without touching them. These lines are called asymptotes. Write down the equation of each of these two lines.

8 a Copy and complete this table of values for $y = \dfrac{4}{x}$

x	0.5	1	2	3	4
y		4			

b Draw the graph of $y = \dfrac{4}{x}$ for values of x from 0.5 to 4

c Use the graph to find an estimate for the solution of $\dfrac{4}{x} = 6 - x$

d Use the method of trial and improvement to find this value correct to 2 decimal places.

9 The equation $h = 15t - 5t^2$ gives the height, in metres, of a ball moving through the air t seconds after it was projected from ground level by a machine.

a Draw the graph of $h = 15t - 5t^2$ for $t = 0$ to $t = 3$, using values of t every 0.5 seconds.

b Use your graph to find out how long it takes the ball to reach its maximum height.

c The ball is caught at a height of 2 m as it falls back to the ground.

How long has the ball been in the air?

10 A boy throws a ball through the air.

The equation $y = 2 + x - \dfrac{1}{80}x^2$ describes the path of the ball where x represents the horizontal distance, in metres, of the ball from the boy and y represents the height, in metres, of the ball above the ground.

a Draw a graph to show the path of the ball for values of x from 0 to 80.

b Use the graph to find

  i the maximum height reached by the ball

  ii the horizontal distances of the ball from the boy when it is at a height of 10 m above the ground

  iii the horizontal distance of the ball from the boy when it hits the ground.

11 The diagram shows a rectangle.

All the measurements are in cm.

The width is x and the length is 3 cm more than the width.

The area of the rectangle is 20 cm².

a Draw a suitable graph

b Find an estimate for the value of x

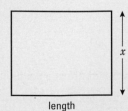

length

12 The diagram shows a cuboid.
The base of the cuboid is a square of side x cm.
The height of the cuboid is $(x + 4)$ cm.
The volume of the cuboid is 100 cm³.
Find the height of the cuboid.

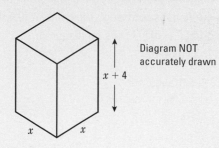

Diagram NOT accurately drawn

$x + 4$

x x

June 2005

13 Match each of the equations with its graph.

i $y = \dfrac{1}{x}$ **ii** $y = 3^x$ **iii** $y = x^2 - 4$

iv $y = x^3$ **v** $y = x^3 - x^2 - 6x$

A

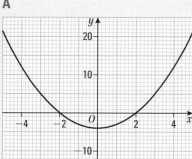

B

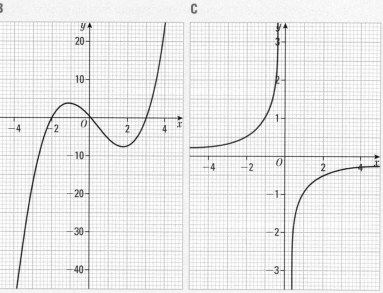

C

D

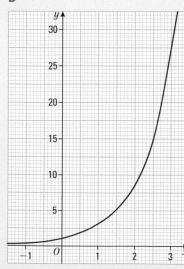

E

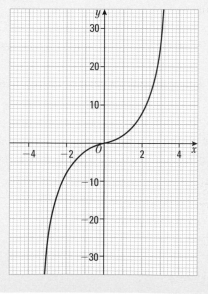

A★
A02

14 The diagram shows a sketch of the graph of $y = ab^x$

The curve passes through the points A (0.5, 1) and B (2, 8).

The point C (−0.5, k) lies on the curve.

Find the value of k.

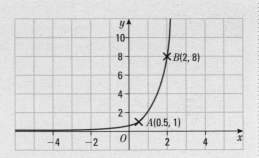

June 2006

7 QUADRATIC AND SIMULTANEOUS EQUATIONS

The graph of a quadratic function is a curve called a parabola. A javelin would follow the path of a perfect parabola if the effects of air resistance, wind and rotation didn't affect it. In the Ancient Olympics, athletes had to throw the javelin from the back of a galloping horse.

◎ Objectives

In this chapter you will:
- set up and solve a pair of simultaneous equations in two unknowns
- solve a pair of simultaneous equations using a graphical approach
- solve quadratic equations by factorisation, completing the square, and using the formula
- solve algebraic fraction equations and quadratic equations
- construct graphs of simple loci
- solve a pair of simultaneous equations in two unknowns when one equation is linear and the other is quadratic
- solve a pair of simultaneous equations in two unknowns when one equation is linear and the other is a circle.

◆ Before you start

You should already know how to:
- factorise algebraic expressions
- solve a simple linear equation
- construct linear and quadratic graphs
- substitute into a quadratic expression
- use surds.

7.1 Solving simultaneous equations

◉ Objective

● You can solve a pair of simultaneous equations in two unknowns.

◈ Why do this?

Economists use simultaneous equations when considering the changes in the price of a commodity, and the resulting change in market demand.

◈ Get Ready

1. Solve
   a $2x + 7 = 4$
   b $4y - 3 = 9 + y$
2. Work out the value of $6 - 5m$ when
   a $m = 2$
   b $m = -3$

🔍 Key Points

● When there are two unknowns you need two equations. These are called **simultaneous equations**.
● Simultaneous equations can be solved using elimination or substitution.
● To **eliminate** an unknown, multiply the equations so that the coefficients of that unknown are the same. Add or subtract the equations to eliminate the chosen unknown.
● To substitute an unknown, rearrange one of the equations to make the unknown the subject, then substitute its value (in terms of the second unknown) into the other equation.
● Once you know one unknown, you can use substitution to find the other.

Example 1 Solve the simultaneous equations
$$4x - y = 3$$
$$x + y = 7$$

Method 1

$4x - y = 3$ (1) ← Label the equations (1) and (2).
$x + y = 7$ (2)
$5x + 0 = 10$ ← Since $-y$ and $+y$ are of different sign, add equations (1) and (2) to eliminate terms in y.

$5x = 10$ so $x = 2$ ← Divide both sides by 5.

When $x = 2$, $2 + y = 7$ ← Substitute $x = 2$ into equation (2) and solve to find the value of y.
$y = 7 - 2 = 5$
So the solution is $x = 2$, $y = 5$.
Check: $4 \times 2 - 5 = 8 - 5 = 3$ ✓ ← Check your solution by substituting into equation (1).

 ResultsPlus
Examiner's Tip

When deciding which unknown to eliminate, if possible choose the unknown where the signs are different. You can then eliminate the unknown by adding the equations.

Method 2

$4x - y = 3$ (1) ← Label the equations (1) and (2).

$x + y = 7$ (2)

$y = 7 - x$ ← Rearrange equation (2) to make y the subject.

ResultsPlus
Examiner's Tip

If a fraction is introduced when making y the subject of an equation, use an alternative method since the fraction will complicate your working.

$4y - (7 - x) = 3$ ← Substitute $y = 7 - x$ into equation (1).

$4x - 7 + x = 3$

$5x - 7 = 3$ ← Expand the bracket and solve by the balance method.

$5x = 3 + 7 = 10$

$5x = 10$ so $x = 2$ ← Divide both sides by 5.

When $x = 2, 2 + y = 7$ ← Substitute $x = 2$ into equation (2) and solve to find the value of y.

$y = 7 - 2 = 5$

So the solution is $x = 2, y = 5$. ← Check your solution by substituting into equation (1).

Check: $4 \times 2 - 5 = 8 - 5 = 3$ ✓

Example 2   Solve the simultaneous equations

$$5x - 6y = 13$$
$$3x - 4y = 8$$

$5x - 6y = 13$ (1)

$3x - 4y = 8$ (2)

$15x - 18y = 39$ (3) ← Multiply (1) by 3 and (2) by 5 to make the coefficients of x equal. Label the new equations (3) and (4).

$15x - 20y = 40$ (4)

$0 + 2y = -1$ ← Subtract equation (4) from equation (3) to eliminate the terms in x. $-18y - (-20y) = -18y + 20y = +2y$

$y = -\frac{1}{2}$

$5x - (6 \times -\frac{1}{2}) = 13$ ← Substitute $y = -\frac{1}{2}$ into equation (1).

$5x - (-3) = 13$

$5x + 3 = 13$

$5x = 10$

$x = 2$

So the solution is $x = 2, y = -\frac{1}{2}$.

Check: $3 \times 2 - (4 \times -\frac{1}{2}) = 6 + 2 = 8$ ← Check your solution by substituting into equation (2).

Exercise 7A

Questions in this chapter are targeted at the grades indicated.

Solve these simultaneous equations.

1 $2x + y = 9$
$x + y = 5$

2 $3x - y = 12$
$2x + y = 13$

3 $5x - 2y = 9$
$3x - 2y = 7$

4 $x + 4y = 6$
$3x - 2y = 4$

5 $x + 2y = 9$
$y = x + 3$

6 $2x + 5y = 12$
$y = 3 - x$

7 $5x - y = -4$
$y = 2x + 1$

8 $3x - 4y = -2$
$y = x + 1$

9 $8x - 3y = -2$
$y = 3 - 2x$

10 $4x - 3y = 14$
$2x + 2y = -7$

11 $3x + 2y = 11$
$2x - 5y = 1$

12 $4x + 6y = 5$
$3x + 4y = 4$

13 $5x + 4y = 5$
$3x - 5y = -34$

14 $7x - 2y = 13$
$4x - 3y = 13$

15 $4x - 3y = 5$
$2x + 2y = -1$

7.2 Setting up equations in two unknowns

◎ Objective

● You can set up and solve a pair of simultaneous equations in two unknowns.

❓ Why do this?

You can set up equations in two unknowns to explain practical situations. For example, 2 adult cinema tickets and 1 child ticket costs £23.50, but 1 adult and 3 children would cost £25.50.

◈ Get Ready

Solve these simultaneous equations.

1. $x + 2y = 8$
$4y - x = 10$

2. $3x + y = 5$
$5y + 4x = 14$

3. $4x + y = 13$
$3x + 2y = 11$

🔍 Key Point

◉ When setting up your simultaneous equations, clearly define the unknowns used.

🔍 Example 3

Zach has some ten pence coins and some twenty pence coins in his piggy bank.
In his piggy bank he has a total of 18 coins which amounts to £2.30. Work out the number of ten pence coins and the number of twenty pence coins in Zach's piggy bank.

Let x be the number of 10p coins in the piggy bank. ← Define the unknowns.
Let y be the number of 20p coins in the piggy bank.

$10x + 20y = 230$ (1) ← The sum of the 10p and 20p coins is £2.30.

$x + y = 18$ (2) ← There are 18 coins altogether.

Results Plus
Watch Out!

Make sure that both sides of each equation have consistent units.

$$10x + 20y = 230 \quad (1)$$
$$10x + 10y = 180 \quad (3)$$

Multiply equation (2) by 10 to give equation (3).

$$10y = 50$$

Subtract to eliminate x.

$$y = 5$$
$$x + 5 = 18$$

Substitute $y = 5$ in (2) to find the value of x.

$$x = 13$$

So Zach has 13 ten pence coins and 5 twenty pence coins.

Check: $13 \times 10p + 5 \times 20p = £2.30$

Exercise 7B

1 The sum of two numbers is 19 and their difference is 5. Find the value of each of the numbers.

2 The total cost of a meal and a bottle of wine is £28.10.
The meal cost £8.90 more than the bottle of wine. Find the cost of the meal.

3 A taxi company charges a fixed amount of £f plus x pence for each mile of a journey.
A journey of 10 miles costs £10.20.
A journey of 6 miles costs £7.40.
Work out the cost of a journey of 8 miles.

4 Cinema tickets for 1 adult and 3 children cost £13.50.
The cost for 2 adults and 5 children is £24. Find the cost of one adult ticket and the cost of one child ticket.

5 Three nuts and six bolts have a combined mass of 72 g.
Four nuts and five bolts have a combined mass of 66 g.
Find the combined mass of one nut and one bolt.

6 The diagram shows a rectangle.
All sides are measured in centimetres.
  a Write down a pair of simultaneous equations in a and b.
  b Solve your pair of simultaneous equations to find a and b.

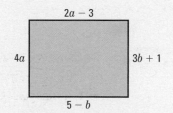

7 Mary and Ann together receive a total of £306 for baby-sitting.
Ann is paid for 14 days' work and Mary is paid for 15 days' work.
Ann's pay for 6 days' work is £6 more than Mary gets for 4 days.
Work out how much they each earn per day.

8 Atif walks for x hours at 5 km/h and runs for y hours at 10 km/h. He travels a total of 35 km and his average speed is 7 km/h. Find the value of x and y.

7.3 Using graphs to solve simultaneous equations

◎ Objective

● You can solve a pair of simultaneous equations by using a graph.

◈ Why do this?

It can be easier to find solutions to some simultaneous equations if they are plotted on a graph.

◈ Get Ready

The sides of a rectangle, in clockwise order, are $4a$, $24 - 3b$, $3b + 4$ and $3a$.
1. Set up two simultaneous equations in a and b.
2. Find a and b.
3. Find the perimeter of the rectangle.

◉ Key Point

● Simultaneous equations can be solved graphically, by drawing the graphs of the two equations and finding the coordinates of their point of **intersection**.

◎ Example 4

Solve the simultaneous equations
$$2x - y = 5$$
$$x + y = 4$$

$2x - y = 5$ ←   Rearrange $2x - y = 5$ to make y the subject.
   $2x = y + 5$
     $y = 2x - 5$

For $y = 2x - 5$,   ←   Find and plot any three points on $y = 2x - 5$.
when $x = 0, y = -5$
when $x = 2, y = -1$
when $x = 4, y = 3$.

For $x + y = 4$,   ←   Find and plot the points where $x + y = 4$ crosses the axes and one other point.
when $x = 0, y = 4$
when $y = 0, x = 4$
when $x = 2, y = 2$.

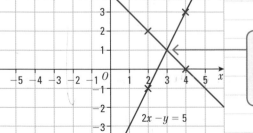

Draw the lines by joining the plots.

Find the coordinates of the point where the lines cross. This gives the solution of the simultaneous equations.

ResultsPlus
Examiner's Tip

Check your solution by substitution into both equations.

So the solution is $x = 3, y = 1$
Check: $2 \times 3 - 1 = 5$
and $3 + 1 = 4$.

Exercise 7C

1. The diagram shows three lines **A**, **B** and **C**.
   a Match the three lines to these equations.

   $y = 2x$ B

   $x + y = 3$ A

   $x - 2y = 3$ C

   b Use the diagram to solve these simultaneous equations.

     i $y = 2x$ **ii** $x + y = 3$ **iii** $y = 2x$

       $x + y = 3$ $x - 2y = 3$ $x - 2y = 3$

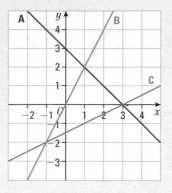

2. For each of these pairs of simultaneous equations, draw two linear graphs on the same grid and use them to solve the equations. Use a scale of -10 to $+10$ on each axis.

   a $4x + y = 8$ **b** $2x + 3y = 12$ **c** $x + y = 10$

      $x - y = 2$ $y = 2x - 3$ $y = 3x + 2$

7.4 Solving quadratic equations by factorisation

Objective

○ You can solve quadratic equations by factorising, when possible.

Why do this?

Quadratic equations can be used to work out car stopping distances.

Get Ready

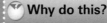

Factorise

1. a $x^2 - 4x$ **b** $2y^2 + 5y$ **c** $2x^2 + 5x - 3$ **d** $6y^2 - 7y - 20$

Key Points

● A **quadratic equation** can always be written in the form $ax^2 + bx + c = 0$ where a ($\neq 0$), b and c represent numbers.

● To factorise a quadratic equation you need to find two numbers whose sum is b and whose product is c. See Unit 2 Section 8.4 for more on factorising quadratic expressions.

● If the product of two numbers is 0, then at least one of these numbers must be 0. For example, if $cd = 0$ then either $c = 0$ or $d = 0$ or they are both 0.

● A quadratic equation has two solutions (or roots). Sometimes these solutions may be equal.

Example 5   Solve

   a $2x^2 = 6x$ b $y^2 - y - 20 = 0$

a $2x^2 = 6x$

   $2x^2 - 6x = 0$ ← Rearrange into the form $ax^2 + bx + c = 0$.

   $2x(x - 6) = 0$ ← Factorise.

   So either $2x = 0$ or $(x - 6) = 0$ ← Solve the linear equations.
   giving the two solutions $x = 0$ and $x = 6$.

b $y^2 - y - 20 = 0$ ← Factorise into two bracketed terms.
   $(y - 5)(y + 4) = 0$
   So either $(y - 5) = 0$ or $(y + 4) = 0$.   ← Remember: You are looking for two numbers whose
   The two solutions are $y = 5$ and $y = -4$.   product is -20 and whose sum is -1 (i.e. -5 and $+4$).

Example 6   Solve $q(q + 4) + 4 = 6q + 3$.

$q(q + 4) + 4 = 6q + 3$
$q^2 + 4q + 4 = 6q + 3$ ← Expand the brackets and rearrange into the form $ax^2 + bx + c = 0$.
$q^2 + 4q + 4 - 6q - 3 = 0$
$q^2 - 2q + 1 = 0$
$(q - 1)(q - 1) = 0$ ← Factorise.

So either $q - 1 = 0$ or $q - 1 = 0$, giving
the two equal solutions $q = 1$ and $q = 1$.   ← The solutions are both the same.
We say the solution is $q = 1$.

Example 7   Solve $4x^2 - 25 = 0$.

Method 1
$4x^2 - 25 = 0$
$4x^2 = 25$
$x^2 = 25 \div 4 = 6.25$ ← Take the square root of both sides.
$x = \pm\sqrt{6.25}$
So the two solutions are $x = 2.5$ or $x = -2.5$.

Method 2
$4x^2 - 25 = 0$
$(2x - 5)(2x + 5) = 0$ ← Factorise by the difference of two squares
So either $(2x - 5) = 0$ or $(2x + 5) = 0$.   method (see Section 7.6).
So the two solutions are $x = 2.5$ or $x = -2.5$.

Exercise 7D

1 Solve
  a $x(x - 4) = 0$
  b $(a + 5)(a - 3) = 0$
  c $(2m - 1)(4m - 9) = 0$
  d $y^2 + 2y = 0$
  e $t^2 - t = 0$
  f $4p^2 - 7p = 0$

2 Solve
  a $x^2 - 6x + 8 = 0$
  b $x^2 + 7x + 6 = 0$
  c $x^2 + x - 12 = 0$
  d $x^2 - 6x + 9 = 0$
  e $x^2 - 5x - 36 = 0$
  f $x^2 - 16 = 0$
  g $x^2 + 10x + 25 = 0$
  h $x^2 - 100 = 0$

3 Solve
  a $5x^2 + 26x + 5 = 0$
  b $3x^2 - 11x + 6 = 0$
  c $2x^2 + 7x - 4 = 0$
  d $5x^2 + 14x - 3 = 0$

4 Solve
  a $x^2 - x = 6$
  b $x^2 - 10 = 3x$
  c $x(x - 3) = x + 21$
  d $x^2 - 36 = 2x - 1$
  e $x(3x - 1) = x^2 + 15$
  f $6(x^2 + 3) = 31x$
  g $(x + 4)(x - 3) = 4(2x - 1)$
  h $(4x - 1)^2 = 10 - x$

B

A

A*

7.5 Completing the square

◎ Objective

○ You can complete the square for a quadratic expression.

⬥ Why do this?

Completing the square can help you to find the coordinates of the minimum (or maximum) point on a quadratic curve, for example, the maximum height of a bouncing ball.

⬆ Get Ready

Expand and simplify
1. $(x + 3)^2$
2. $(x - 5)^2$
3. $(x + a)^2$

🌐 Key Points

◉ Expressions such as $(x + 1)^2$, $(x + 4)^2$ and $(x + \frac{1}{2})^2$ are all called **perfect squares**.

◉ Expressions like $x^2 + bx + c$ can be written in the form $\left(x + \frac{b}{2}\right)^2 - \left(\frac{b}{2}\right)^2 + c$.
  This process is called **completing the square**.

◉ Expressions like $ax^2 + bx + c$ are rewritten as $a\left(x^2 + \frac{b}{a}x\right) + c$ before completing the square for the expression inside the brackets.

Example 8 Write $x^2 + 4x + 5$ in the form $(x + p)^2 + q$, stating the values of p and q.

$x^2 + 4x = (x + 2)^2 - 4$ ← Ignore the constant term. Find the perfect square which will give the correct terms in x^2 and x, then subtract 4 to make the identity true.

So
$x^2 + 4x + 5 = (x + 2)^2 - 4 + 5$ ← Add 5 to obtain $x^2 + 4x + 5$.

$= (x + 2)^2 + 1$ ← Simplify the expression.

$p = 2, q = 1$ ← Compare $(x + 2)^2 + 1$ with $(x + p)^2 + q$ and write down the values of p and q.

Example 9 a Write the expression $2y^2 - 12y - 5$ in the form $p(y + q)^2 + r$.
b Hence write down the minimum possible value of $2y^2 - 12y - 5$.
c Find the value of y for which $2y^2 - 12y - 5$ has its minimum value.

a $2y^2 - 12y - 5 = 2(y^2 - 6y) - 5$ ← Take out the coefficient of y^2 for the y^2 and y terms. Leave the constant term separate.

$= 2[(y - 3)^2 - 9] - 5$ ← Complete the square for $y^2 - 6y$.

$= 2(y - 3)^2 - 18 - 5$ ← Multiply out the square brackets.

$= 2(y - 3)^2 - 23$ ← Simplify the expression so that it is in the required form.

b The minimum possible value of any square number is zero so $(y - 3)^2 \geqslant 0$ for any value of y.

The minimum possible value of
$2(y - 3)^2 - 23 = 2 \times 0 - 23$. ← Substitute $(y - 3)^2 = 0$ into the answer to part a.
The minimum value of
$2y^2 - 12y - 5$ is therefore -23.

c The minimum value of $2y^2 - 12y - 5$ occurs when $y = 3$. ← Find the value of y which makes $(y - 3)^2 = 0$.

Exercise 7E

A

1 Write the following in the form $(x + p)^2 + q$.
    a $x^2 + 4x$
    b $x^2 + 10x$
    c $x^2 + 12x$
    d $x^2 - 2x$
    e $x^2 - 14x$
    f $x^2 - 24x$
    g $x^2 + x$
    h $x^2 - 3x$
    i $x^2 + 4x + 7$
    j $x^2 + 8x + 17$
    k $x^2 + 10x - 20$
    l $x^2 - 6x + 11$
    m $x^2 - 20x + 80$
    n $x^2 - 26x - 1$
    o $x^2 - x + 1$
    p $x^2 + 5x - 5$

2 Write the following in the form $a(x + p)^2 + q$.

a $2x^2 + 12x$

b $2x^2 - 4x + 5$

c $3x^2 - 12x + 10$

d $5x^2 + 50x + 100$

3 For all values of x, $x^2 + 8x + 24 = (x + p)^2 + q$.

a Find the value of the constants p and q.

b Write down the minimum value of $x^2 + 8x + 24$.

4 The diagram shows a sketch of the curve with equation $y = x^2 + 6x + 10$.

a Write down the coordinates of the point A, at which the curve crosses the y-axis.

b By completing the square for $x^2 + 6x + 10$, find the coordinates of the minimum point B.

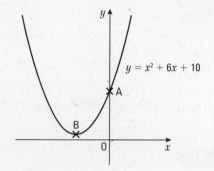

5 a By writing $1 + 4x - x^2$ as $-(x^2 - 4x - 1)$ find the value of r and the value of s for which $1 + 4x - x^2 = r - (x - s)^2$.

b Use your answer to part **a** to write down the maximum value of $1 + 4x - x^2$.

c For what value of x does this occur?

7.6 Solving quadratic equations by completing the square

Objective

● You can solve quadratic equations by completing the square.

Why do this?

The path of a cricket ball can be modelled using a quadratic equation.

Get Ready

Solve

1. $x^2 - 4 = 0$

2. $x^2 - x - 6 = 0$

3. $x^2 - 3x - 28 = 0$

4. Simplify

a $\sqrt{12}$

b $\sqrt{32}$

Key Points

● By completing the square, any quadratic expression can be written in the form $p(x + q)^2 + r$.

● Similarly, any quadratic equation can be written in the form $p(x + q)^2 + r = 0$.

Example 10

Solve $x^2 - 12x + 9 = 0$.

Give your solutions: **a** in surd form

b correct to 3 significant figures.

$x^2 - 12x + 9$ will not factorise into two brackets since no two integers have a product of 9 and a sum of -12.

$$x^2 - 12x + 9 = 0$$
$$(x - 6)^2 - 36 + 9 = 0$$
$$(x - 6)^2 - 27 = 0$$

Complete the square for $x^2 - 12x$.
Comparing this with $p(x + q)^2 + r = 0$ gives, $p = 1$, $q = -6$ and $r = -27$.

$$(x - 6)^2 = 27$$
$$x - 6 = \pm\sqrt{27}$$
$$x - 6 = \pm 3\sqrt{3}$$
$$x = 6 \pm 3\sqrt{3}$$

Take the square root of both sides.

Add 6 to both sides.

a The two solutions are $x = 6 + 3\sqrt{3}$ and $x = 6 - 3\sqrt{3}$.

b The two solutions are $x = 11.2$ and $x = 0.804$.

Example 11

Solve $2x^2 - 5x + 1 = 0$.

Give your solutions correct to 2 decimal places.

$$2x^2 - 5x + 1 = 0$$
$$x^2 - 2.5x + 0.5 = 0$$

Divide both sides by 2, the coefficient of x^2.

$$(x - 1.25)^2 - 1.25^2 + 0.5 = 0$$
$$(x - 1.25)^2 - 1.0625 = 0$$
$$(x - 1.25)^2 = 1.0625$$
$$x - 1.25 = \pm\sqrt{1.0625}$$
$$x - 1.25 = \pm 1.030\,776\,406$$

Complete the square for $x^2 - 2.5x$.
Evaluate $-1.25^2 + 0.5$

Take the square root of both sides.

The two solutions are $x = 2.28$ and $x = 0.22$.

Results Plus

Examiner's Tip

Write down more digits than are required from your calculator display before you do any rounding.

Exercise 7F

A

1 Solve these quadratic equations, giving your solutions in surd form.

    a $x^2 - 6x - 2 = 0$ **b** $x^2 + 4x + 1 = 0$ **c** $x^2 + 10x - 12 = 0$

    d $x^2 - 2x - 7 = 0$ **e** $2x^2 - 6x - 3 = 0$ **f** $5x^2 + 12x + 3 = 0$

2 Solve these quadratic equations, giving your solutions correct to 2 decimal places.

    a $x^2 + 8x + 5 = 0$ **b** $x^2 - 9x + 6 = 0$ **c** $x^2 + x - 8 = 0$

    d $2x^2 + 4x - 5 = 0$ **e** $6x^2 - 3x - 2 = 0$ **f** $10x^2 - 5x - 4 = 0$

7.7 Solving quadratic equations using the formula

⊙ Objective

- You can use the formula $x = \dfrac{-b \pm \sqrt{b^2 - 4ac}}{2a}$ to solve quadratic equations.

⦾ Why do this?

Many physical situations can be modelled by quadratic equations. For example, the time it takes a high diver to dive into a pool from a particular springboard is the solution of $7t^2 - t - 4 = 0$.

⬆ Get Ready

Solve these equations by completing the square.

1. $x^2 + 4x - 10 = 0$ **2.** $x^2 + 6x - 5 = 0$ **3.** $x^2 - 2x - 7 = 0$

4. Work out the value of $b^2 - 4ac$, when

   a $a = 1, b = 4, c = 2$ **b** $a = 2, b = -5, c = 3$ **c** $a = 5, b = -7, c = -5$

🕐 Key Points

- You can use the method of solving the general quadratic equation $ax^2 + bx + c = 0$ by completing the square (see Sections 7.5 and 7.6) to develop a formula which can be used to solve all quadratic equations. This is called a **quadratic formula**.

$$ax^2 + bx + c = 0$$

$$x^2 + \frac{b}{a}x + \frac{c}{a} = 0 \qquad \leftarrow \boxed{\text{Divide both sides by } a.}$$

$$\left(x + \frac{b}{2a}\right)^2 - \left(\frac{b}{2a}\right)^2 + \frac{c}{a} = 0 \qquad \leftarrow \boxed{\text{Complete the square on } x^2 + \frac{b}{a}x.}$$

$$\left(x + \frac{b}{2a}\right)^2 = \left(\frac{b}{2a}\right)^2 - \frac{c}{a} \qquad \leftarrow \boxed{\text{Rearrange.}}$$

$$\left(x + \frac{b}{2a}\right)^2 = \frac{b^2}{4a^2} - \frac{c}{a} = \frac{b^2 - 4ac}{4a^2}$$

$$x + \frac{b}{2a} = \pm\sqrt{\frac{b^2 - 4ac}{4a^2}} = \frac{\sqrt{b^2 - 4ac}}{2a} \qquad \leftarrow \boxed{\text{Take the square root of both sides.}}$$

$$x = -\frac{b}{2a} \pm \frac{\sqrt{b^2 - 4ac}}{2a} \qquad \leftarrow \boxed{\text{Make } x \text{ the subject of the formula and simplify.}}$$

$$x = \frac{-b \pm \sqrt{b^2 - 4ac}}{2a}$$

- If the value of $b^2 - 4ac$ is negative, the quadratic equation does not have any real solutions.

Example 12   Solve $x^2 - 5x + 3 = 0$.

Give your solutions correct to 2 decimal places.

Results**Plus**
Examiner's Tip

$x^2 - 5x + 3 = 0$

$a = 1, b = -5, c = 3$

> Compare with $ax^2 + bx + c = 0$ and write down the values of a, b and c.

> In equations like $x^2 + bx + c = 0$ it is helpful to write it as $1x^2 + bx + c = 0$ so that the value of a is clearly 1.

$x = -(-5) \pm \dfrac{\sqrt{(-5)^2 - 4 \times 1 \times 3}}{2 \times 1}$

> Substitute a, b and c into the quadratic formula.

$x = \dfrac{5 \pm \sqrt{25 - 12}}{2}$

$x = \dfrac{5 + \sqrt{13}}{2}$

or $x = \dfrac{5 - \sqrt{13}}{2}$

The solutions are $x = 4.30$ or $x = 0.70$

Example 13   Solve $5x^2 + x - 3 = 0$.

Give your solutions correct to 2 decimal places.

$5x^2 + x - 3 = 0$

$a = 5, b = 1, c = -3$

> Compare with $ax^2 + bx + c = 0$ and write down the values of a, b and c.

$x = \dfrac{-1 \pm \sqrt{1^2 - 4 \times 5 \times -3}}{2 \times 5}$

> Substitute a, b and c into the quadratic formula.

$x = \dfrac{-1 \pm \sqrt{1 + 60}}{10}$

$x = \dfrac{-1 + \sqrt{61}}{10}$ or $x = \dfrac{-1 - \sqrt{61}}{10}$

The solutions are $x = 0.68$ or $x = -0.88$.

Exercise 7G

Solve these quadratic equations. Give your solutions correct to 3 significant figures.

1 $x^2 + 4x + 2 = 0$ **2** $x^2 + 7x + 5 = 0$ **3** $x^2 + 6x - 4 = 0$

4 $x^2 + x - 10 = 0$ **5** $x^2 - 4x - 7 = 0$ **6** $x^2 - 5x + 3 = 0$

7 $2x^2 + 4x + 1 = 0$ **8** $5x^2 - 9x + 2 = 0$ **9** $6x^2 - 5x - 8 = 0$

10 $10x^2 + 3x - 2 = 0$ **11** $4x^2 - 7x + 2 = 0$ **12** $x(x - 1) = x + 5$

13 $x(2x + 3) = 4 - 7x$ **14** $(x + 2)(x - 3) = 15$ **15** $3x + 5 = 5x^2 + 2(x - 4)$

A

7.8 Solving algebraic fraction equations leading to quadratic equations

◉ Objective

◉ You can solve equations involving both algebraic fractions and quadratic equations.

⬧ Get Ready

Multiply out:

1. $(x + 6)(2x - 3)$ **2.** $(2x - 2)(3x - 6)$ **3.** $(x + 6)(x + 1) \times \dfrac{2}{(x + 1)}$

🔍 Key Points

◉ Equations with algebraic fractions often occur in mathematics.

◉ These sometimes lead to quadratic equations which you can solve using one of the methods already described.

Example 14   Solve $\dfrac{5}{2x + 1} + \dfrac{6}{x + 1} = 3$.

$(2x + 1)(x + 1) \times \dfrac{5}{2x + 1} +$ ←   Multiply both sides by $(2x + 1)(x + 1)$ and cancel.

$(2x + 1)(x + 1) \times \dfrac{6}{x + 1} =$

$(2x + 1)(x + 1) \times 3$

$5(x + 1) + 6(2x + 1) = 3(2x + 1)(x + 1)$

$5x + 5 + 12x + 6 = 3(2x^2 + 3x + 1)$ ←   Expand the brackets. Simplify both sides.

$17x + 11 = 6x^2 + 9x + 3$

$6x^2 - 8x - 8 = 0$ ←   Rearrange into the form $ax^2 + bx + c = 0$.

$3x^2 - 4x - 4 = 0$

$(3x + 2)(x - 2) = 0$ ←   Solve by factorisation.

So either $3x + 2 = 0$ or $x - 2 = 0$.

The solutions are $x = -\frac{2}{3}$ and $x = 2$.

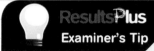

ResultsPlus

Examiner's Tip

When the denominators have no common factor, multiply by the product of the denominators.

⚙ Exercise 7H

Solve these quadratic equations.

1 $\dfrac{2}{x - 1} + \dfrac{3}{x + 1} = 1$ **2** $\dfrac{6}{x} = \dfrac{5x - 1}{3}$ **3** $\dfrac{6}{x} - \dfrac{5}{x + 1} = 2$

4 $\dfrac{2}{x + 1} + \dfrac{4}{3x - 1} = 3$ **5** $\dfrac{3}{x} - \dfrac{2}{2x + 1} = 5$ **6** $\dfrac{3}{2x - 1} + \dfrac{4}{x + 2} = 2$

A*

In questions 7–12, give your solutions correct to 3 significant figures.

A★

7 $\dfrac{3}{x} - \dfrac{1}{1+x} = 1$

8 $\dfrac{1}{x+4} - \dfrac{2}{x-5} = 4$

9 $\dfrac{2}{x-1} + \dfrac{5}{x+4} = 1$

10 $\dfrac{3}{x-1} - \dfrac{2}{x+3} = 1$

11 $\dfrac{2x+1}{x} = \dfrac{1-x}{6}$

12 $\dfrac{2}{x-1} - \dfrac{1}{1-2x} = 1$

7.9 Setting up and solving quadratic equations

◎ Objective

● You can solve practical problems which involve quadratic equations.

？ Why do this?

You can find the best positioning for a satellite dish by setting up and solving a quadratic equation involving the diameter and depth of the dish.

⬆ Get Ready

Multiply out:

1. $2x(2-x)$

2. $(x+1)(1+x)$

3. $(3x+2)(4-x)$

🔍 Key Points

◎ To find the equation to represent a problem:
  ◉ where relevant, draw a diagram and put all of the information you are given on it
  ◉ use x to represent the unknown which you have been asked to find
  ◉ use other letters to identify any other relevant unknowns
  ◉ look for information in the question which links these letters to x and write them down
  ◉ try simple numbers for the unknowns and see if this helps you to find a method
  ◉ make sure that the units on both sides of your equation are the same.

Example 15 The diagram shows a rectangular lawn surrounded on three sides by flower beds.
Each flower bed is 2 m wide.
The area of the lawn is 14 m².
Find the length of the lawn.

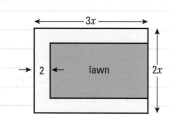

Length of lawn $= 3x - 2$ ← Write down expressions for the length and width of the lawn.
Width of lawn $= 2x - 4$
Area of lawn $= (3x - 2)(2x - 4)$ ← Using area of rectangle = length × width.
$\qquad = 6x^2 - 16x + 8.$
But the area of the lawn is given as 14 m².
So $6x^2 - 16x + 8 = 14$ ← Set up a quadratic equation.
$\quad 6x^2 - 16x - 6 = 0$
$\quad 3x^2 - 8x - 3 = 0$
$\quad (3x + 1)(x - 3) = 0$
So either $x = -\frac{1}{3}$ or $x = 3$. ← Solve using the method of factorisation.

The only acceptable solution is $x = 3$.
Length of lawn $= 3x - 2 = 3 \times 3 - 2$
$\qquad\qquad\quad = 7$ m.

> The solution cannot be negative since x is a measurement of length.
> Substitute $x = 3$ into the expression for the length of the lawn.

Example 16

Angela drove 300 km to the seaside.
Her average speed was 10 km/h less than she expected.
The journey therefore took 1 hour longer than she had planned.
Find Angela's actual average speed.

Let Angela's actual average speed be x km/h.

> Define the letters to be used.

Her expected average speed was $(x + 10)$ km/h.

The time taken for the journey $= \dfrac{300}{x}$ hours.

> Time $= \dfrac{\text{distance}}{\text{speed}}$

Her expected time for the journey $= \dfrac{300}{x + 10}$ hours.

$\dfrac{300}{x} - \dfrac{300}{x + 10} = 1$

> Set up an equation using the information that the difference in the times is 1 hour.

$x(x + 10) \times \dfrac{300}{x} - x(x + 10) \times \dfrac{300}{x + 10} = x(x + 10) \times 1$

> Multiply by $x(x + 10)$ and cancel.

$300(x + 10) - 300x = x(x + 10)$
$300x + 3000 - 300x = x^2 + 10x$

> Expand brackets and rearrange.

$\qquad\quad x^2 + 10x - 3000 = 0$
$(x + 60)(x - 50) = 0$

> Solve by factorising.

Since x cannot be negative, $x = 50$.
Angela's actual average speed is 50 km/h.

Exercise 7I

1   The sum of the square of an integer and 2 times itself is 24.
Find the two possible values of the integer.

2   The product of three numbers 5, $2x$ and $x - 8$ is -160.
Find the value of the integer x.

3   A man is four times as old as his son, and 8 years ago the product of their ages was 160.
Find their present ages as integers.

4   The length of a rectangular wall is 5 m greater than the height of the wall.
The area of the wall is 16 m². Work out the length of the wall. Give your answer correct to 3 significant figures.

5   The sum of the squares of two consecutive integers is 41.
a   If x is one of the integers, show that $x^2 + x - 20 = 0$.
b   Solve $x^2 + x - 20 = 0$ to find the two consecutive integers.

A02
A03

A*

A02

A03

A★ A02
A03

6   Find the length of each side of this right-angled triangle.
    The measurements are given in cm.
    Give your answers correct to 2 d.p.

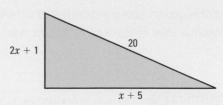

A02
A03

7   192 square tiles are needed to tile a kitchen wall. If the tiles had measured 2 cm less each way, 300 tiles
    would have been needed. Find the size of the larger tiles.

8   A farmer uses 60 m of fencing to make three sides of a rectangular sheep pen. The fourth side of the
    pen is a wall. Work out the length of the shorter sides of the pen if the area enclosed is 448 m².

A03

9   The diameters of two circles are $4x$ cm and $(x + 3)$ cm.

    The area of the shaded region is 84π cm².
    Work out the value of x.

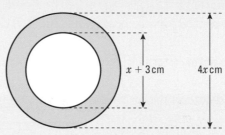

A02
A03

10 On a journey of 420 km, a train driver calculates that the journey would take 40 minutes less if he
    increased his average speed by 5 km/h.
    Work out his present average speed. Give your answer correct to the nearest whole number.

7.10 Constucting graphs of simple loci

◉ Objectives

- You can construct the graphs of simple loci including the circle
  $x^2 + y^2 = r^2$ for a circle of radius r centred at the origin of the
  coordinate plane.
- You can select and apply construction techniques and use your
  understanding of loci to draw graphs based on circles and
  perpendiculars of lines.

⊘ Why do this?

Planets and stars move in paths that
can be described by simple loci.
Scientists model these paths using
these techniques.

◈ Get Ready

1. Which of the following are Pythagorean triples?
   a 3, 4, 5   **b** 4, 5, 6   **c** 6, 8, 12   **d** 5, 12, 13   **e** 8, 15, 17
2. Construct a circle of radius 5 cm.
3. On your circle in question 2, draw **a** a tangent to the circle, **b** a perpendicular from the centre of the circle to
   the tangent in **a**.

◉ Key Points

- The locus of a circle is the path of all points equidistant from a given fixed point.

● To construct the graph of $x^2 + y^2 = r^2$ consider any point P with coordinates (x, y) at a distance r from the origin O $(0, 0)$.
Draw the lines PQ and OQ to form the right-angled triangle PQO.

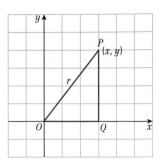

● By Pythagoras, $x^2 + y^2 = r^2$.
This is true for any point P, at a distance r from the fixed point O.
This is the definition of a circle.

● Joining all of the points A, B, C, D, E, etc gives a circle of radius r.

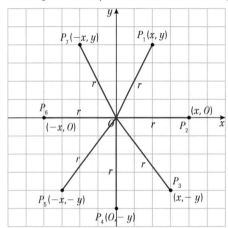

So $x^2 + y^2 = r^2$ is the equation of a circle of radius r, centre $(0, 0)$.

Example 17

a Construct the graph of the locus of all points distance 3 units from the line $y = x$.

b Find the equation of this locus.

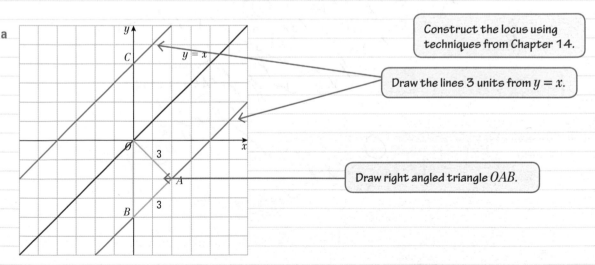

Construct the locus using techniques from Chapter 14.

Draw the lines 3 units from $y = x$.

Draw right angled triangle OAB.

b In triangle OAB, $OA = AB = 3$ units.
 By Pythagoras, $OB^2 = 3^2 + 3^2 = 9 + 9 = 18$
  $OB = \sqrt{18} = 4.2426\ldots$
 Equation of locus through B is $y = x - 4.24\ldots$
 Equation of locus through C is $y = x + 4.24\ldots$

Apply Pythagoras to triangle OAB to find the length of OB.

Example 18
a Construct the graph of $x^2 + y^2 = 25$.

b Find the equation of the tangent to the curve at the point $(3, -4)$.

$r = \sqrt{25} = 5$

> Compare $x^2 + y^2 = 25$ with $x^2 + y^2 = r^2$.

a

> Draw a circle, centre O, with compasses set to 5 units.

$(3, -4)$

> Draw tangent through $(3, -4)$ and read off y-intercept (-6.5).

4.1

> Estimate the gradient: $4.1 \div 5$.

-6.5

5

b Gradient of tangent $= 4.1 \div 5$

$= 0.82$

> Using $y = mx + c$

y-intercept $= -6.5$

The equation is $y = 0.82x - 6.5$

Exercise 7J

A

1 On graph paper, draw the graphs of the following equations.

a $x^2 + y^2 = 4$ b $x^2 + y^2 = 16$ c $x^2 + y^2 = 36$ d $x^2 + y^2 = 64$ e $x^2 + y^2 = 100$

A03

2 Using your graph of $x^2 + y^2 = 16$, construct a tangent parallel to the line $y = x$.
Write down the coordinates of the point where this tangent touches the graph.

A*
A03

3 Find the equation of the locus of points 5 units from the following lines.

a $y = 6$ b $y = -4$ c $x = 3$ d $x = -5$

A03

4 Find the equation of the locus of points 6 units from the line with equation $y = x + 4$.

7.11 Solving simultaneous equations when one is linear and the other is quadratic

⊚ Objective

⊙ You can solve a pair of simultaneous equations in two unknowns when one equation is linear and the other is quadratic.

⬥ Get Ready

Use a graphical method to find the solution of these pairs of simultaneous equations.

1. $x + y = 10$
$y = 2x + 1$

2. $y = -3x + 3$
$y = 2x - 7$

🔍 Key Points

⊙ You can solve a pair of simultaneous equations where one equation is linear and the other quadratic by an algebraic method or a graphical approach.

⊙ The solution of a pair of simultaneous equations where one is linear and one is quadratic is represented by the points of intersection of a straight line and a quadratic curve.

🔍 Example 19 Solve the simultaneous equations **a** graphically **b** algebraically.

$x^2 + 2y = 1$
$y = x - 1$

a $y = \dfrac{1}{2} - \dfrac{x^2}{2}$ ⟵ [Make y the subject of $x^2 + 2y = 1$.]

For $y = x - 1$,
when $x = 3, y = 2$
when $x = 0, y = -1$
when $x = -4, y = -5$

x	-3	-2	-1	0	1	2	3
y	-4	-1.5	0	0.5	0	-1.5	-4

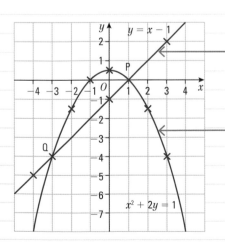

[Find and plot any three points then draw the graph.]

[Construct a table of values, plot the points and draw the graph.]

ResultsPlus
Examiner's Tip

The graph of the linear equation crosses the graph of the quadratic at **two** points. So the pair of simultaneous equations has **two** solutions.

The solutions are $x = 1, y = 0$ and $x = -3, y = -4$. ⟵ [The points of intersection of the two graphs are P(1, 0) and Q(-3, -4).]

b $x^2 + 2y = 1$ (1)   ←   Label the equations (1) and (2).
           $y = x - 1$ (2)

$x^2 + 2(x - 1) = 1$ ←   Substitute (2) into (1) and rearrange.
   $x^2 + 2x - 2 = 1$
    $x^2 + 2x - 3 = 0$
   $(x + 3)(x - 1) = 0$ ←   Solve the quadratic equation using the method of factorisation. See Section 22.4.

So either $(x + 3) = 0$ or $(x - 1) = 0$
$x = -3$ or $x = 1$
When $x = -3$,   ←   Substitute values of x into (2).
$y = -3 - 1$
$y = -4$.
When $x = 1$,
$y = 1 - 1$
$y = 0$.
So the solutions are $x = -3, y = -4$ and $x = 1, y = 0$.

Exercise 7K

For each of these pairs of simultaneous equations:
a draw a quadratic graph and a linear graph on the same grid and use them to solve the simultaneous equations (use a scale of -10 to $+10$ on each axis)
b solve them using an algebraic method. You must show all of your working.

A

1 $x^2 + y = 6$
     $y = x$

2 $x^2 - 2y = 2$
     $y = x + 3$

3 $x^2 + 4y = 7$
     $2y + x = 2$

4 $y = 3x^2 - 2$
     $y = 3 - 2x$

5 $y = 3 - x^2$
     $y = 5 - 3x$

6 $2y = 4x^2 - 7$
     $y = 6x$

7.12 Solving simultaneous equations when one is linear and one is a circle

Objective

- You can solve a pair of simultaneous equations in two unknowns when one equation is linear and the other is of the form $x^2 + y^2 = r^2$ (i.e. a circle).

Get Ready

On graph paper, draw graphs of the following equations:
1. $x^2 + y^2 = 16$
2. $x^2 + y^2 = 64$
3. $x^2 + y^2 = 49$

Key Points

- The **equation of a circle** with centre (0, 0) and radius r can be written as $x^2 + y^2 = r^2$.
- You can solve simultaneous equations where one is linear and one is the equation of a circle graphically and algebraically.
- If the solutions are not integer values, they can only be estimated using the graphical method and can be calculated using the quadratic formula (see Section 7.7).

Example 20 Solve the simultaneous equations   **a** graphically   **b** algebraically.

$$x^2 + y^2 = 25$$
$$y = x + 1$$

a The graph of $x^2 + y^2 = 25$ is a circle,   ← | Compare with $x^2 + y^2 = r^2$.
centre (0, 0) of radius 5 units.

For $y = x + 1$,   ← | Plot the points and draw the line $y = x + 1$.
when $x = 2, y = 3$
when $x = 0, y = 1$
when $x = -2, y = -1$.

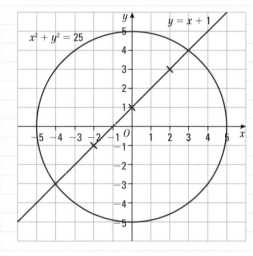

The solutions are the coordinates of the points of intersection.

The solutions are
$$x = 3, y = 4 \text{ and } x = -4, y = -3.$$

b $x^2 + y^2 = 25$ (1)
         $y = x + 1$ (2)
   $x^2 + (x + 1)^2 = 25$ ← | Substitute $y = x + 1$ into (1).

$x^2 + x^2 + 2x + 1 = 25$ ← | Expand and simplify.
$2x^2 + 2x - 24 = 0$
$x^2 + x - 12 = 0$
$(x - 3)(x + 4) = 0$ ← | Solve the quadratic equation by the factorisation method.

So $x = 3$ or $x = -4$
$y = 3 + 1 = 4$ or $y = -4 + 1 = -3$ ← | Substitute values of x to find the corresponding values of y.

The solutions are $x = 3, y = 4$ and $x = -4, y = -3$.

Example 21   Draw suitable graphs to find estimates of the solutions of:
$$x^2 + y^2 = 16$$
$$y = x - 1.$$

The graph of $x^2 + y^2 = 16$ is a circle, centre $(0, 0)$ of radius 4 units. ← Compare with $x^2 + y^2 = r^2$.

For $y = x - 1$, ← Plot the points and draw the line $y = x - 1$.
when $x = 2, y = 1$
when $x = 0, y = -1$
when $x = -2, y = -3$

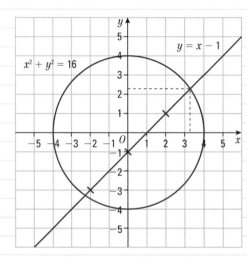

The solutions can only be estimated.

The estimated solutions are
$x = 3.3, y = 2.3$ and $x = -2.3, y = -3.3$.

Example 22   Solve these simultaneous equations
$$x^2 + y^2 = 16$$
$$y = x - 1$$
Give your answers correct to 2 decimal places.

$$x^2 + y^2 = 16 \quad (1)$$
$$y = x - 1 \quad (2)$$
$$x^2 + (x - 1)^2 = 16 \quad ← \quad \text{Substitute (2) in (1) and rearrange to give a quadratic equation.}$$
$$x^2 + x^2 - 2x + 1 = 16$$
$$2x^2 - 2x - 15 = 0$$
$$x = \frac{2 \pm \sqrt{2^2 - 4 \times 2 \times (-15)}}{2 \times 2} \quad ← \quad \text{Solve using the quadratic formula.}$$
$$x = \frac{2 \pm \sqrt{4 + 120}}{2 \times 2} = \frac{2 \pm \sqrt{124}}{4}$$
$$x = 3.28 \text{ or } -2.28$$

The solutions are $x = 3.28, y = 2.28$ ← Write the solutions correct to 2 decimal places.
and $x = -2.28, y = -3.28$ to 2 d.p.

Exercise 7L

1. On graph paper, draw the graph of the circle with equation $x^2 + y^2 = 36$. On the same axes, draw the straight line with equation $y = 2x$.
   Hence find estimates of the solutions of the simultaneous equations
   $x^2 + y^2 = 36$ and $y = 2x$.

2. Draw suitable graphs to find estimates of the solutions of the simultaneous equations
   $x^2 + y^2 = 19$ and $y = x + 3$.

3. Draw suitable graphs to find estimates of the solutions of the simultaneous equations
   $x^2 + y^2 = 49$ and $x + y = 5$.

4. Solve these simultaneous equations.
   a $x^2 + y^2 = 13$
      $y = x + 1$
   b $x^2 + y^2 = 20$
      $y = 2 - x$
   c $x^2 + y^2 = 34$
      $y = 1 + 2x$

5. Solve these simultaneous equations. Give your answers correct to 3 significant figures.
   a $x^2 + y^2 = 20$
      $y = x + 4$
   b $x^2 + y^2 = 32$
      $y = 1 + 3x$
   c $x^2 + y^2 = 100$
      $y = 2x - 3$

Chapter review

- When there are two unknowns you need two equations. These are called **simultaneous equations**.
- Simultaneous equations can be solved using elimination or substitution.
- To **eliminate** an unknown, multiply the equations so that the coefficients of that unknown are the same. Add or subtract the equations to eliminate the chosen unknown.
- To substitute an unknown, rearrange one of the equations to make the unknown the subject, then substitute its value (in terms of the second unknown) into the other equation.
- Once you know one unknown, you can use substitution to find the other.
- When setting up your simultaneous equations, clearly define the unknowns used.
- Simultaneous equations can be solved graphically, by drawing the graphs of the two equations and finding the coordinates of their point of **intersection**.
- A **quadratic equation** can always be written in the form $ax^2 + bx + c = 0$ where a ($\neq 0$), b and c represent numbers.
- To factorise a quadratic equation you need to find two numbers whose sum is b and whose product is c.
- If the product of two numbers is 0, then at least one of these numbers must be 0. For example, if $cd = 0$ then either $c = 0$ or $d = 0$ or they are both 0.
- A quadratic equation always has two solutions (or roots). Sometimes these solutions may be equal.
- Expressions such as $(x + 1)^2$, $(x + 4)^2$ and $(x + \frac{1}{2})^2$ are all called **perfect squares**.
- Expressions like $x^2 + bx + c$ can be written in the form $\left(x + \frac{b}{2}\right)^2 - \left(\frac{b}{2}\right)^2 + c$.
  This process is called **completing the square**.
- Expressions like $ax^2 + bx + c$ are rewritten as $a\left(x^2 + \frac{b}{a}x\right) + c$ before completing the square for the expression inside the brackets.
- By completing the square, any quadratic expression can be written in the form $p(x + q)^2 + r$.

- Similarly, any quadratic equation can be written in the form $p(x + q)^2 + r = 0$.
- All quadratic equations can be solved by the formula

$$x = \frac{-b \pm \sqrt{b^2 - 4ac}}{2a}$$

- If the value of $b^2 - 4ac$ is negative, the quadratic equation does not have any real solutions.
- Equations with algebraic fractions sometimes lead to quadratic equations.
- To find the equation to represent a problem:
  - where relevant, draw a diagram and put all of the information on it
  - use x to represent the unknown which you have been asked to find
  - use other letters to identify any other relevant unknowns
  - look for information in the question which links these letters to x and write them down
  - try simple numbers for the unknowns and see if this helps you to find a method
  - make sure that the units on both sides of your equation are the same.
- The locus of a circle is the path of all points equidistant from a given fixed point.
- You can solve simultaneous equations where one is linear and one is the equation of a circle graphically and algebraically.
- The solution of a pair of simultaneous equations where one is linear and one is quadratic is represented by the points of intersection of a straight line and a quadratic curve.
- The **equation of a circle** with centre (0, 0) and radius r can be written as $x^2 + y^2 = r^2$.
- You can solve a pair of simultaneous equations where one equation is linear and the other quadratic by an algebraic method or a graphical approach.
- If the solutions are not integer values they can only be estimated using the graphical method and can be calculated using the quadratic formula.

Review exercise

1 Solve the following equations **a** $x^2 = 9$ **b** $2x^2 = 72$ **c** $2x^2 - 108 = 0$

2 Solve the following equations **a** $4 - y^2 = 0$ **b** $\dfrac{t^2}{4} = 1$ **c** $\dfrac{p^2}{3} - 3 = 0$

3 The diagram shows a trapezium.
The lengths of three sides of the trapezium are
$x - 5$, $x + 2$ and $x + 6$.
All measurements are given in centimetres.
The area of the trapezium is 36 cm².
Find the length of the shortest side of the trapezium.

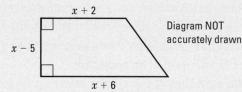
Diagram NOT accurately drawn

4 Solve the following equations
  a $y^2 - 5y - 6 = 0$ **b** $4t^2 - 16 = 0$ **c** $4 - 3p - p^2 = 0$

5 Solve these simultaneous equations.
  a $2x + y = 11$ **b** $3x - y = -4$ **c** $4x - 3y = 7$
   $y = x + 7$ $y = 3 + 2x$ $y = 2x - 1$

6 For each of these pairs of simultaneous equations, draw two linear graphs on the same grid and use them to solve the simultaneous equations. Use a scale of -10 to $+10$ on each axis.
  a $y = 8 - 3x$ **b** $2x + y = 4$
   $x + y = 4$ $3x + 4y = 12$

C

A03

B

7 A rectangular room is 2 metres longer than it is wide.
If its area is 52 m², what is its perimeter? Give your answer to 2 decimal places.

8 The height of a ball above the ground in metres can be
calculated from the formula:

$$h = 30t - 5t^2$$

where t = time in seconds after being thrown.
Find:
a the total time that the ball was in the air
b the maximum height of the ball above the ground
c the time at which the ball was 25 cm above the ground.

9 Solve the following equations
a $2k^2 - 11k + 5 = 0$ b $4m^2 - 4m = 3$ c $(2n - 1)(3n + 2) = 24$

10 $x^2 + 8x + 5$ can be written in the form $(x + p)^2 + q$.
a Find the value of p and the value of q.
b Use your answer to part **a** to solve the equation $x^2 + 8x + 5 = 0$.
Give your solutions to 3 significant figures.

11 Solve this quadratic equation $x^2 - 5x - 8 = 0$.
Give your answers correct to 3 significant figures.   *June 2006*

12 a Solve the equation $x^2 - 2x - 1 = 0$.
Give your answer correct to 3 significant figures.
Hence, or otherwise
b solve the equation $3x^2 - 6x - 3 = 0$.   *June 2009*

13 a $x^2 - 2x + 3$ is to be written in the form $(x + a)^2 + b$.
Find the values of a and b.
b Use your answer to part **a** to find the minimum value of $x^2 - 2x + 3$.
c Write down the value of x for which $x^2 - 2x + 3$ has a minimum value.
d Sketch the graph of $y = x^2 - 2x + 3$.

14 Write $4x^2 + 24x$ in the form $a(x + p)^2 + q$. State the values of a, p and q.

15 Solve these simultaneous equations.
a $2x + 3y = 10$ b $5x + 4y = 8$ c $2x + 3y = 1$
   $3x + 5y = 16$ $2x - 3y = -6$ $7x + 8y = -4$

16 A gas bill consists of a fixed charge (£F) and a charge (g pence) for each unit used.
Mrs Anwar used 350 units and paid £30. Mr White used 450 units and paid £35. Find the fixed charge
and the charge per unit.

A

17 Write down the pair of simultaneous equations that are solved by the coordinates of the point of intersection of the two lines shown in the diagram.

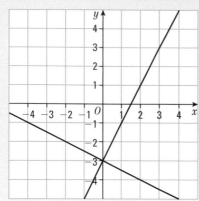

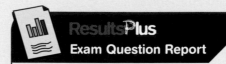

Exam Question Report

58% of students answered this sort of question well. They used all the information in the question.

18 Solve this pair of simultaneous equations
$$4x + 3y = 4$$
$$2y = 1 - 3x$$
a   by drawing two linear graphs on the same grid
    (use a scale of -10 to $+10$ on each axis)
b   by using an algebraic method (you must show all of your working).

A02

19 $x^2 - 8x + 23 = (x - p)^2 + q$ for all values of x.
a   Find the value of p and the value of q.
Here is a sketch of the curve with equation
$$y = x^2 - 8x + 23$$
B is the minimum point on the curve.
b   Find the coordinates of B.

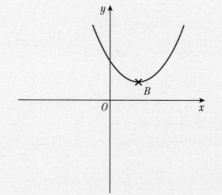

June 2006

A03

20 Kate buys 2 lollies and 5 choc ices for £6.50.
Pete buys 2 lollies and 3 choc ices for £4.30.
Work out the cost of 1 lolly.
Give your answer in pence.

June 2007

A02

21 The diagram shows a sketch of the graph of $y = 3(x^2 - x)$.
The line $y = 4 - 4x$ intersects the curve $y = 3(x^2 - x)$ at the points A and B.
Use an algebraic method to find the coordinates of A and B.

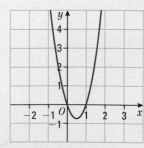

Nov 2005

22 **a** Show that the equation $\dfrac{5}{x + 2} = \dfrac{4 - 3x}{x - 1}$ can be rearranged to give $3x^2 + 7x - 13 = 0$.

b Solve $3x^2 + 7x - 13 = 0$.

Give your solutions correct to 2 decimal places.

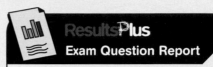

ResultsPlus

Exam Question Report

80% of students answered this question poorly because they did not take care with the brackets in the denominators.

June 2008

23 The diagram shows a 6-sided shape.

All the corners are right angles.

All the measurements are given in centimetres.

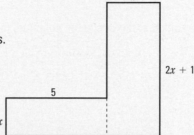

Diagram **NOT** accurately drawn

The area of the shape is 95 cm^2.

a Show that $2x^2 + 6x - 95 = 0$.

b Solve the equation $2x^2 + 6x - 95 = 0$.

Give your answers correct to 3 significant figures.

Nov 2008

24 Sean runs in a 20 km fun run. He runs the first 10 km at a speed of x km per hour. He runs the second 10 km at a speed 1 km per hour less than the first 10 km. His total time for the fun run is 4 hours.

a Show that $\dfrac{10}{x} + \dfrac{10}{x - 1} = 4$.

b Show x satisfies the quadratic equation $4x^2 - 24x + 10 = 0$.

c Find the two solutions of the equation $4x^2 - 24x + 10 = 0$.

Give your answers correct to 3 significant figures.

d What was Sean's speed for the second 10 km?

25 Solve these simultaneous equations. Give your answers correct to 3 significant figures.

a $x^2 + y^2 = 19$
   $y = x + 5$

b $x^2 + y^2 = 45$
   $y = 6 + 2x$

26 Solve the equation $x^2 - 4x + 2 = 0$.

27 Show that any straight line that passes through the point $(1, 2)$ must intersect the curve with equation $x^2 + y^2 = 16$ at two points.

June 2006

A*

A02

A02
A03

A02

A03

A03

28 The diagram shows a circle, radius 5 units, centre the origin.

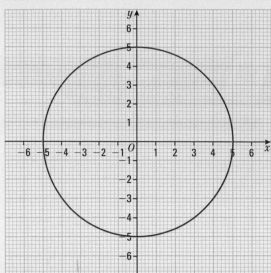

Use the diagram to find estimates of the solutions to the equations:

$x^2 + y^2 = 25$

$y = 2x + 1.$

Nov 2005, adapted

8 PROPORTION

On 6 August 1945, a nuclear bomb known as 'Little Boy' was dropped on Hiroshima, Japan. The energy of the explosion caused was equivalent to around 18 kilotons of TNT, caused by 600 mg of uranium within the weapon converting into energy. Three days later a second bomb, 'Fat Man', was detonated over Nagasaki, Japan. This bomb was 15% larger, and the energy released in the explosion was equivalent to about 21 kilotons of TNT.

Objectives

In this chapter you will:

- learn to solve problems involving direct proportion
- discover how to write down the statement of proportionality and the formula for a variety of problems
- find out how to solve problems involving inverse, square and cubic proportion.

Before you start

You need to be able to:

- understand, use and calculate proportion
- derive simple formulae and simple rules for number sequences
- plot points on a graph
- derive quadratic and cubic formulae.

8.1 Direct proportion

● You can use graphs to solve problems involving direct proportion.

◈ **Why do this?**

Scientists use proportion to work out the amount of energy produced by a nuclear explosion.

◈ **Get Ready**

1. A pen costs 25p. What is the cost of:
   a 2   **b** 3   **c** 4   **d** 5 pens?

2. A photocopier takes 10 seconds to copy 2 letters. How long does it take to copy:
   a 4   **b** 8   **c** 16   **d** 32 letters?

Key Point

◉ When a graph of two quantities is a straight line through the origin, one quantity is directly proportional to the other.

Example 1

The cost of buying 5 litres of fuel is £11.

  a Show that the cost, £C, of buying the fuel is directly proportional to the amount, x litres, of fuel bought.

  b Find a formula for C in terms of x.

a The cost of buying 5 litres of fuel is £11. ←
   So, the cost of buying 10 litres of fuel is £22,
   the cost of buying 15 litres of fuel is £33,
   the cost of buying 20 litres of fuel is £44,
   and so on.

> 10 litres cost twice as much as 5 litres. 5 litres cost £11, so 10 litres cost 2 × £11 = £22.
>
> 15 litres cost three times as much as 5 litres. 5 litres cost £11, so 15 litres cost 3 × £11 = £33 (and so on).
> Summarise the information in a table.

x	5	10	15	20
C	11	22	33	44

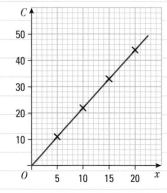

> Plot a graph of C against x.
> Draw a line through the points.
> The graph is a straight line which passes through the point O. So C is directly proportional to x.

The graph shows that the cost, £C, of buying the fuel is directly proportional to the amount, x litres, bought.

b $C = kx$

The point $(11, 5)$ lies on the line, so

$11 = k \times 5$

$k = \frac{11}{5} = 2.2$

The formula is $C = 2.2x$.

> The graph passes through the origin O. The equation of a straight line which passes through the origin has the form $y = mx$. Here $y = C$ and $m = k$ (the constant of proportionality). Find the constant k. Substitute $C = 11$ and $x = 5$ into the formula.
> Write down the formula, substituting $k = 2.2$ into $C = kx$.

Exercise 8A

Questions in this chapter are targeted at the grades indicated.

C

1 The table gives information about the variables x and Y.

x	2	4	6	8	10
Y	3	6	9	12	15

  a Plot the graph of Y against x.

  b Is Y directly proportional to x? Give a reason for your answer.

2 Here is a graph of $Y = kd$. Use the information in the graph to work out the value of k.

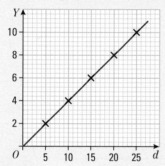

B

3 The table gives information about the variables x and M.

x	4	10	16	20
M	9	22.5	36	45

  a Show that M is directly proportional to x.

  b Given that $M = kx$, work out the value of k.

  c Use your formula to work out the value of M when $x = 32$.

4 Y is directly proportional to t. $Y = 10$ when $t = 20$.

  a Sketch a graph of Y against t.

  b Work out a formula for Y in terms of t.

  c Use your formula to work out the value of Y when $t = 100$.

A

5 The cost (£C) of a bottle of white correcting fluid is directly proportional to the volume (v cm^3) of fluid in the bottle. A bottle containing 4 cm^3 of fluid costs £1.80.

  a Work out the cost of **i** 8, **ii** 12, **iii** 16 cm^3 of the liquid.

  b Work out a formula for C in terms of v.

  c The cost of a bottle of correcting fluid is £5.85. Work out the volume of fluid in the bottle.

8.2 Further direct proportion

⊙ **Objective**

○ You can use formulae to solve problems involving direct proportion.

⊘ **Why do this?**

If you know the currency exchange rate for the country you're on holiday in then you can convert local prices into pounds, to see roughly how expensive goods are.

⊕ **Get Ready**

1. Solve for x

   a $x = \frac{6}{7}a$ when $a = 21$ **b** $\frac{x}{3} = \frac{6}{7}a$ when $a = 28$ **c** $\frac{2}{5}x = \frac{6}{7}a$ when $a = 35$

🔍 **Key Points**

◉ The symbol $\propto$ means 'is proportional to'.

◉ When y is directly proportional to x:

   ◉ $y \propto x$ is the statement of proportionality.

   ◉ $y = kx$ is the formula for direct proportion, where k is the **constant of proportionality**.

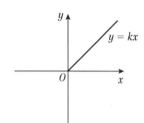

🔍 **Example 2** W is directly proportional to x. $W = 18$ when $x = 1.5$.
Work out the value of W when $x = 7$.

$W \propto x$ ← Write down the statement of proportionality. W is 'directly proportional' to x.

$W = kx$ ← Replace $\propto$ with '$= k$'.

$18 = 1.5k$ ← Work out the value of k. Substitute $W = 18$ and $x = 1.5$.

$k = \frac{18}{1.5} = 12$ ← Divide both sides by 1.5.

So, $W = 12x$ ← Write down the formula, putting in the value of k.

When $x = 7$
$W = 12 \times 7 = 84$ ← Put $x = 7$ into $W = 12x$.

⚙ **Exercise 8B**

B

1   y is directly proportional to x so that $y = kx$. $y = 12$ when $x = 8$. Work out the value of k.

2   B is directly proportional to t so that $B = kt$. $B = 1.75$ when $t = 2.5$. Work out the value of k.

3 P is directly proportional to h. $P = 40.5$ when $h = 18$.

    a   Show that $P = \frac{9}{4}h$.

    b   Work out the value of P when $h = 32$.

    c   Work out the value of h when $P = 27$.

4   The voltage V across a resistor (in volts) is directly proportional to the current c flowing through it (in amps).

    a   Show that $V = 250c$.

    b   Work out the value of c when $V = 9$ volts.

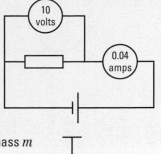

A02

5   The extension E of an elastic string (in mm) is directly proportional to the mass m on the string (in grams).

    a   Find a formula for E in terms of m.

    b   Find the extension of the string when the mass is 450 g.

    c   Find the mass that will extend the string by 52.5 mm.

A02

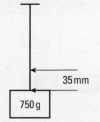

6   The volume, $V\,\text{cm}^3$, of mercury in a tube is directly proportional to the height, h cm, of the tube. When the height of the tube is 12 cm, the volume of mercury is 40 cm³. Work out the volume of mercury in the tube when the height of the tube is 20 cm.

A02

8.3 Writing statements of proportionality and formulae

Objective

● You can write down the statement of proportionality and the formula for a variety of problems.

Why do this?

When scientists are trying to establish a formula, they will often write a statement of proportionality before they perform an experiment and work out the exact results.

Get Ready

1.   **a**   The volume, V, of a gas is directly proportional to its temperature, T.
     Find a formula for V in terms of T.

    b   The pressure, P, is directly proportional to its temperature, T. Find a formula for P in terms of T.

Key Points

● Sometimes quantities are proportional to the square, cube or other power of another quantity.

    ● 'y is proportional to the square of x' so $y \propto x^2$ means $y = k \times x^2$ or $y = kx^2$, where k is the constant of proportionality.

    ● 'y is proportional to the cube of x' so $y \propto x^3$ means $y = k \times x^3$ or $y = kx^3$, where k is the constant of proportionality.

    ● 'y is proportional to the square root of x' so $y \propto \sqrt{x}$ means $y = k \times \sqrt{x}$ or $y = k\sqrt{x}$, where k is the constant of proportionality.

◉ Quantities can also be **inversely proportional** to each other.

 ◉ 'y is inversely proportional to x' so $y \propto \dfrac{1}{x}$ means $y = k \times \dfrac{1}{x}$ or $y = \dfrac{k}{x}$, where k is the constant of proportionality.

◉ Some common proportional graphs are shown below.

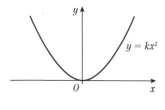
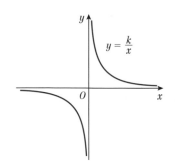
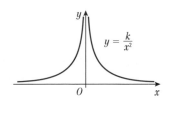

🔍 **Example 3**　Write down **i** the statement of proportionality, **ii** the formula, for each of the following.

    a　H is proportional to the square of d.　　　　**b**　T is proportional to the cube of f.

    c　G is proportional to the square root of p.　　**d**　R is inversely proportional to x.

    e　L is inversely proportional to the square of n.

a　**i**　$H \propto d^2$　　　　**ii**　$H = kd^2$　←   The symbol $\propto$ means 'is proportional to', so replace the words with the symbol $\propto$. The 'square of d' is the same as 'd squared', i.e. d^2.
Replace the symbol $\propto$ with '$= k$'. So $H \propto d^2$ becomes $H = kd^2$.

b　**i**　$T \propto f^3$　　　　**ii**　$T = kf^3$

c　**i**　$G \propto \sqrt{p}$　　　**ii**　$G = k\sqrt{p}$

d　**i**　$R \propto \dfrac{1}{x}$　　　**ii**　$R = \dfrac{k}{x}$　←   You write an inverse proportion as a reciprocal (i.e. 'one over...'). The reciprocal of x is $\dfrac{1}{x}$.

e　**i**　$L \propto \dfrac{1}{n^2}$　　　**ii**　$L = \dfrac{k}{n^2}$　←   The reciprocal of n^2 is $\dfrac{1}{n^2}$.

🔍 **Example 4**　Write these statements of proportionality in words.

    a　$P \propto w^2$　　**b**　$A \propto g^3$　　**c**　$F \propto \sqrt[3]{t}$　　**d**　$Y \propto \dfrac{1}{d}$　　**e**　$B \propto \dfrac{1}{r^3}$

a　P is proportional to the square of w.　←   This could also be written as 'P is proportional to w squared'.

b　A is proportional to the cube of g.

c　F is proportional to the cube root of t.

d　Y is inversely proportional to d.　← $\dfrac{1}{d}$ is the reciprocal of d. So Y is inversely proportional to d.

e　B is inversely proportional to the cube of r.　←   This could also be written as B is inversely proportional to r cubed.

A

Exercise 8C

1 Write down **i** the statement of proportionality, **ii** the formula, for each of the following.
Use the symbol $\propto$.

   a M is directly proportional to n.

   b L is proportional to the square of h.

   c P is proportional to the cube of t.

   d Q is proportional to the square root of y.

   e W is proportional to the cube root of x.

   f A is inversely proportional to b.

   g H is inversely proportional to the square of g.

   h U is inversely proportional to the square of f.

   i E is inversely proportional to the cube of w.

   j V is inversely proportional to the square root of r

2 The number of ants in an ants' nest, N, is proportional to the square of the diameter, d cm, of the ants' nest.

   a Write down the statement of proportionality.

   b Write down a formula for N in terms of d and k (the constant of proportionality).

3 The height, in metres, of a tree is proportional to the square of the circumference, of the tree in metres. Write down a formula for H in terms of c.

A02

4 The quantity of fuel, Q tonnes, used by a rocket is proportional to the cube of the time, t seconds, that the fuel burns. Write down a formula for Q in terms of t.

5 A formula for the force of attraction (F newtons) between two objects is given by $F = \dfrac{k}{r^2}$ where r is the distance between the objects (in metres) and k is a constant. What is the relationship between F and r?

6 A formula for the period (T seconds) of a pendulum is given by $T = k\sqrt{l}$, where l is the length of the pendulum (in metres) and k is a constant. Tom says that T is inversely proportional to the square of l. He is wrong. Explain why.

8.4 Problems involving square and cubic proportionality

◉ Objective

○ You can solve problems involving square and cubic proportions.

◈ Why do this?

The power output of a wind turbine is proportional to the cube of the wind speed.

◈ Get Ready

1. Solve

   a 15^2

   b 8^3

   c 11^3

   d $\sqrt{81}$

   e $\sqrt[3]{512}$

   f $\sqrt{4096}$

◓ Key Points

◉ To solve problems involving square or cubic proportion, first write the statement of proportionality, then the formula.

◉ Substitute given values into the formula to find the solution.

🔍 **Example 5** — The resistance (R newtons) to the motion of a racing car is directly proportional to the square of the speed (v m/s) of the racing car. Given that $R = 5000$ when $v = 10$, work out the value of R when $v = 30$.

$R \propto v^2$ ← Write down the statement of proportionality. R is proportional to the square of v.

$R = kv^2$ ← Write down the formula. Replace the symbol $\propto$ with '$= k$'.

$5000 = k \times 10^2$
$5000 = k \times 100$ ← Work out the value of k. Substitute $R = 5000$ and $v = 10$ into $R = kv^2$.

$k = \dfrac{5000}{100} = 50$ ← Divide both sides of the equation by 100.

So, $R = 50v^2$
When $v = 30$ ← Work out the value of R when $v = 30$. Substitute $v = 30$ into $R = 50v^2$.
$R = 50 \times 30^2$
$R = 50 \times 900 = 45\,000\,\text{N}$

🔍 **Example 6** — $T \propto m^3$. $T = 10$ when $m = 2$. Find the value of m when $T = 80$.

$T \propto m^3$
$T = km^3$ ← Replace the symbol $\propto$ with '$= k$'.

$T = 10$ when $m = 2$, so ← Work out the value of k. Substitute $T = 10$ and $m = 2$ into $T = km^3$.
$10 = k \times 2^3$
$10 = 8k$
$k = 1.25$
So, $T = 1.25m^3$ ← Write down the formula. Substitute $k = 1.25$ into $T = km^3$.

When $T = 80$, ← You need to find m when $T = 80$, so substitute $T = 80$ into $T = 1.25m^3$.
$1.25m^3 = 80$
$m^3 = \dfrac{80}{1.25} = 64$ ← Divide both sides by 1.25.

$m = \sqrt[3]{64} = 4$ ← Take the cube root of both sides.

⚙️ **Exercise 8D**

A

1. A is proportional to the square of x so that $A = kx^2$. $A = 20$ when $x = 5$. Work out the value of k.

2. G is proportional to the cube of f so that $G = kf^3$. $G = 54$ when $f = 1.5$. Work out the value of k.

3. $Z \propto b^2$. $Z = 32$ when $b = 5$. Find a formula for Z in terms of b.

4. $L \propto v^3$. $L = 120$ when $v = 4$. Find a formula for L in terms of v.

5. W is proportional to the square of p. $W = 8$ when $p = 4$. Work out the value of W when $p = 3$.

6 A stone is thrown vertically upwards with a speed of v m/s. The height, H metres, reached by the stone is proportional to the square of v. When $v = 20$ m/s, $H = 20$ m. Work out the value of:

 a H when $v = 15$ m/s

 b v when $H = 10$ m.

7 The volume, V mm³, of a raindrop is proportional to the cube of its radius, r mm. When the radius of a raindrop is 2 mm its volume is 33 mm³. Work out the radius of a raindrop which has a volume of 65 mm³. Give your answer to 2 decimal places.

8 The mass of an earthworm (in grams) is proportional to the cube of its length (in cm). Copy and complete this table.

Mass (grams)	Length (cm)
20	5
	6
43.94	
50	

8.5 Problems involving inverse proportion

⊙ Objective

⊙ You can solve problems involving inverse proportion.

⊙ Why do this?

You can use inverse proportion to work out how much each person that holds a winning ticket in a lottery draw actually wins.

⊙ Get Ready

1. If $V \propto T$ and $P \propto \dfrac{1}{V}$, write P in terms of T.

Key Points

⊙ To solve problems involving inverse proportion, first write the statement of proportionality, then the formula.

⊙ Substitute given values into the formula to find the solution.

Example 7 The number of hours (H) needed to dig a certain hole is inversely proportional to the number of men (x) available to dig the hole. If it takes 5 men 8 hours to dig the hole, how long will it take 6 men to dig the hole?

$H \propto \dfrac{1}{x}$ ← Write down the statement of proportionality. H is inversely proportional to x.

$H = \dfrac{k}{x}$ ← Write down the formula. Replace the symbol $\propto$ with '$= k$'.

$8 = \dfrac{k}{5}$ ← Work out the value of k. Substitute $H = 8$ and $x = 5$ into $H = \dfrac{k}{x}$.

$k = 40$ ← Multiply both sides of the equation by 5.

So, $H = \dfrac{40}{x}$ ← Substitute the value of k into the formula.

When $x = 6$, $H = \dfrac{40}{6} = 6\frac{2}{3}$ hours. ← Work out the value of H when $x = 6$. Substitute $x = 6$ into $H = \dfrac{40}{x}$.

Example 8

The force of attraction (F newtons) between two magnets is inversely proportional to the square of the distance (d cm) between them. When the magnets are 1.5 cm apart, the force of attraction is 32 newtons.

a Find a formula for F in terms of d.

b Work out the distance between the magnets when the force of attraction is 1.125 newtons.

a $F \propto \dfrac{1}{d^2}$

$F = \dfrac{k}{d^2}$

When $d = 1.5, F = 32$

So, $32 = \dfrac{k}{1.5^2}$ ← Multiply both sides by 1.5^2.

$k = 1.5^2 \times 32$

$= 2.25 \times 32$

$= 72$

The formula is $F = \dfrac{72}{d^2}$.

b $F = 1.125$

So $1.125 = \dfrac{72}{d^2}$ ← Multiply both sides by d^2.

$d^2 \times 1.125 = 72$ ← Divide both sides by 1.125.

$d^2 = \dfrac{72}{1.125}$

$d^2 = 64$

$d = \sqrt{64} = 8$ cm ← Take the square root of both sides.

Exercise 8E

1 L is inversely proportional to d so that $L = \dfrac{k}{d}$.

$L = 2.25$ when $d = 20$. Work out the value of k.

2 $A \propto \dfrac{1}{x}$

$A = 3.75$ when $x = 8$.

a Work out the value of A when $x = 12$.

b Work out the value of x when $A = 5$.

3 $G \propto \dfrac{1}{d}$

$G = 4.8$ when $d = 6$.

a Work out the value of G when $d = 8$.

b Work out the value of d when $G = 10.8$.

4 The volume V (m³) of a gas is inversely proportional to the pressure P (N/m²). $V = 4$ m³ when $P = 500$ N/m². Work out the volume of the gas when the pressure is 750 N/m².

5 The frequency F of sound (in hertz) is inversely proportional to the wavelength w (in metres). The musical note of middle C has a frequency of 256 hertz and a wavelength of 1.29 m.

a Work out the frequency of a note with a wavelength of 0.86 m.

b Work out the wavelength of a note with frequency 344 hertz.

6 $M \propto \dfrac{1}{f^2}$

$M = 0.625$ when $f = 8$.

a Work out the value of M when $f = 3.5$.

b Work out the value of f when $M = 1$.

7 The shutter speed (S) of a camera is inversely proportional to the square of the aperture setting (f). When $f = 8$, $S = 125$.

    a Find a formula for S in terms of f.

    b Work out the value of S when $f = 4$.

A02   A

Chapter review

◉ When a graph of two quantities is a straight line through the origin, one quantity is directly proportional to the other.

◉ The symbol $\propto$ means 'is proportional to'.

◉ $y \propto x$ means 'y is directly proportional to x'.

◉ When y is directly proportional to x:

    ◉ $y \propto x$ is the statement of proportionality.

    ◉ $y = kx$ is the formula, where k is the **constant of proportionality**.

◉ Where k is the constant of proportionality:

    ◉ $y = kx^2 = y \propto x^2$ means y is proportional to the square of x.

    ◉ $y - kx^3 = y \propto x^3$ means y is proportional to the cube of x.

    ◉ $y = k\sqrt{x} = y \propto \sqrt{x}$ means y is proportional to the square root of x.

◉ When y is inversely proportional to x:

    ◉ $y \propto \dfrac{1}{x}$ is the statement of proportionality.

    ◉ $y = k \times \dfrac{1}{x}$ or $y = \dfrac{k}{x}$ are ways of writing the formula, where k is the constant of inverse proportionality.

◉ To solve problems involving proportion:

    ◉ write the statement of proportionality

    ◉ write the formula

    ◉ substitute given values into the formula to find the solution.

Review exercise

1 In an experiment, the value of V was measured for different values of d. The results are given in this table. Show that V is directly proportional to d.

A02   B

d	0.5	1.5	3.5	4.5	6.0
V	0.75	2.25	5.25	6.75	9.0

2 The time, T seconds, it takes a water heater to boil some water is directly proportional to the mass of water, m kg, in the water heater.

When $m = 250$, $T = 600$.

    a Find T when $m = 400$.

The time, T seconds, it takes a water heater to boil a costant mass of water is inversely proportional to the power, P watts, of the water heater.

When $P = 1400$, $T = 360$.

    b Find the value of T when $P = 900$.

June 2006

A

3 Copy and complete this table.

a	T is directly proportional to b	
b		$R \propto a$
c	P is proportional to the square of m	
d	Z is proportional to the cube of g	
e		$H \propto \dfrac{1}{y}$

4 $S \propto p.\ S = 12$ when $p = 8$.

a Find a formula for S in terms of p.

b Sketch a graph of S against p.

ResultsPlus
Exam Question Report

79% of students answered this sort of question well. They knew how to move from a statement of proportionality to a formula.

A03 **5** For an oil company, the cost (£C) of drilling a hole is directly proportional to the depth (d metres) of the hole. The cost of drilling a hole to a depth of 100 metres is £8500. Work out the cost of drilling a hole to a depth of 825 metres.

A03 * **6** The pressure P of water on a diver (in bars) is directly proportional to his depth d (in metres). When the diver is at a depth of 5 metres the pressure on the diver is 0.5 bars. For safety reasons the pressure on a particular diver must not exceed 6.5 bars. The diver wants to dive to a depth of 60 m. Can the diver do this safely? Give a reason for your answer.

A03 **7** For batteries having the same length, the energy stored in a battery is proportional to the square of the circumference of the battery. When the circumference is 3.5 cm the energy stored in the battery is 5 units. The radius of a battery is 5 cm. Work out the energy stored in the battery.

8 When a stone is dropped from a cliff it travels a distance D (in metres) after a time t (in seconds). This table gives information about the stone.

t	0	1	2	3	4	5
D	0	5	20			

a Show that $D \propto t^2$, and complete the table.

b Work out the time taken for the stone to travel 15 metres. Give your answer correct to 2 decimal places.

9 The resistance, R ohms, of a particular cable is inversely proportional to the square of its radius, r mm. Copy and complete this table. Give your answers correct to 2 decimal places.

Radius (r mm)	Resistance (R ohms)
10	500
15	
17.5	
	250

10 X is proportional to the square root of h. $X = 3$ when $h = 16$.

a Show that $X = 0.75\sqrt{h}$.

b Work out the value of X when $h = 25$.

c Work out the value of h when $X = 6$.

11 In a particular industrial process, Germalex is added to increase the speed of a chemical reaction The time taken, T seconds, for the reaction is inversely proportional to the square root of the mass, m grams, of Germalex added. When 50 grams of Germalex is added the time taken for the reaction is 20 seconds.

   a Show that $T = 100\sqrt{\dfrac{2}{m}}$.

   b It is required that the time taken for a particular reaction should be 15 seconds. Work out the mass of Germalex that needs to be added to achieve this reaction time. Give your answer to 3 significant figures.

12 Here are 4 sketch graphs.

   A 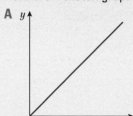   **B** 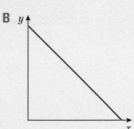   **C** 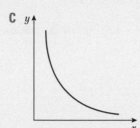   **D**

   a Write down the letter of the graph which shows y is directly proportional to x.

   b Write down the letter of the graph which shows y is inversely proportional to x.

13 M is directly proportional to L^3.

   When $L = 2$, $M = 160$.

   a Find the value of M when $L = 3$.

   b Find the value of L when $M = 120$.

June 2009, adapted

14 q is inversely proportional to the square of t.

   When $t = 4$, $q = 8.5$.

   a Find a formula for q in terms of t.

   b Calculate the value of q when $t = 5$.

ResultsPlus

Exam Question Report

78% of students answered this question poorly because they did not read the details of the question properly.

June 2008

15 The time, T seconds, for a hot sphere to cool is proportional to the square root of the surface area, $A\ m^2$, of the sphere.

   When $A = 100$, $T = 40$.

   Find the value of T when $A = 60$.

   Give your answer correct to 3 significant figures.

June 2009

16 y is directly proportional to the square of x.

   x is directly proportional to the square root of z.

   a Find a formula for y in terms of z and a constant of proportionality.

   u is directly proportional to the square of v.

   v is inversly proportional to the square root of w.

   b Show that the product of u and w is constant.

117

9 TRANSFORMATIONS OF FUNCTIONS

It can take just a transformation to make a new song successful. Music producers use sound mixers to change the pitch and volume of different sounds, as well as to add different layers of sounds and other effects.

◎ Objectives

In this chapter you will:
- use function notation
- learn the relationship between simple transformations of curves and their effect on the equations of curves.

◈ Before you start

You should be able to:
- identify transformations
- solve an equation in x for a given value of x.

9.1 Using function notation

⊙ Objective

⊙ You can use function notation.

? Why do this?

Computer programmers use function notation as a means of shorthand for long lines of code.

⬆ Get Ready

1. Work out the value of $2x^2$ when $x = 3$.
2. Find the value of x if $4x - 3 = 8$.
3. If $y = x^2$ and $x = t + 1$, express y in terms of t.

🔍 Key Point

⊙ A **function** $y = f(x)$ is a rule for working out values of y when given values of x.

Example 1

$y = f(x) = 2x + 3$
Find the values of

a $f(5)$ b $f(-4)$ c a where $f(a) = 5$.

a $f(5) = 2 \times 5 + 3 = 13$ ⟵   Substitute $x = 5$ in $2x + 3$.
b $f(-4) = 2 \times (-4) + 3 = -5$
c $f(a) = 2a + 3 = 5$ ⟵   Solve the equation to find the value of a.
   so $a = 1$

Example 2

$y = g(x) - 2x^2 + 1$
a Find the value of $g(-3)$.
b Find the value of $2g(1)$.
c What is the algebraic expression for $3g(x - 1)$?

a $g(-3) = 2 \times (-3)^2 + 1 = 19$
b $2g(1) = 2 \times (2 \times 1^2 + 1) = 6$ ⟵   Work out $g(1)$ first and then multiply by 2.

c $g(x - 1) = 2(x - 1)^2 + 1 = 2x^2 - 4x + 3$ ⟵   Replace x by $(x - 1)$ in the expression $2x^2 + 1$. Then multiply by 3.
   $3g(x - 1) = 3(2x^2 - 4x + 3)$
   $= 6x^2 - 12x + 9$

⚙ Exercise 9A

Questions in this chapter are targeted at the grades indicated.

1 $f(x) = 3x^2$, $g(x) = \frac{4}{x}$
   Find the values of
   a $f(2)$ b $f(0)$ c $f(-4)$ d $g(4)$
   e $g(-1)$ f $g\left(\frac{1}{2}\right)$

B

B

2 $f(x) = 2x^3$, $g(x) = x^2 + x$

Find the values of

    a $f(1) + g(1)$ **b** $f(2) + g(3)$ **c** $f(2) \times g(2)$ **d** $\dfrac{f(4)}{g(4)}$

A

3 $f(x) = 2x + 2$

    a Find the value of $f(3)$.

    b $f(a) = 6$ Find the value of a.

A*
A02

4 $g(x) = x^2 - 4$

    a Find the values of   **i** $g(0)$ **ii** $g(1)$ **iii** $g(-2)$.

    b $g(k) = 12$ Find the values of k.

A02

5 $g(x) = (x - 3)(x + 4)$

    a Find the values of   **i** $g(5)$ **ii** $g(0)$ **iii** $3g(-2)$.

    b $g(a) = 0$ Find the values of a.

A02
A03

6 $f(x) = x(x - 4)$

    a Find the values of   **i** $f(1)$ **ii** $f(2)$ **iii** $2f(-1)$.

    b $f(k) = 0$ Find the values of k.

    c $f(m) = 5$ Find the values of m.

A02

7 $f(x) = x^2$

    a Find $f(4)$.   **b** Write out in full $f(x) - 4$.   **c** Write out in full $f(x - 4)$.

A02

8 $g(x) = 4(x + 1)$

    a Find $g(-3)$.   **b** Write out in full $g(2x)$.   **c** Write out in full $3g(x)$.

9.2 Translation of a curve parallel to the axes

◎ Objective

● You understand the relationship between the translation of a curve parallel to an axis and the change in its function form.

�ʔ Why do this?

You could draw the graphs of the paths of two balls being juggled on the same axes. The graph of the second ball is a translation along the x-axis of the graph of the first ball.

⬆ Get Ready

1. The point P $(2, 3)$ is translated by 2 units parallel to the y-axis. Find the new coordinates.
2. The point P $(2, 3)$ is translated by 2 units parallel to the x-axis. Find the new coordinates.
3. The point P $(2, 3)$ is translated by $\begin{pmatrix} -3 \\ 0 \end{pmatrix}$. Find the new coordinates.

🔑 Key Points

● The relationship between the curves given by $y = f(x)$ and $y = f(x) + a$ is a translation of a units parallel to the y-axis or a translation by $\begin{pmatrix} 0 \\ a \end{pmatrix}$.

● The relationship between the curves given by $y = f(x)$ and $y = f(x + a)$ is a translation of $-a$ units parallel to the x-axis or a translation of $-\begin{pmatrix} a \\ 0 \end{pmatrix}$.

Example 3

Here is the graph of $y = f(x)$ where
$$f(x) = x^2$$

a Draw the graph with equation $y = f(x) + 3$.

b Describe the transformation that maps
$y = f(x)$ to $y = f(x) + 3$.

c Write the algebraic form for $y = f(x) + 3$.

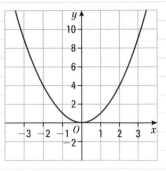

a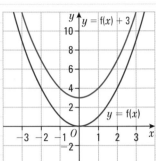

Each point on the curve $y = f(x)$ is moved up by 3 units.

b The transformation is a translation by $\begin{pmatrix} 0 \\ 3 \end{pmatrix}$.

c $f(x) = x^2$ so $f(x) + 3 = x^2 + 3$
The equation of the transformed curve is $y = x^2 + 3$.

Example 4

Here is a sketch of the graph of $y = f(x)$ where
$f(x) = x^2$.

a Describe the transformation which maps
$y = f(x)$ to $y = f(x - 2)$.

b Sketch the curve with equation $y = f(x - 2)$.
The coordinates of the minimum point of
$y = f(x)$ are $(0, 0)$.

c Write down the coordinates of the minimum
point of $y = f(x - 2)$.

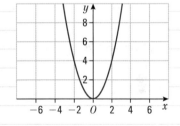

a Translation by $+2$ units parallel to the x-axis or translation by $\begin{pmatrix} 2 \\ 0 \end{pmatrix}$.

b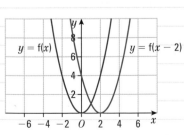

c $(0, 0)$ is mapped to $(0, 2)$.

Exercise 9B

A*
A03

1 Here is the graph of $y = f(x) = x^2$.

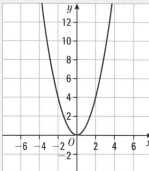

Draw the graphs of:

a $y = f(x) + 3$ **b** $y = f(x) - 4$

c $y = f(x + 2)$ **d** $y = f(x - 1)$.

A03

2 Here is a sketch of the graph of $y = f(x) = x^3$.

a Draw sketches of the graphs of: **i** $y = f(x) + 3$ **ii** $y = f(x - 1)$.

b Write down the coordinates of the point to which the point $(0, 0)$ is mapped in each case.

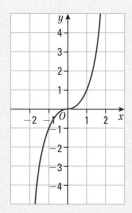

A03

3 Here is a sketch of the graph of $y = f(x) = \dfrac{1}{x}$.

The curve $y = f(x)$ is translated by $\begin{pmatrix} 0 \\ 2 \end{pmatrix}$.

a Sketch the graph of the new curve.

b Write down the coordinates to which the point $(2, 0.5)$ is mapped.

c Write down the equation of the translated curve:

   i in function form

   ii in algebraic form.

The curve $y = f(x)$ is now translated by $\begin{pmatrix} -2 \\ 0 \end{pmatrix}$.

d Sketch the transformed curve.

e The point $(1, 4)$ is mapped to the point (p, q).

   Write down the values of p and q.

f Write down the equation of the translated curve:

   i in function form

   ii in algebraic form.

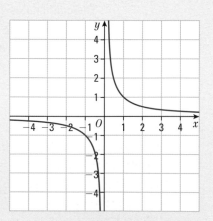

4

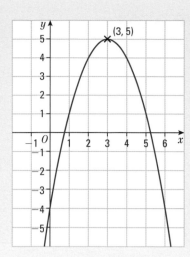

Here is a curve with equation $y = f(x)$.
The maximum point of the curve is (3, 5).

a Write down the coordinates of the maximum point of $y = f(x) - 3$.

b Write down the coordinates of the maximum point of $y = f(x + 2)$.

A03 A*

5 Here are two curves, C_1 and C_2. The equation of the curve C_1 is $y = f(x)$.
The curve C_1 can be mapped to the curve C_2 by a translation.
The maximum point of C_1 is (2, 4) and the maximum point of C_2 is (2, 7).

a Describe the translation.

b Write down the equation of the curve C_2 in function form.

The algebraic equation of the curve C_1 is $y = 4x - x^2$.

c Write down the algebraic equation of the curve C_2.

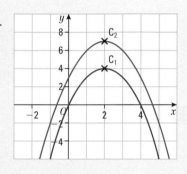

A03

6 Here are two curves, C_1 and C_2. The equation of the curve C_1 is $y = f(x)$.
The curve C_1 can be mapped to the curve C_2 by a translation.

a Describe the translation.

b Write down the equation of the curve C_2 in function form.

The algebraic equation of the curve C_1 is $y = x^2$.

c Write down the algebraic equation of the curve C_2.

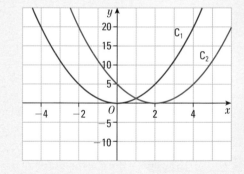

A03

7 The expression $x^2 + 4x + 9$ can be written in the form $(x + a)^2 + b$ for all values of x.

a Find the value of a and the value of b.

The graph of $x^2 + 4x + 9$ can be obtained from the graph of $y = x^2$ by a translation.

b Describe this translation.

c Sketch the graph of $y = x^2$.

d Sketch the graph of $x^2 + 4x + 9$ on the same axes.

A03

8 Describe fully the transformation that will map the curve with equation $y = x^2$ to the curve with equation $y = x^2 - 6$.

A02
A03

9.3 Stretching a curve parallel to the axes

Objective

You understand the effect that stretching a curve parallel to one of the axes has on its function form.

Why do this?

If you were designing a bridge in the style of the Golden Gate Bridge, you could experiment with the length of the support struts. Taller struts would stretch the shape of the curved cable.

Get Ready

1. Draw the triangles with coordinates (1, 1), (3, 1), (3, 2) and (2, 1), (6, 1), (6, 2). What is the relationship between the two triangles?
2. Draw the triangles with coordinates (1, 1), (3, 1), (3, 2) and (1, 3), (3, 3), (3, 6). What is the relationship between the two triangles?

Key Points

- The relationship between the curves $y = f(x)$ and $y = af(x)$ (where a is a constant) is that of a **stretch** of magnitude a parallel to the y-axis.
- The relationship between the curves $y = f(x)$ and $y = f(ax)$ (where a is a constant) is that of a stretch of magnitude $\frac{1}{a}$ parallel to the x-axis.

Example 5

Here is the graph of $y = f(x)$.

a Sketch the graph of $y = 2f(x)$.

b To what point is the point (2, 1) mapped?

a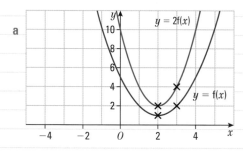

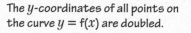

The transformation is a stretch parallel to the y-axis of magnitude 2.

The y-coordinates of all points on the curve $y = f(x)$ are doubled.

b The minimum point (2, 1) of $y = f(x)$ is mapped to the minimum point (2, 2) of $y = f(2x)$.

Example 6

Here is the graph of $y = f(x)$.

a Sketch the graph of $y = f(2x)$.

b To what point is the point (2, 1) mapped?

a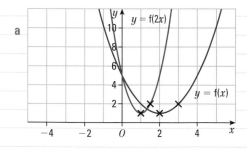

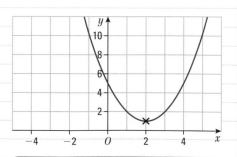

The x-coordinates of all points on the curve $y = f(x)$ are halved.

b The point (2, 1) of $y = f(x)$ is mapped to the point (0.5, 1) of $y = f(2x)$.

Example 7

Here is the graph of $y = f(x) = \sin x°$.

a Draw the graph of $y = 2f(x)$.

b Write down the algebraic equation of $y = 2f(x)$.

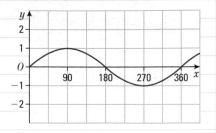

a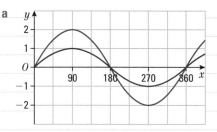

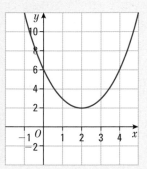

The graph is stretched parallel to the y-axis.

b $y = 2\sin x°$

Exercise 9C

1 Here is the graph of $y = f(x)$. It has a minimum point at $(2, 2)$.
Draw the graph of $y = f(2x)$.
To which point is the minimum point of $y = f(x)$ mapped?

A03 A*

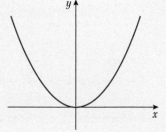

2 Here is a sketch of the curve $C_1\ y = f(x) = x^2$.

a Sketch the curve C_2 with equation $y = 4f(x)$.

b Write down the equation of the curve C_2 in algebraic form.

c Give two different transformations that will each map the curve C_1 to the curve C_2.

A02 A03

3 Here is the graph of $y = f(x) = \cos x°$.

a Draw the graph of $y = 2f(x)$.
Write the equation of $y = 2f(x)$ in algebraic form.

b Draw the graph of $y = f(2x)$.
Write the equation of $y = f(2x)$ in algebraic form.

A02

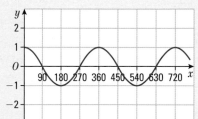

4 Here is the graph of $y = f(x)$.
The graph crosses the y-axis at $(0, 4)$ and the x-axis at $\left(-\frac{2}{3}, 0\right)$.

a Sketch the graph with equation $y = 3f(x)$.

b Write down the coordinates of the points to which $(0, 4)$ and $\left(-\frac{2}{3}, 0\right)$ are mapped.

A03

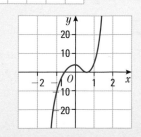

A*

5   Here is the graph of $y = f(x) = 1 + \sin x°$.
On separate graphs, sketch the curves with equations: a $y = 2f(x)$
                                                          b $y = f(2x)$
                                                          c $y = f(x) + 2$.

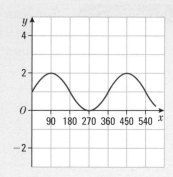

A03

6   Here is the graph of $y = f(x) = \cos x°$.
a Sketch the graph with equation $y = f\left(\dfrac{x}{2}\right)$.
b How many solutions does the equation
   $f\left(\dfrac{x}{2}\right) = 0.5$ have in the range $0 < x < 360$?

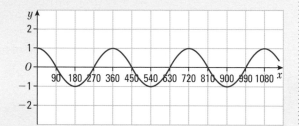

A03

7   Here is a sketch of the curve $C_1, y = f(x) = x(x - 4)$.
The curve C_1 has a minimum point at $(2, -4)$.
The curve C_1 is mapped to the curve C_2 by a stretch.
The minimum point on C_2 is $(4, -4)$.
The minimum point on C_1 is mapped to the minimum point of C_2.
a Describe the stretch fully.
b Draw a sketch of C_2.
c Write the equation of C_2:   i using functional form
                                  ii in algebraic form.

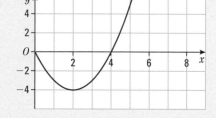

A02
A03

8   The expression $x^2 - 8x + 5$ can be written in the form $(x - p)^2 + q$.
a Find the values of p and q.
b Write down the coordinates of the point P where the curve $y = f(x) = x^2 - 8x + 5$ crosses the y-axis.
The curve $y = f(x)$ is mapped by a stretch parallel to the y-axis, so that the point P is mapped to the point $(0, 3)$.
c Describe the stretch and write down the equation of the new curve.

Enlargement

Key Points

◉ The relationship between the curves $y = f(x)$ and $y = af\left(\dfrac{x}{a}\right)$ (where a is a constant) is that of an enlargement, centre the origin and scale factor a.

Example 8

The graph of the curve C_1 $y = f(x) = x^2 + 1$ is shown on the grid.

On the same grid,

a sketch the curve C_2 with equation $y = 2f(x)$

b sketch the curve C_3 with equation $y = 2f\left(\dfrac{x}{2}\right)$

c Describe fully the transformation that maps

i C_1 to C_2 ii C_2 to C_3 iii C_1 to C_3

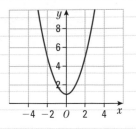

a C_2 is in black.

b C_3 is in blue.

c i Stretch parallel to the y-axis of magnitude 2.

  ii Stretch parallel to the x-axis of magnitude 2.

iii Enlargement with scale factor 2 and centre the origin.

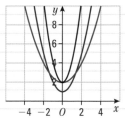

Exercise 9D

1 Here is a graph of the curve with equation $y = \sin x°$.

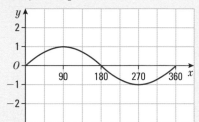

a Copy the graph and sketch the curve C_2,
the enlargement of C_1 with scale factor 2.

b Write down the equation of the curve C_2.

A03   A★

2 a Draw a sketch of the U-shaped curve C_1, with equation $y = x(x + 2)$.

The curve C_1 is enlarged with a scale factor 2, centre O, to give the curve C_2.

b Find the equation of the curve C_2 in function form.

c Sketch the curve C_2.

A03

3 a Sketch the curve with equation $y = f(x) = \cos x°$ for values of x from 0 to 1080.

b On the same axes, sketch the graph of the curve with equation $y = 3f\left(\dfrac{x}{3}\right)$.

A03

4 Write down the algebraic equation of the given curve after it has been enlarged by scale factor 4 and centre O.

a $y = x^2 + 3$ b $y = \dfrac{1}{x} + 1$ c $y = 2^x$ d $y = 2\sin(2x)$

A03

9.4 Rotation about the origin and reflection in the axes

⊙ Objectives

● You understand the effect that a reflection of a curve in one of its axes has on its function form.
● You understand the effect that a rotation of a curve by 180° about the origin has on its function form.

⟐ Why do this?

If two cars race between two towns, but one starts in one town and the other car starts in another, the graph of one car's displacement against time will be a reflection of the other car's graph.

⬦ Get Ready

1. Draw the line joining the origin to the point (4, 5). Reflect this line in the x-axis.
2. Draw the line joining (1, 0) to the point (4, 3). Reflect this line in the y-axis.
3. Draw the line joining (1, 0) to the point (4, 3). Rotate this line by 180° about the origin.

◎ Key Points

● The curve $y = f(-x)$ is a reflection in the y-axis of the curve $y = f(x)$.
● The curve $y = -f(x)$ is a reflection in the x-axis of the curve $y = f(x)$.
● The curve $y = -f(-x)$ is a rotation by 180° about the origin of the curve $y = f(x)$.

◉ Example 9

Here is the graph of $y = f(x)$.
Sketch the curves with equations:

a $y = f(-x)$ **b** $y = -f(x)$ **c** $y = -f(-x)$.

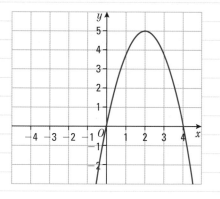

a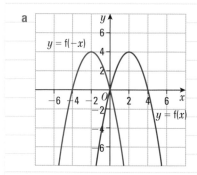

b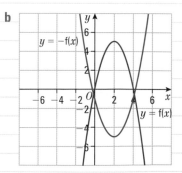

c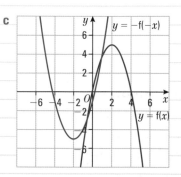

Exercise 9E

1 The graph of $y = f(x) = x^2 + 3$ has been drawn.
  a Sketch the graph of $y = -f(x)$.
  b Write down the equation of the new graph in algebraic form.

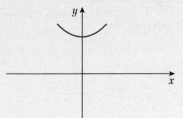

A03

2 The graph of $y = f(x) = x^3$ has been drawn.
  a Sketch the graph of $y = f(-x)$.
  The point $(2, 8)$ on $y = f(x) = x^3$ has been mapped to
  the point (p, q).
  b Write down the values of p and q.
  c Write down the equation of the new curve.
    Give your answer in algebraic form.

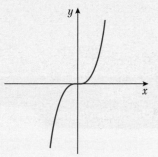

A03

3 The graph of $y = f(x) = 2^x$ has been drawn.
  a Sketch the graph of $y = -f(-x)$.
  The point $(1, 2)$ has been mapped to the point (r, t).
  b Write down the values of r and t.
  c Write down the equation of the new curve.
    Give your answer in algebraic form.

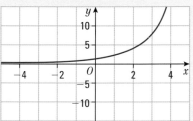

A03

Chapter review

⊚ A **function** $y = f(x)$ is a rule for working out values of y when given values of x.
⊚ You should know the following transformations:
  ⊚ $y = f(x) + a$ is a translation by $+a$ units, parallel to the y-axis of $y = f(x)$
  ⊚ $y = f(x + a)$ is a translation by $-a$ units, parallel to the x-axis of $y = f(x)$
  ⊚ $y = af(x)$ is a **stretch** of magnitude a units parallel to the y-axis of $y = f(x)$
  ⊚ $y = f(ax)$ is a stretch of magnitude $\frac{1}{a}$ units parallel to the x-axis of $y = f(x)$
  ⊚ $y = af\left(\frac{x}{a}\right)$ is an enlargement, centre the origin and scale factor a, of $y = f(x)$
  ⊚ $y = f(-x)$ is a reflection in the y-axis of $y = f(x)$
  ⊚ $y = -f(x)$ is a reflection in the x-axis of $y = f(x)$
  ⊚ $y = -f(-x)$ is a rotation by $180°$ about the origin of $y = f(x)$.

Review exercise

1 $f(x) = x^2 + 2$
  Work out   **a** $f(2)$ **b** $f(-3)$ **c** a where $f(a) = 2$.

B

2 $f(x) = x^2 + 3x$ Show that $f(x-1) = (x + 2)(x - 1)$.

A03 A

3 Sketch the graph of $y = x^2$. Hence sketch the graph of $y = -x^2$.

A★ A03

4 The graph of $y = f(x)$ is shown on the grid.

a Copy the graph and then draw the graph of $y = f(x) + 2$ on the same axes.

b On another copy of the graph, draw the graph of $y = -f(x)$.

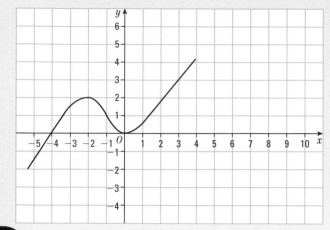

ResultsPlus
Exam Question Report

55% of students answered this sort of question well. They understood how translation affects the shape of a graph.

June 2007, adapted

A03

5 The curve with equation $y = f(x)$ is translated so that the point at (0, 0) is mapped onto the point (4, 0).

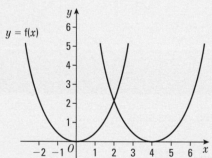

ResultsPlus
Exam Question Report

68% of students answered this question poorly because they did not check their answers by substituting in key values of x.

Find an equation of the translated curve.

Nov 2007, adapted

A02 A03

6 The diagram shows a sketch of the graph of $y = x^2 - x$.

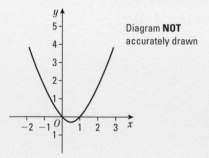

Diagram **NOT** accurately drawn

a On the same diagram, sketch and label the graph of
$y = (x - 1)^2 - (x - 1)$.
Show clearly where this graph crosses the x-axis and where it crosses the y-axis.

b On the same diagram sketch and label the graph of $y = 3(x^2 - x)$.

c Write down the solutions of the equation $(x - 1)^2 - (x - 1) = 0$.

Nov 2005

7 The diagram shows a sketch of part of the curve $y = \sin x°$.
  a Write down the coordinates of the point A.
  b On the same diagram, sketch the graph of $y = \sin 2x°$.

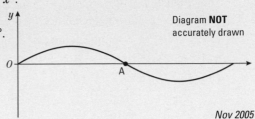

Diagram **NOT** accurately drawn

Nov 2005

A03 **A***

8 The diagram shows a sketch of part of the curve $y = \sin x°$.

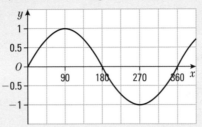

On a suitable grid, draw the graph of $y = 4\sin 2x°$

A03

9 Describe the transformation that will map the curve $y = x^2$ to the curve $y = x^2 - 8x + 11$.

A02
A03

10 The curve $y = x^2$ can be mapped to the curve $y = 4x^2$ using a single transformation.
  a Describe one possible transformation.
  b Describe another possible transformation.

A02
A03

11 The equation of the curve C_1 is $y = f(x) = 8 + 4x - x^2$.
  a Write $8 + 4x - x^2$ in the form $q - (x - p)^2$
     where p and q are numbers to be found.
  Here is a sketch of the curve $y = 8 + 4x - x^2$.
  b Write down the coordinates of the maximum point of the curve.
  The curve C_1 is stretched to the curve C_2 so that the maximum point of C_1 is mapped to $(2, 24)$.
  c Describe the stretch.
  d Write down the equation of C_2 in function form.

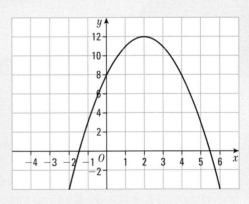

A02
A03

The average price of a wedding cake is between £250 and £500, with the most spectacular costing up to an amazing £2000. Cake designers often wrap ribbon around the edge of the cake but how could they work out how much ribbon they need? They could try and measure it using a piece of string or they could use maths. If they knew they had baked the cake in a 25 cm cake tin, then they would know that the diameter of the cake is 25 cm and they can easily calculate the circumference.

Objectives

In this chapter you will:

- know and use the formulae for the circumference and area of a circle
- find the length of an arc, the area of a sector of a circle, and answer problems involving circles in terms of π
- convert between units of area.

Before you start

You need to:

- be able to solve problems involving perimeters and areas
- know what a circle, semicircle and quarter circle are, and be able to name the parts of a circle and related terms
- know how to draw circles and arcs to a given radius.

10.1 Circumference and area of a circle

◉ Objectives

- ◉ You can work out the circumference of a circle.
- ◉ You can work out the area of a circle.
- ◉ You can solve problems involving circles, including semicircles and quarter circles.

❔ Why do this?

To fit a new tyre on the wheel of your bike, you may need to know the circumference of the wheel to find the correct size.

⬆ Get Ready

1. Draw a circle of radius 5 cm. For this circle, draw and label clearly:

   a a radius **b** a diameter **c** a chord **d** a sector **e** an arc **f** a segment **g** a tangent.

◉ Key Points

◉ For all circles $\dfrac{\text{circumference of circle}}{\text{diameter of circle}} = \pi$ (pi).

This value cannot be found exactly.

To 3 decimal places, $\pi = 3.142$.

circumference of circle $= \pi \times$ diameter of circle

$C = 2\pi r$

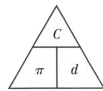

$C = \pi \times d$

$d = C \div \pi$

Results**Plus**
Examiner's Tip

Calculator exam papers have the following instruction about π.
'If your calculator does not have a π button, take the value of π to be 3.142 unless the question instructs otherwise.'

Results**Plus**
Watch Out!

It is important not to confuse the diameter with the radius.

🔍 Example 1

Work out the circumference of a circle with:

a diameter 8.7 cm

b radius 3.1 m.

Give your answers correct to 3 significant figures.

Results**Plus**
Examiner's Tip

Remember that the circumference is approximately 3 times the diameter or 6 times the radius.

a $C = \pi \times 8.7$ ← Use $C = \pi d$ with $d = 8.7$ cm.

   $= 27.3318\ldots$

   Circumference $= 27.3$ cm ←

   Use the π button or 3.142.
   Write down at least 4 figures of the calculator display.
   Give the answer correct to 3 significant figures. The units are the same as the diameter (cm).

b $C = 2 \times \pi \times 3.1$ ←

   $= 19.47787\ldots$

   Circumference $= 19.5$ m

   The diameter can be worked out from $d = 2r$ so $d = 2 \times 3.1 = 6.2$ and then use $C = \pi d$.
   Or use $C = 2\pi r$ with $r = 3.1$ m.
   The units are the same as the radius (m).

Example 2   The circumference of a circle is 84.3 cm. Work out the radius of the circle.
Give your answer correct to 3 significant figures.

$84.3 = 2 \times \pi \times r = 2\pi \times r$ ← Use $C = 2\pi r$ with $C = 84.3$ cm as the radius is given in the question.
Divide both sides by 2π and write down at least 4 figures of the calculator display.

$r = 84.3 \div (2\pi)$
$= 13.4167\ldots$

Radius $= 13.4$ cm ← Give the answer correct to 3 significant figures. The units are the same as the circumference (cm).

Watch Out!

Be careful when dividing by 2π on a calculator. It is best to use brackets.

Exercise 10A

Questions in this chapter are targeted at the grades indicated.

In this exercise, if your calculator does not have a π button, take the value of π to be 3.142. Give answers correct to 3 significant figures unless a question says differently.

D

1   Work out the circumference of a circle with diameter:
   a 7 cm   **b** 12.9 mm   **c** 5.6 cm   **d** 40 cm   **e** 21.9 m

2   The radius of a basketball net hoop is 23 cm.
   a Work out the circumference of a basketball net hoop.

   A netball hoop has a radius of 19 cm.
   b Work out how much longer the circumference of a basketball net hoop is than the circumference of a netball hoop.

C

3   The circumference of a CD is 37.7 cm. Work out the radius of the CD.

4   The diameter of the front wheel of Michael's bicycle is 668 mm.
   a Work out the circumference of the wheel.
   Give your answer in cm correct to the nearest cm.

   Michael rides his bicycle.
   b Work out the distance cycled when the wheel makes 1000 complete turns.
   Give your answer in km correct to 2 significant figures.

   The distance Michael rides his bicycle is 6 km.
   c Work out the number of complete turns made by this wheel.

A02
A03

A02
A03

5   The length of the minute hand of a watch is 1.2 cm.
   a Work out the distance moved by the point end of the hand in 1 hour.
   b Work out the distance moved by the point end of the hand in: **i** 6 hours **ii** 20 minutes.

A02

6 A circular table has a radius of 95 cm.
a Work out the circumference of the table.
The circumference of a circular tablecloth is 1.5 m.
The tablecloth is put symmetrically on the table so that the distance from the table to the edge of the tablecloth is the same all around the table.
b Work out the distance from the table to the edge of the tablecloth.

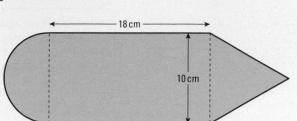

7 The diagram shows a shape made from a semicircle, a rectangle and an equilateral triangle.
The rectangle has length 18 cm and width 10 cm.
Work out the perimeter of the shape.

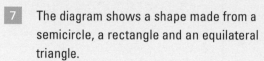

Area of a circle

Key Points

To find the area of a circle means to find the area enclosed by the circle.
Here is a circle that has been divided into four equal wedges or sectors. The sectors are then arranged as shown to form a parallelogram-like shape.

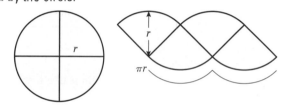

The length shown as πr is half the circumference, $2\pi r$, of the circle.
The area of the circle is the same as the area of the shape.
Here is what happens when the circle is divided into more sectors.

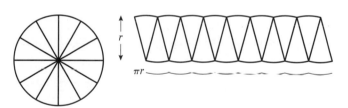

The shape looks more like a parallelogram and as the number of sectors increases the parallelogram becomes more like a rectangle.

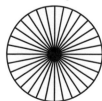 

The width of this rectangle is equal to half of the circumference of the original circle and the height of the rectangle is equal to the radius of the circle.

Area of circle = area of rectangle = $\pi r \times r = \pi r^2$

Taking A as the area of a circle and r as the radius of the circle, $A = \pi r^2$
That is Area = $\pi \times$ radius $\times$ radius

Example 3 Work out the area of a circle with: **a** a radius of 9 cm **b** a diameter of 12.8 m.
Give your answers correct to 3 significant figures.

a $A = \pi \times 9^2$ ← Use $A = \pi r^2$ with $r = 9$ cm.
   $= 254.4690\ldots$ Write down at least 4 figures of the calculator display.

   Area $= 254$ cm^2 ← Give the answer correct to 3 significant figures.
   As the units of the radius are cm, the units of the area are cm^2.

b Radius $= 12.8 \div 2$ m ← Divide the diameter by 2 to get the radius.
   $= 6.4$ m Write down at least 4 figures of the calculator display.
   $A = \pi \times 6.4^2$ Give the answer correct to 3 significant figures.
   $= 128.6796\ldots$

   Area $= 129$ m^2 ← As the units of the radius are m, the units of the area are m^2.

 Results**Plus**
Examiner's Tip

When the diameter of a circle is given, to work out the area of the circle first find the radius by dividing the diameter by 2.

Example 4 Work out the radius of a circle with area 46 cm^2.

   $46 = \pi \times r^2$ ← Use $A = \pi r^2$ with $A = 46$ cm^2.

   $r^2 = 46 \div \pi = 14.64225\ldots$ ← Work out the value of r^2 by dividing both sides by π.

   $r = \sqrt{14.64225\ldots} = 3.8265\ldots$ ← Take the square root to find the value of r.

   Radius $= 3.83$ cm

Exercise 10B

In this exercise, if your calculator does not have a π button, take the value of π to be 3.142.
Give answers correct to 3 significant figures unless the question says differently.

1 Work out the area of a circle with radius:
   a 8 cm   **b** 12.7 cm   **c** 28.5 mm   **d** 9.72 cm   **e** 12.6 m

2 Work out the area of a circle with diameter:
   a 24 cm   **b** 8.3 cm   **c** 0.95 m   **d** 58.4 mm   **e** 18.26 cm

D

3 The diagram shows a pond surrounded by a path.
a Work out the area of the blue region of the pond.
b Work out the area of the path.
c The path is made of shingle that costs £1.95 per square metre of path. Work out the cost of the shingle to make the path.

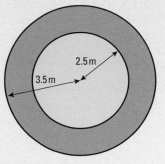

4 The diagram represents the plan of a sports field. The field is a rectangle with semicircular ends. The rectangle has length 100 m and width 70 m. The semicircles have diameter 70 m.
a Work out the area of the field.

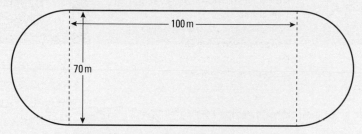

The field is to be covered in fertiliser that costs 23p per square metre.
b Work out the cost of the fertiliser for the field.

5 A circle of diameter 8 cm is cut from a piece of yellow card.

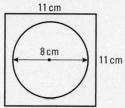

The card is in the shape of a square of side 11 cm.
The card shown yellow in the diagram is thrown away.
Work out the area of the card thrown away.

6 A, B and C are three circles. Circle A has radius 5 cm and circle B has radius 12 cm. Circle C is such that area of circle C = area of circle A + area of circle B. Work out the radius of circle C.

7 The diagram shows a star made by removing four identical quarter circles from the corners of a square of side 30 cm.

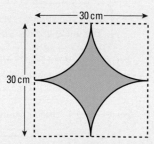

Work out the area of the star.

10.2 **Sectors of circles**

◎ **Objectives**

● You can work out the length of an arc of a circle.

● You can work out the area of a sector of a circle.

● You can solve problems involving arc lengths and areas of sectors of circles, including finding the area of a segment of a circle.

❓ **Why do this?**

Astronomers often use the arc length of a circle when working out distances in space.

⬧ **Get Ready**

1. Here is a circle and a sector with an angle of 60°.

   a How many of these sectors will fill the circle without overlapping?

   b What fraction of the circle is the sector?

   c What fraction of the circle is a sector with an angle of 40°?

🔍 **Key Points**

◉ For a sector with angle $x°$ of a circle with radius r:

sector $= \dfrac{x}{360}$ of the circle so

area of sector $= \dfrac{x}{360} \times \pi r^2$

and arc length $= \dfrac{x}{360} \times 2\pi r$

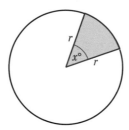

🔍 **Example 5**

For this sector of a circle, work out:

a the arc length

b the perimeter.

Give your answers correct to 3 significant figures.

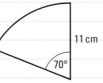

11 cm

70°

a arc length $= \dfrac{70}{360} \times 2 \times \pi \times 11$ ← Use arc length $= \dfrac{x}{360} \times 2\pi r$ with $x = 70$ and $r = 11$ cm.

arc length $= 13.4390...$ ← Write down at least 4 figures of the calculator display.

arc length $= 13.4$ cm ← Give the answer correct to 3 significant figures. The units are the same as the radius (cm).

b perimeter $= 13.4390 + 2 \times 11$ ← perimeter $=$ arc length $+$ two radii.

perimeter $= 35.4390...$ ← Use the unrounded value for the arc length.

perimeter $= 35.4$ cm ← Give the answer correct to 3 significant figures.

Example 6

Calculate the area of this sector.
Give your answer correct to 3 significant figures.

130°
9 m

$\text{area} = \frac{130}{360} \times \pi \times 9^2$ ←

area of sector $= \frac{x}{360} \times \pi r^2$
with $x = 130$ and $r = 9$ m.

area = 91.8915...

area of sector = 91.9 m² ← r is in metres so area is in m².

ResultsPlus
Examiner's Tip

Always write down the angle at the centre of the circle as a fraction of 360.

Exercise 10C

In this exercise, if your calculator does not have a π button, take the value of π to be 3.142.
Give answers correct to 3 significant figures unless the question says differently.

1 Calculate the arc length of each of these sectors.

a
20° 5 cm

b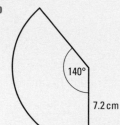
140°
7.2 cm

c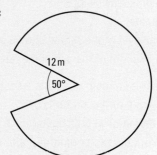
12 m
50°

2 Calculate the area of each of these sectors.

a
7.6 cm 110°

b
4.7 cm
48°

c
10 cm 96°

3 Calculate the perimeter of each of these sectors.

a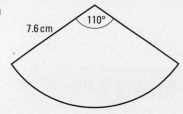
6 cm 70°

b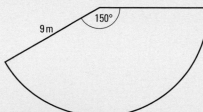
150°
9 m

A

A* A02 A03

4 The diagram shows a sector of a circle of radius 5.5 cm.
The length of the arc of the sector is 6.72 cm.
Work out the size of the angle, x, of the sector.
Give your answer to the nearest degree.

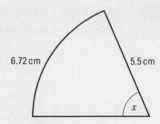

6.72 cm   5.5 cm x

A02 A03

5 The diagram shows a sector of a circle.
The area of the sector is 17.453 cm².
Work out the radius of the circle.
Give your answer correct to the nearest cm.

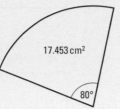

17.453 cm²   80°

A02 A03

6 The diagram shows information about the throwing circle and the landing area for a
discus competition.
The radius of the throwing circle is 1.25 m.
Distances are measured from the front of the circle as shown.
The discus must land in the sector shown green in the
diagram. The angle of the sector is 40°.
The winning throw in the men's discus in the 2008
Olympics was 68.82 m.
Calculate:
a the area of the region shown green
b the length of the arc of the landing area sector.

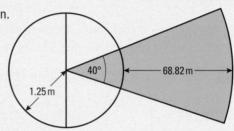

40°   68.82 m   1.25 m

A02 A03

7 Here is a shape made from a sector of a circle of radius 16 cm and a semicircle.
The angle of the sector is 60°.
The diameter of the semicircle is a radius of the sector.
a Work out the perimeter of the shape.
b Work out the area of the shape.

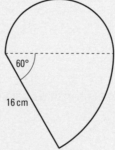

60°   16 cm

10.3 Problems involving circles in terms of π

◉ Objectives

- You can give answers in terms of π.
- You can solve problems when the information is given in terms of π.

❔ Why do this?

It is impossible to calculate the exact value of π so when scientists or architects need an exact answer it has to be given in terms of π.

◈ Get Ready

1. Simplify **a** $2 \times a \times 5$ **b** $\dfrac{12 \times b}{4}$

2. Find the value of x when $4ax = 18a$

Example 7 A circle has a radius of 6 cm. Find: **a** the circumference of the circle **b** the area of the circle. Give your answers in terms of π.

a $C = 2 \times \pi \times 6 = 12\pi$ ← Use $C = 2\pi r$ with $r = 6$ cm.

Circumference $= 12\pi$ cm ← $2 \times 6 = 12$
Do not forget the units.

b $A = \pi \times 6^2 = \pi \times 36 = 36\pi$ ← Use $A = \pi r^2$ with $r = 6$ cm.

Area $= 36\pi$ cm^2 ← $6^2 = 36$

Example 8 The perimeter of this sector is $(2r + 12\pi)$ m. Find the radius, r m, of the sector.

r m

arc length $= \frac{90}{360} \times 2 \times \pi \times r = \frac{1}{4} \times 2\pi \times r$ ← Find the arc length of the sector.

arc length $= \frac{1}{2}\pi r$

perimeter $= 2r + \frac{1}{2}\pi r$ ← Add $2r$ to the arc length to find the perimeter of the sector.

$2r + \frac{1}{2}\pi r = 2r + 12\pi$ ← Use the given expression $2r + 12\pi$.

$\frac{1}{2}\pi r = 12\pi$ ← Solve for r.

$\frac{1}{2}r - 12$

$r = 2 \times 12 = 24$
Radius of sector $= 24$ m

A02
A03

Exercise 10D

1 Giving your answer in terms of π, find the circumference of a circle:
   a with diameter 9 cm   **b** with diameter 1 m   **c** with radius 26 mm

2 Giving your answer in terms of π, find the area of a circle:
   a with radius 2 cm   **b** with radius 10 m   **c** with diameter 40 m

3 The diagram shows two circles with the same centre.
   a Show that the area of the coloured region between the two circles is 64π cm^2.
   b Find, in terms of π, the circumference of a circle whose area is the same as the area of the coloured region.

A02
A03

D

C

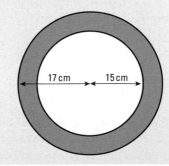

17 cm   15 cm

B

4 The blue arc and the red arc make a complete circle.

The length of the blue arc is 6π cm and the length of the red arc is 10π cm.

a Find the radius of the circle.

b Find, in terms of π, the area of the circle.

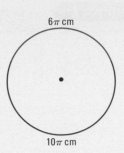

6π cm

10π cm

A
A03

5 Find, in terms of π:

a the area of this sector

b the arc length of the sector

c the perimeter of the sector.

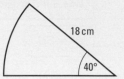

18 cm

40°

A02

6 A shape is made by drawing three semicircles on the line AB as shown. Two of the semicircles have diameter X cm and Y cm as shown.

Show that the perimeter of the shape is the same as the circumference of a circle with diameter $(X + Y)$ cm.

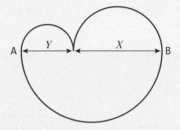

A   Y   X   B

10.4 Units of area

◎ Objective

● You can change between m^2 and cm^2, between cm^2 and mm^2 and between km^2 and m^2.

❓ Why do this?

It is sometimes necessary to change from one unit to another, for example, a garden design may have measurements in both metres and centimetres.

◈ Get Ready

1. Work out   **a** 10×10 **b** 100×100 **c** 1000×1000

2. Work out   **a** 5×100 **b** $13 \times 10\,000$ **c** $7.6 \times 1\,000\,000$

3. Work out   **a** $7000 \div 100$ **b** $3500 \div 10\,000$ **c** $49\,000 \div 1\,000\,000$

◍ Key Points

◉ The diagram shows two identical squares.

The sides of square A are measured in metres and the sides of square B are measured in centimetres.

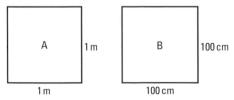

A   1 m

1 m

B   100 cm

100 cm

Square A is 1 m by 1 m so the area of square A is 1×1 m^2 = 1 m^2.

Square B is 100 cm by 100 cm so the area of square B is 100×100 cm^2 = 10 000 cm^2.

The squares have the same area so 1 m^2 = 100×100 cm^2 = 10 000 cm^2.

There are similar results for other units.

Length	Area
1 cm = 10 mm	$1\,cm^2 = 10 \times 10 = 100\,mm^2$
1 m = 100 cm	$1\,m^2 = 100 \times 100 = 10\,000\,cm^2$
1 km = 1000 m	$1\,km^2 = 1000 \times 1000 = 1\,000\,000\,m^2$

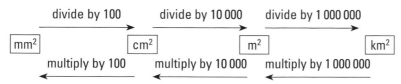

divide by 100   divide by 10 000   divide by 1 000 000

mm^2 cm^2 m^2 km^2

multiply by 100   multiply by 10 000   multiply by 1 000 000

Example 9   Convert 4.6 m² to cm².

$$4.6\,m^2 = 4.6 \times 10\,000\,cm^2$$
$$= 46\,000\,cm^2$$

$1\,m^2 = 10\,000\,cm^2$
Multiply the number of m² by 10 000.

Remember that to change from a larger unit to a smaller unit, you multiply.

Example 10   Convert 870 mm² to cm².

$$870\,mm^2 = 870 \div 100\,cm^2 = 8.7\,cm^2$$

$1\,cm^2 = 100\,mm^2$
So $1\,mm^2 = \frac{1}{100}\,cm^2$
Divide the number of mm² by 100.

Exercise 10E

1 Work out the area of this circle in:
    a cm²   **b** m².

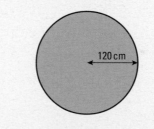

120 cm

2 Work out the area of this triangle in:
    a cm²   **b** mm².

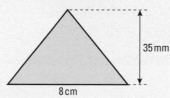

35 mm

8 cm

3 Convert to cm².
    a 4 m²   **b** 6.9 m²   **c** 600 mm²   **d** 47 mm²

4 Convert to m².
    a 5 km²   **b** 0.3 km²   **c** 40 000 cm²   **d** 560 cm²

5 **a** How many mm are there in 1 m?   **b** How many mm² are there in 1 m²?
    c Convert 8.3 m² to mm².

6 Find, in cm², the area of a rectangle:
    a 3.2 m by 1.4 m   **b** 45 mm by 8 mm.

D

C

Chapter review

- For all circles, $\dfrac{\text{circumference of circle}}{\text{diameter of circle}} = \pi$ (pi).

- To 3 decimal places, $\pi = 3.142$.

- Circumference of a circle $= \pi d = 2\pi r$ where d is the diameter of the circle, and r is the radius of the circle.
  Area of a circle $= \pi r^2$ where r is the radius of the circle.

- For a sector with angle $x°$ of a circle with radius r:

  area of sector $= \dfrac{x}{360} \times \pi r^2$

  and arc length $= \dfrac{x}{360} \times 2\pi r$

Length	Area
1 cm = 10 mm	1 cm^2 = 10 × 10 = 100 mm^2
1 m = 100 cm	1 m^2 = 100 × 100 = 10 000 cm^2
1 km = 1000 m	1 km^2 = 1000 × 1000 = 1 000 000 m^2

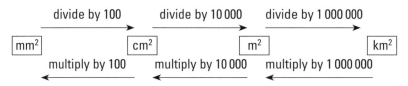

Review exercise

1 Convert to cm^2.
  a 450 mm^2 **b** 6 m^2

2 The area of a large farm is 6 540 000 m^2.
  Convert 6 540 000 m^2 to km^2.

3 A ring-shaped flowerbed is to be created around a circular lawn of radius 2.55 m.

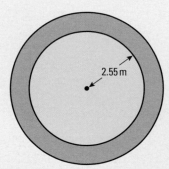

2.55 m

Roses costing £4.20 are to be planted approximately every 50 cm around this flowerbed.
How much money will be needed for roses?

4 The diagram shows a garden that includes a lawn, a vegetable patch, a circular pond and a flowerbed.
All measurements are shown in metres.
The lawn is going to be relaid with turf costing £4.60 per square metre.
How much will this cost?

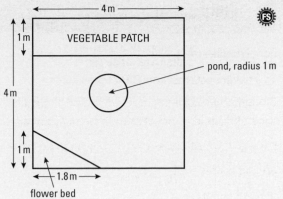

5 A running track consists of two 60 m straights and two semicircular bends of diameter 60 m.

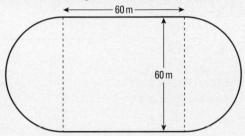

a Find the length of one lap of this running track.
b The owners of the track wish to stage athletics meetings and need it to be exactly 400 m long.
This can be done by altering the straights or widening the bends.
Calculate what adjustments would need to be made.

6 Discs of diameter 2 cm are cut from a metal strip that is 2 cm by 100 cm.

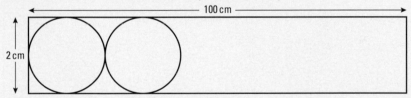

What is the minimum amount of waste material?

7 The diagram shows an equilateral triangle ABC with sides of length 6 cm.
P is the midpoint of AB.
Q is the midpoint of AC.
APQ is a sector of a circle, centre A.

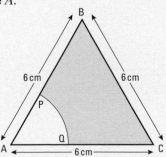

Calculate the area of the shaded region.
Give your answer correct to 3 significant figures.

June 2009

A

8 The diagram shows a sector of a circle, centre O.
The radius of the circle is 6 cm.
Angle $AOB = 120°$.

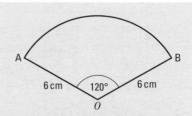

Work out the perimeter of the sector.
Give your answer in terms of π in its simplest
form.

ResultsPlus
Exam Question Report

82% of students answered this question poorly
because they forgot to include all the parts of the
perimeter.

May 2009

A02 A03

9 OAD is a sector of a circle centre O radius 6 cm.
OBC is a sector of a circle centre O radius 8 cm.
OAB and ODC are straight lines.
Angle $COB = 45°$.
Find, in terms of π:
a the area of $ABCD$
b the perimeter of $ABCD$.

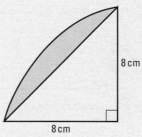

A* **A02 A03**

10 Calculate the area of the shaded segment of this quarter circle.

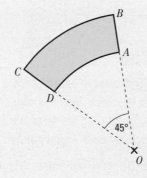

8 cm

8 cm

11 VOLUME

Hot air balloons rely on the buoyancy of hot air in order to lift their load. The volume of air required to lift a certain load depends on many things including the temperature of the air inside and outside the balloon, the altitude above sea level and the humidity of the atmosphere. However, approximately 3.91 m² of air is required within the balloon to lift each 1 kg. So, to lift a 60 kg adult, a balloon containing 234.6 m² of air would be required.

◉ Objectives

In this chapter you will:
- work out the volumes and surface areas of 3D shapes
- convert between units of volume
- solve problems with the compound measure density.

◈ Before you start

You need to know how to:
- work out the volumes and surface areas of cuboids and prisms
- use the formulae for the circumference and area of a circle
- convert between units of area.

11.1 Volume of a cylinder

⊙ Objective

● You can work out the volume of a cylinder.

? Why do this?

You could work out the volume of liquid that your mug can hold if you wanted to boil only that exact amount of water, to save energy.

⊕ Get Ready

1. Find the area of these circles:

   a radius 3 cm   **b** diameter 5 cm   **c** radius 10 cm.

🕐 Key Point

● Volume of **cylinder** = area of cross-section × length
$$= \pi r^2 h$$
where r is the radius and h is the height.

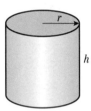

🔍 Example 1

Work out the volume of this cylinder.
Give your answer in terms of π and to 3 significant figures.

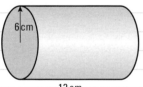

12 cm

> The cross-section of the cylinder is a circle with radius 6 cm. Remember: area of circle = π × radius². Take π as 3.142.

Area of cross-section = $\pi \times 6^2$
$$= 36\pi$$

Volume of cylinder = $3.142 \times 6 \times 6 \times 12$ ⟵ Use volume of cylinder = area of cross-section × length.

$$= 1357.344 \text{ cm}^3$$ ⟵ Do not round your answer at this stage. Write down all the digits on your calculator display.

$$= 1360 \text{ cm}^3 \text{ (3 s.f.)}$$ ⟵ Give your final answer correct to 3 significant figures.

Exercise 11A

Questions in this chapter are targeted at the grades indicated.

C

1   Work out the volumes of these cylinders.
Give your answers correct to 3 significant figures.

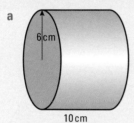

a   4 cm   5 cm

b   240 mm   300 mm

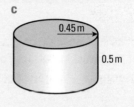

c   30 mm   5 cm

d   12 cm   79 cm

2   Work out the volumes of these cylinders. Give your answers in terms of π.

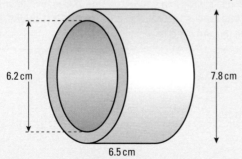

a   6 cm   10 cm

b   20 cm   6.5 cm

c   0.45 m   0.5 m

3   An aircraft hangar has a semicircular cross-section of diameter 20 m.
The length of the hangar is 32 m.
Work out the volume of the hangar. Give your answer in terms of π.

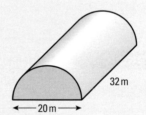

32 m   20 m

A03

4   An annulus has an external diameter of 7.8 cm, an internal diameter of 6.2 cm and a length of 6.5 cm.
Work out the volume of the annulus. Give your answer correct to 1 decimal place.

A03   **B**

6.2 cm   7.8 cm   6.5 cm

5   A gold coin has a height of 2.5 mm and a volume of 2000 mm^3. Work out the diameter of the gold coin.
Give your answer correct to 2 decimal places.

A03

6   An oil drum has a radius of 0.9 m and a height of 1.4 m. The oil drum is completely filled with oil.
Work out the volume of the oil in the oil drum.
Give your answer correct to 3 significant figures.

**A02
A03**

11.2 Volume of a pyramid and a cone

◎ Objective

● You can work out the volume of a pyramid and a cone.

? Why do this?

A manufacturer of ice-cream cones would use the volume of a cone formula to work out the volume of ice-cream that each different-sized cone requires.

⟡ Get Ready

1. Work out the area of:
   a a rectangle with sides 1.1×2.01 cm
   b a triangle of height 3.2 cm and base 9.1 mm
   c a circle with diameter 9.1 mm.

Key Points

● Volume of **pyramid** $= \frac{1}{3} \times$ area of base $\times$ vertical height
● Volume of **cone** $= \frac{1}{3} \times$ area of base $\times$ vertical height
  $= \frac{1}{3}\pi r^2 h$

where r is the radius and h is the height.

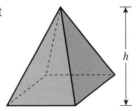
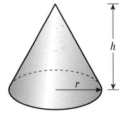

Example 2

A pyramid has a square base of side 3.6 m and a vertical height of 5 m.
Work out the volume of the pyramid.

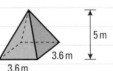

Volume of pyramid $= \frac{1}{3} \times 3.6 \times 3.6 \times 5$ ⟵ | Use volume of pyramid $= \frac{1}{3} \times$ area of base $\times$ vertical height.

$= \frac{1}{3} \times 64.8 = 21.6 \text{ m}^3$ ⟵ | Here the base is a square of side 3.6 m, so the area of the base $= 3.6 \times 3.6 \text{ m}^2$. The vertical height of the pyramid is 5 m.

Example 3

A cone has a circular base of radius 7 cm and a vertical height of 16.2 cm.
Work out the volume of the cone.
Give your answer correct to 1 decimal place.

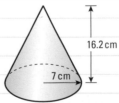

Volume of cone $= \frac{1}{3} \times \pi \times 7^2 \times 16.2$

$= \frac{1}{3} \times 3.142 \times 7 \times 7 \times 16.2$

A cone is a pyramid with a circular base.
Use volume of pyramid $= \frac{1}{3} \times$ area of base $\times$ vertical height.

Here area of base $= \pi \times 7^2 \text{ cm}^2$ and vertical height $= 16.2$ cm.

$= \frac{1}{3} \times 2493.796248\ldots$
$= 831.2654161\ldots$
$= 831.3 \text{ cm}^3 \text{ (1 d.p.)}$

A

Exercise 11B

1   Work out the volumes of these pyramids.

   a

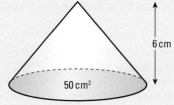

5 m

6 m

6 m

   b

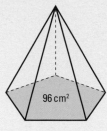

96 cm²

   c

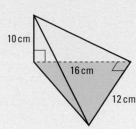

15 cm

10 cm

16 cm

12 cm

2   The Great Pyramid at Giza has height of 262 m and a square base of side 434 m.
       Work out the volume of the Great Pyramid.

3   Work out the volumes of these cones. Use $\pi = 3.142$.
       Give your answers correct to 3 significant figures.

   a

6 cm

50 cm²

   b

5.2 cm

4.8 cm

   c

0.32 cm

0.75 cm

4   The diagram shows a shape made from a cone with
       base radius 4 cm and height 4 cm joined to a cone
       with base radius 4 cm and height 6 cm.
       Work out the total volume of the shape.
       Give your answer in terms of π.

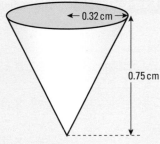

4 cm

4 cm   6 cm

A03

5   The diagram shows a cone cut into two parts,
       part A and part B. Work out the volume of part B.

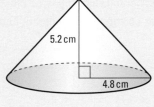

A

10 cm

16 cm

B

5 cm

24 cm

A02
A03

A★

11.3 Volume of a sphere

⑦ Why do this?

A factory making footballs would have to work out the volumes of their different types of footballs to know how much to inflate them.

⊕ Get Ready

1. Work out the area of a circle with diameter 6.4cm:
   a in terms of π **b** with $\pi = 3.142$
2. Calculate
   a 4^3 **b** 6^3 **c** 11^3

Key Point

⊙ Volume of **sphere** $= \frac{4}{3}\pi r^3$, where r is the radius.

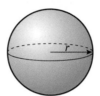

Example 4   The radius of a spherical raindrop is 1.8 mm. Work out the volume of the raindrop. Give your answer correct to 3 significant figures.

Volume of raindrop $= \frac{4}{3} \times \pi \times 1.8^3$

← Use volume of sphere $= \frac{4}{3}\pi r^3$.
Here $r = 1.8$ mm.
Remember 1.8^3 means $1.8 \times 1.8 \times 1.8$.

$= \frac{4}{3} \times 3.142 \times 1.8 \times 1.8 \times 1.8$

$= \frac{4}{3} \times 18.324144$
$= 24.432192$
$= 24.4 \text{ mm}^3 \ (3 \text{ s.f.})$

Example 5   The volume of a spherical ball is 0.86 m³. Work out the radius of the ball. Give your answer correct to 3 significant figures. Use $\pi = 3.142$.

Let the radius of the ball $= r$ metres. ← Write down an equation in terms of r, the radius of the spherical ball.

ResultsPlus
Examiner's Tip

Show all stages in your working.

So, $\frac{4}{3}\pi r^3 = 0.86$ ← The volume of the ball is 0.86 m³, so $\frac{4}{3}\pi r^3 = 0.86$.

$3 \times \frac{4}{3}\pi r^3 = 2.58$ ← To make r the subject of the equation, multiply both sides by 3, divide both sides by 4, divide both sides by 3.142 and then take the cube root of both sides.
$\pi r^3 = 0.645$
$r^3 = 0.2052832591$
$r = \sqrt[3]{0.2052832591}$
$= 0.5899083062$
radius $= 0.590 \text{ m} \ (3 \text{ s.f.})$

In the following questions, use the value of π on your calculator and give your answers correct to 3 significant figures.

1 A spherical soap bubble has a radius of 4 cm. Work out the volume of the bubble.

2 A hemispherical dome has a diameter of 25 m. Work out the volume of the dome.

3 The volume of a sphere is 2400 cm³. Work out the radius of the sphere.

4 The dimensions of a cuboid are 30 cm × 20 cm × 40 cm. The volume of a sphere has the same volume as the cuboid. Work out the radius of the sphere.

5 A spherical ball is made from plastic. The external and internal diameters of the ball are 50 cm and 49.5 cm, respectively. Work out the volume of plastic used to make the ball.

6 The volume of a sphere of radius r metres is twice the volume of a sphere of radius 4 metres. Work out the value of r.

A

A03

A03 **A***

A03

11.4 Further volumes of shapes

⊙ Objective

○ You can work out the volumes of harder shapes made from cuboids, cylinders, cones, pyramids and spheres.

⊘ Why do this?

Many toys, such as a wooden toy trains, are made up of combinations of simple shapes, such as cuboids and cylinders.

⊙ Get Ready

1. Calculate the volume of:

a a cylinder of radius 2 cm and length 5 mm

b a triangular pyramid where the length of the base of the triangle is 12 cm, the height of the base triangle is 6 cm and the height of the pyramid is 24 cm.

🔍 Key Point

◉ To work out the volume of a composite shape, work out the volumes of the shapes it is made from and add the volumes together.

🔍 Example 6

Work out the volume of this shape.

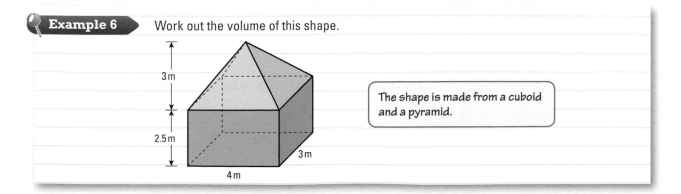

The shape is made from a cuboid and a pyramid.

3 m

2.5 m

3 m

4 m

3 m

Volume of cuboid $= 4 \times 3 \times 2.5 = 30 \, m^3$

> Work out the volume of the cuboid.
> Use volume of cuboid $= a \times b \times c$.
> Here $a = 4$ m, $b = 3$ m and $c = 2.5$ m.

Volume of pyramid $= \frac{1}{3} \times 4 \times 3 \times 3 = 12 \, m^3$

> Work out the volume of the pyramid.
> Use volume of pyramid $= \frac{1}{3} \times$ area of base $\times$ vertical height.
> Here area of the base $= 4 \times 3 \, m^2$ and vertical height is 3 m.

Total volume $= 30 + 12 = 42 \, m^3$ ←

> Work out the total volume. Add the volume of the cuboid and the volume of the pyramid.

Example 7 Work out the volume of this shape. Leave your answer in terms of π.

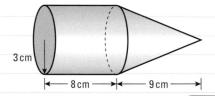

> The shape is made from a cylinder and a cone.

3 cm

8 cm   9 cm

Volume of cylinder $= \pi \times 3^2 \times 8 = 72\pi \, cm^3$

> Work out the volume of the cylinder.
> Use volume of cylinder $= \pi r^2 h$. Here $r = 3$ cm and $h = 8$ cm.

Volume of cone $= \frac{1}{3} \times \pi \times 3^2 \times 9 = 27\pi \, cm^3$

> Work out the volume of the cone.
> Use volume of cone $= \frac{1}{3} \times$ area of base $\times$ vertical height $= \frac{1}{3}\pi r^2 h$.
> Here $r = 3$ cm and $h = 9$ cm.

Total volume $= 72\pi + 27\pi = 99\pi \, cm^3$ ←

> Work out the total volume in terms of π. Add the volume of the cylinder and the volume of the cone.

Exercise 11D

A A02 A03

1 Work out the volumes of these shapes. Give your answers correct to 3 significant figures.

a
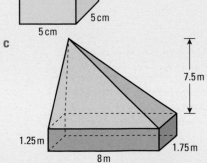
5 cm
5 cm
5 cm
5 cm
5 cm

b
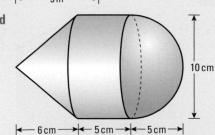
6 m
7 m
5 m

c
7.5 m
1.25 m
1.75 m
8 m

d
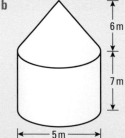
10 cm
6 cm   5 cm   5 cm

2 A cone is joined to a cylinder, as shown in the diagram.
Work out the total volume of the shape. Give your answer in terms of π.

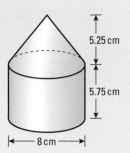

5.25 cm

5.75 cm

8 cm

3 A hemisphere is joined to a cylinder, as shown in the diagram.
Work out the volume of the shape.
Give your answer in terms of π.

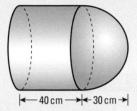

40 cm   30 cm

4 A wooden cube of side 15 cm is used to make a spherical ball.
Work out the volume of wood that must be cut from the cube
to make a ball with the largest possible radius.

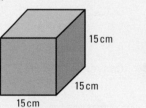

15 cm

15 cm

15 cm

5 A conical hole is cut into a cylinder to make the shape shown in the diagram.
Work out the volume of the shape. Give your answer in terms of π.

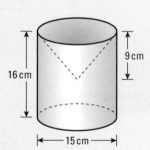

16 cm

9 cm

15 cm

6 A shape is made by joining a hemisphere of radius r cm to a cone of radius r cm.
The height of the cone is $2r$ cm. Find an expression, in terms of r and π, for
the volume of the shape.

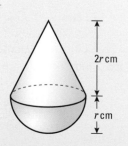

2r cm

r cm

A02
A03

A02
A03

A02 A*

A02
A03

A02
A03

A

11.5 Units of volume

⊙ Objectives

- You can convert between volume measures in metric units
- You can convert between units of volume and units of capacity.

❓ Why do this?

Factories need to convert between units of volume when they pack small boxes, like DVDs, into larger boxes for transporting.

⊕ Get Ready

1. Calculate:
   a $1000 \times 1000 \times 1000$
   b 4.62×1000
   c $35\,m \times 125\,cm \times 2.1\,m$

🔑 Key Points

- $1\,m^3 = 1\,000\,000\,cm^3$ $1\,cm^3 = \frac{1}{1\,000\,000}\,m^3$
- $1\,cm^3 = 1000\,mm^3$ $1\,mm^3 = \frac{1}{1000}\,cm^3$
- Litres are often used to measure the capacity or amount a container can hold.
- $1\,litre = 1000\,cm^3$
- $1\,cm^3 = 1\,ml$

🔍 Example 8

Convert $4\,m^3$ to cm^3.

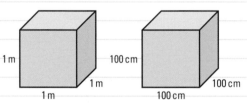

> Convert m^3 to cm^3.
> Draw a cube of side 1 m.
> Now 1 m = 100 cm,
> so 1 m × 1 m × 1 m =
> 100 cm × 100 cm × 100 cm.

$1\,m^3 = 100\,cm \times 100\,cm \times 100\,cm = 1\,000\,000\,cm^3$ ← Replace $1\,m^3$ with $1\,000\,000\,cm^3$.

So $4\,m^3 = 4 \times 1\,000\,000\,cm^3 = 4\,000\,000\,cm^3$

🔍 Example 9

Convert $358\,mm^3$ to cm^3.

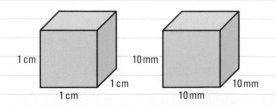

> Convert cm^3 to mm^3.
> Draw a cube of side 1 cm.
> Now 1 cm = 10 mm,
> so 1 cm × 1 cm × 1 cm =
> 10 mm × 10 mm × 10 mm.

$1\,cm^3 = 10\,mm \times 10\,mm \times 10\,mm = 1000\,mm^3$ ←

So $1\,mm^3 = \frac{1}{1000}\,cm^3$

$358\,mm^3 = 358 \times \frac{1}{1000}\,cm^3$

$= 0.358\,cm^3$

> Convert mm^3 to cm^3.
> $1\,cm^3 = 1000\,mm^3$, so $1\,mm^3 = \frac{1}{1000}\,cm^3$.
> Replace $1\,mm^3$ with $\frac{1}{1000}\,cm^3$,
> so $358\,mm^3 = 358 \times \frac{1}{1000}\,cm^3$.

Example 10
   a Convert 3.5 litres to cm^3.
   b Convert 17 000 cm^3 to litres.

a $3.5 \times 1000 = 3500\,cm^3$
b $17\,000 \div 1000 = 17\,litres$

Exercise 11E

1 Convert these to cm^3.
   a $2\,m^3$ **b** $6.75\,m^3$ **c** $450\,mm^3$ **d** $6.8\,mm^3$

2 Convert these to mm^3.
   a $7\,cm^3$ **b** $3.75\,cm^3$ **c** $0.025\,cm^3$

3 Convert these to m^3.
   a $75\,000\,cm^3$ **b** $800\,cm^3$ **c** $125\,000\,mm^3$

4 Convert to litres.
   a $830\,ml$ **b** $5600\,cm^3$ **c** $1\,m^3$ **d** $3540\,mm^3$

5 A swimming pool has length 50 m, width 9 m and depth 1.6 m.
How much water does it hold?
Give your answer in litres.

6 A cylinder hold 34.5 litres of molten metal.
The metal is to be made into cubes of side 3 cm.
How many cubes can be made?

7 The table gives the volumes of three shapes.
Which shape has the greatest volume?
Give a reason for your answer.

Shape	Volume
A	$1.25 \times 10^7\,cm^3$
B	$2.45\,m^3$
C	$3.75 \times 10^8\,mm^3$

11.6 Further surface area of shapes

◉ Objective

○ You can work out the surface area of further shapes made from cylinders, cones and spheres.

⑦ Why do this?

If you have an oddly shaped present that you need to wrap, you can work out the surface area of the present so that you know how much wrapping paper you will need.

⬦ Get Ready

1. What is the radius of a circle if the circumference is $6.4\,\pi\,cm$?
2. What is the diameter of a circle if the area is $12.96\,\pi\,cm^2$?
3. What is the radius of a cylinder which has a volume of $44.064\,mm^3$ and length 3.4 mm?

Key Points

● Total surface area of cylinder $= 2\pi rh + 2\pi r^2$, where r is the radius and h is the height.

● Total surface area of cone $= \pi r^2 + \pi rl$, where r is the radius and l is the slant height.

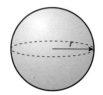

● Surface area of sphere $= 4\pi r^2$, where r is the radius.

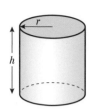

Example 11 A cylindrical can has a radius of 4 cm and a height of 12 cm. Work out the surface area of the can. Use $\pi = 3.142$. Give your answer correct to 3 significant figures.

Surface area $= 2 \times \pi \times 4 \times 12 + 2 \times \pi \times 4^2$
$= 2 \times 3.142 \times 4 \times 12 +$
$\quad 2 \times 3.142 \times 4 \times 4$
$= 301.632 + 100.544$
$= 402.176$
$= 402 \text{ cm}^2 \text{ (3 s.f.)}$

> Use total surface area of cylinder $= 2\pi rh + 2\pi r^2$.
> Here $r = 4$ cm and $h = 12$ cm.
> Replace π with 3.142.
> Remember $4^2 = 4 \times 4$.
> Write down all the figures on your calculator display.
> Give your final answer correct to 3 significant figures.
> Remember to give the units with your answer.

Example 12 A cone has a radius of 3 cm and a vertical height of 4 cm.
Work out the total surface area of the cone. Give your answer in terms of π.

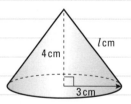

> To work out the total surface area of a cone you need to find the slant height of the cone. Label the slant height, l, in the diagram.

Let the slant height $= l$ cm.
Using Pythagoras' Theorem,
$l^2 = 4^2 + 3^2$
$\quad = 16 + 9 = 25$
$l = \sqrt{25} = 5 \text{ cm}$

> Remember Pythagoras' theorem: $c^2 = a^2 + b^2$.
> Here $c = l$, $a = 4$ cm and $b = 3$ cm.

Total surface area of cone
$= \pi \times 3^2 + \pi \times 3 \times 5$
$= 9\pi + 15\pi$
$= 24\pi \text{ cm}^2$

> Use total surface area of cone $= \pi r^2 + \pi l$.
> Here $r = 3$ cm and $l = 5$ cm.

Example 13

A hemisphere has a radius r cm.

Show that the total surface area of the hemisphere is $3\pi r^2$ cm^2.

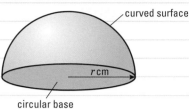
curved surface

r cm

circular base

> A hemisphere is half a sphere, so the area of the curved surface $= \frac{1}{2} \times 4\pi r^2$.
> The area of the circular base $= \pi r^2$, so the total surface area of the hemisphere $= \frac{1}{2} \times 4\pi r^2 + \pi r^2$.

$$\text{Total surface area} = \frac{1}{2} \times 4\pi r^2 + \pi r^2$$
$$= 2\pi r^2 + \pi r^2$$
$$= 3\pi r^2 \text{ cm}^2$$

Exercise 11F

1 Work out the total surface areas of these cylinders. Give your answers correct to 1 decimal place.

a
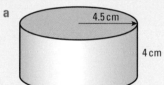
4.5 cm

4 cm

b

0.78 m

0.95 m

2 Work out the total surface areas of these cones. Give your answers correct to 3 significant figures.

a

12 cm

8 cm

b
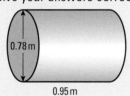
50 mm

50 mm

c

6 cm

8 cm

3 Work out the total surface areas of these shapes. Give your answers in terms of π.

a
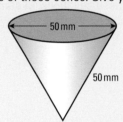
5 cm

b
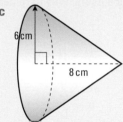
4 cm

8 cm

c
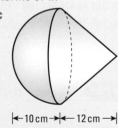
|←10 cm→|←12 cm→|

4 Work out the total surface areas of these prisms.

a
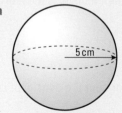
3 cm

3 cm

5 cm

5 cm

5 cm

b
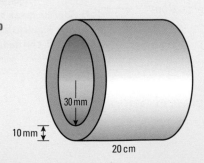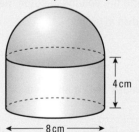
30 mm

10 mm

20 cm

C

A

A02

A02 A★

11.7 Density

◎ Objective

○ You can solve problems with density.

⦿ Why do this?

A submarine's ability to submerge and surface depends on its total density. It changes this by adjusting the amount of water in its ballast tanks.

⦿ Get Ready

1. Work out these calculations, giving your answers to one decimal place.
   a $50 \div 6$ b $4 \div 0.3$ c $400 \div 15.4$ d $347.1 \div 27$

◉ Key Points

◉ Density is a compound measure (see Unit 2 Sections 9.7 and 9.8). To solve density-related problems, we can use the following equations:

$$\text{density} = \frac{\text{mass}}{\text{volume}}$$

$$\text{mass} = \text{density} \times \text{volume}$$

$$\text{volume} = \frac{\text{mass}}{\text{density}}$$

◉ The diagram below is a useful way to remember these equations: M stands for mass, D stands for density and V stands for volume.

$$M = D \times V$$
$$D = \frac{M}{V}$$
$$V = \frac{M}{D}$$

◉ When the mass is measured in kilograms and the volume in cubic metres or m³, then density is measured in kg per m³ or kg/m³. Density can also be measured in g/cm³.

🔍 Example 14

A piece of silver has a mass of 42 g and a volume of 4 cm³. Work out the density of silver.

$Density = \frac{42}{4}$

$= 10.5 \, g/cm^3$ ◀

> Density = $\frac{mass}{volume}$
> Divide the mass by the volume.
> As the mass is in g and the volume is in cm³, the density is in g/cm³.

ResultsPlus
Examiner's Tip

Usually in GCSE mathematics the term weight is used as it is easier to understand. However in problems involving density, the correct term, mass, is used.

Example 15 The density of steel is 7700 kg/m³.

    a A steel bar has a volume of 2.5 m³. Work out the mass of the bar.

    b A block of steel has a mass of 1540 kg. Work out the volume of the block.

a $Mass = 7700 \times 2.5$ ← Mass = density × volume
Multiply the density by the volume.

    $= 19\,250\,kg$ ← As the density is in kg/m³ and the volume is in m³, the mass is in kg.

b $Volume = \dfrac{1540}{7700}$ ← $Volume = \dfrac{mass}{density}$
Divide the mass by the density.

    $= 0.2\,m^3$ ← As the mass is in kg and the density is in kg/m³ the volume is in m³.

Exercise 11G

1 A slab of concrete has a volume of 60 cm³ and a mass of 150 g. Work out the density of the concrete.

2 Gold has a density of 19.3 g/cm³. The gold in a ring has a mass of 15 g. Work out the volume of gold in the ring.

3 14.7 g of sulphur has a volume of 7.5 cm³. Work out the density of sulphur.

4 The density of aluminium is 2590 kg/m³. The density of lead is 11 400 kg/m³.
A block of aluminium has a volume of 0.5 m³. A block of lead has a volume of 0.1 m³.
Which of the two blocks has the greater mass and by how many kilograms?

C

A02
A03 B

Chapter review

● Volume of **cylinder** = area of cross-section × length
       = $\pi r^2 h$
where r is the radius and h is the height.

● Volume of **pyramid** = $\frac{1}{3}$ × area of base × vertical height
● Volume of **cone** = $\frac{1}{3}$ × area of base × vertical height
       = $\frac{1}{3}\pi r^2 h$
where r is the radius and h is the height.

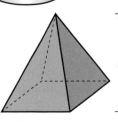
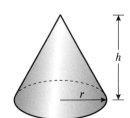

● Volume of **sphere** = $\frac{4}{3}\pi r^3$, where r is the radius.

- To work out the volume of a composite shape, work out the volumes of the shapes it is made from and add the volumes together.
- $1 \text{ m}^3 = 1\,000\,000 \text{ cm}^3$ $1 \text{ cm}^3 = \frac{1}{1\,000\,000} \text{ m}^3$
- $1 \text{ cm}^3 = 1000 \text{ mm}^3$ $1 \text{ mm}^3 = \frac{1}{1000} \text{ cm}^3$
- Litres are often used to measure the capacity or amount a container can hold.
- $1 \text{ litre} = 1000 \text{ cm}^3$
- $1 \text{ cm}^3 = 1 \text{ m}l$
- Total surface area of cylinder $= 2\pi rh + 2\pi r^2$, where r is the radius and h is the height.

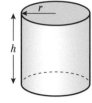

- Total surface area of cone $= \pi r^2 + \pi rl$, where r is the radius and l is the slant height.

- Surface area of sphere $= 4\pi r^2$, where r is the radius.

- **Density** is a compound measure. The following diagram is a useful way to remember the relationships between **mass**, density and **volume**: M stands for mass, D stands for density and V stands for volume.

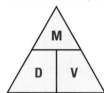

$$M = D \times V$$

$$D = \frac{M}{V}$$

$$V = \frac{M}{D}$$

- When the mass is measured in kilograms and the volume in cubic metres or m³, then density is measured in kg per m³ or kg/m³. Density can also be measured in g/cm³.

Review exercise

C AO3

1   You are planning a party for 30 children.
You buy some concentrated orange squash and some plastic cups.

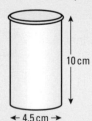

Each plastic cup will have 150 ml of drink in it. (150 ml = 150 cm³)

a   Check that the plastic cup shown can hold 150 ml of drink. Use the formula:

volume $5 \, \pi \times h \times \dfrac{d^2}{4}$

Each of the 30 children at the party will have a maximum of three drinks of orange squash.
Each plastic cup is to be filled with 150 ml of drink.
The squash needs to be diluted as shown on the bottle label.
A bottle of concentrated orange squash contains 0.8 litres of squash and costs £1.25.

b   How many bottles of concentrated orange squash do you need for the party?

c   How much will they cost in total?

2 The volume of this cube is 8 m³
Convert 8 m³ to cm³.

2 m
2 m
2 m

June 2007

3 The density of juice is 4 grams per cm³.
The density of water is 1 gram per cm³.
315 cm³ of drink is made by mixing 15 cm³
of juice with 300 cm³ of water.
Work out the density of the drink.

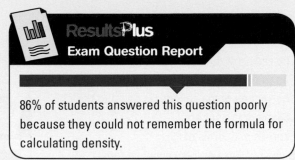

Results Plus
Exam Question Report

86% of students answered this question poorly
because they could not remember the formula for
calculating density.

June 2009

4 The volume of a gold bar is 100 cm³.
The density of gold is 19.3 grams per cm³.
Work out the mass of the gold bar.

Nov 2008

5 A cylindrical oil tank has a radius 60 cm and a length of 180 cm.
It is made from reinforced steel and is full of oil.
The oil has a density of 4.3 g/cm³.
The reinforced steel has a mass of 2.8 g/cm².
Find the total mass of the tank and the oil in kg.

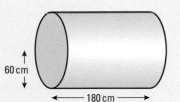

60 cm

180 cm

A02
A03 **B**

6 The diagram shows a storage tank.
The storage tank consists of a hemisphere on top of a cylinder.
The height of the cylinder is 30 metres.
The radius of the cylinder is 3 metres.
The radius of the hemisphere is 3 metres.
  a Calculate the total volume of the storage tank.
     Give your answer correct to 3 significant figures.
A sphere has a volume of 500 m³.
  b Calculate the radius of the sphere.
     Give your answer correct to 3 significant figures.

3 m
3 m
30 m
3 m

Nov 2008

7 Rainfall on a flat rectangular roof 10 m by 5.5 m flows into a cylindrical tub of diameter 3 m.
Find, in cm, the increase in depth of the water in the tub caused by a rainfall of 1.2 cm.
Give your answer correct to 2 significant figures.

A02
A03

8 The volume of a cone with base radius $2x$ cm and height $5x$ cm is equal to the total surface area of a
cylinder with radius $3x$ cm and height h cm. Find an expression for h in terms of x.

A03 **A**

9 A cylindrical bowl has a radius of 15 cm.
It is filled with water to a depth of 12 cm.
Work out the volume of water in the bowl.
Give your answer in litres as a multiple of π.

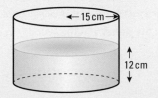

15 cm
12 cm

A*
A03

10 The diagram shows a cylinder and a sphere.

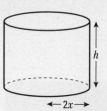

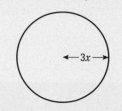

Diagram **NOT** accurately drawn

The radius of the base of the cylinder is $2x$ cm and the height of the cylinder is h cm.
The radius of the sphere is $3x$ cm.
The volume of the cylinder is equal to the volume of the sphere.
Express h in terms of x.
Give your answer in its simplest form.

Results Plus
Exam Question Report

91% of students answered this question poorly because they did not apply the power to all the parts of the expression.

June 2007

A02
A03

11 The diagram represents a large cone of height 30 cm and base diameter 15 cm. The large cone is made by placing a small cone A of height 10 cm and base diameter 5 cm on top of a frustum B.
Calculate the volume of the frustum B.
Give your answer correct to 3 significant figures.

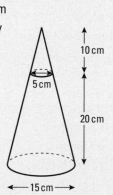

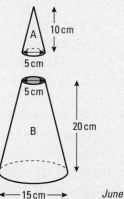

June 2003

A02
A03

* 12 The diagram shows a pyramid and a cone. The volumes are equal. Hamud says that the heights of the solids are equal.
Is he right? Give a reason for your answer.

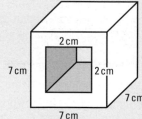

13 The solid shape, shown in the diagram, is made by cutting a hole all the way through a wooden cube.
The cube has edges of length 7 cm.
The hole has a square cross-section of side 2 cm.
a Work out the volume of wood in the solid shape.
The mass of the solid shape is 189 grams.
b Work out the density of the wood.

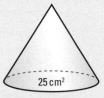

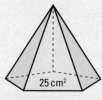

March 2009, adapted

12 CONGRUENCE AND SIMILARITY

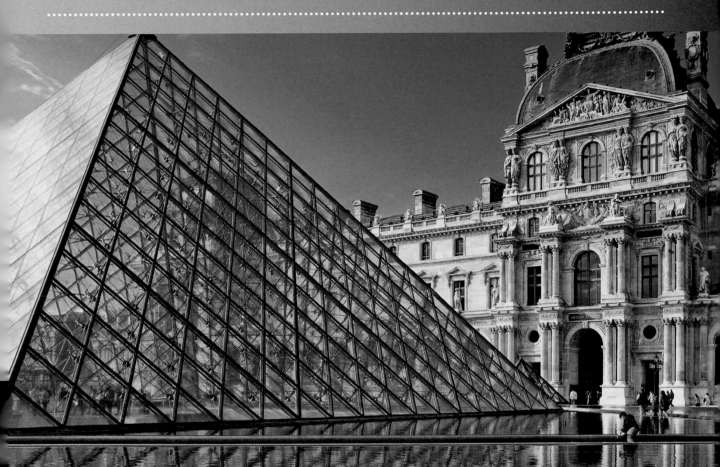

The photo shows the Pyramide du Louvre in Paris. There are actually five pyramids, the large one, three smaller ones and an inverted pyramid which provides the entrance to the Louvre museum. The larger pyramid is made up of 603 diamond-shaped panes of glass with 70 triangular-shaped panes along the base of the pyramid.

◎ Objectives

In this chapter you will:
- prove two triangles are congruent
- recognise similar shapes and use scale factors to find missing sides in similar triangles
- formally prove that triangles are similar
- use similar shapes to solve problems involving lengths, areas and volumes
- understand and use the relationship between length, area and volume scale.

◖ Before you start

You need to:
- know the angle properties of triangles and quadrilaterals
- know what a vertex and a diagonal of a shape are
- use the angle properties associated with parallel lines
- use ratios to compare lengths.

12.1 Congruent triangles

◎ Objective

● You will understand how to prove that two triangles are congruent.

⦿ Why do this?

Designers, engineers and map makers often use scale drawings and plans. Using congruent and similar triangles enables them to find measurements for inaccessible lengths and angles.

⬆ Get Ready

1. If two triangles have the same angles, are the triangles the same?
2. Given the lengths of all three sides, is it possible to draw two different triangles?
3. Given lengths of two sides and the size of the included angle, is it possible to draw two different triangles?

🕐 Key Points

● Two triangles are **congruent** if they have exactly the same shape and size.
● For two triangles to be congruent one of the following **conditions of congruence** must be true.
  ● The three sides of each triangle are equal (SSS).
  ● Two sides and the **included angle** are equal (SAS).
  ● Two angles and a corresponding side are equal (AAS).
  ● Each triangle contains a right angle, and the hypotenuses and another side are equal (RHS).

🔍 Example 1

ABCD is a quadrilateral.
AD = BC.
AD is parallel to BC.
Prove that triangle ADC is congruent to triangle ABC.

ResultsPlus
Watch Out!

The only properties that can be used to prove congruence are those given in the question.

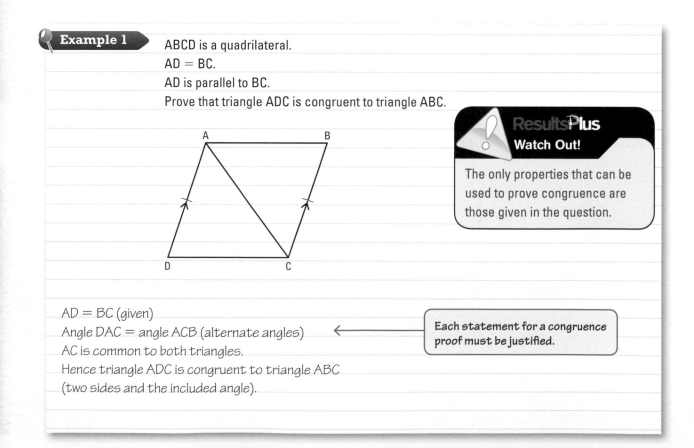

AD = BC (given)
Angle DAC = angle ACB (alternate angles) ← Each statement for a congruence proof must be justified.
AC is common to both triangles.
Hence triangle ADC is congruent to triangle ABC (two sides and the included angle).

congruent   conditions of congruence   included angle

Exercise 12A

Questions in this chapter are targeted at the grades indicated.

* **1** Prove that triangles PQS and QRS are congruent.

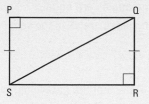

A03

* **2** Prove that triangles XYZ and XVW are congruent.
Hence prove that X is the midpoint of YW.

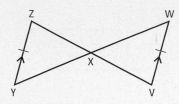

A03

* **3** PQR is an isosceles triangle. S and T are points on QR.
PQ = PR, QS = TR.
Prove that triangle PST is isosceles.

* **4** LMN is an isosceles triangle with LM = LN. Use congruent triangles to prove that the line from L which cuts the base MN of the triangle at right angles also bisects the base.

* **5** ABC is a triangle. D is the midpoint of AB. The line through D drawn parallel to the side BC meets the side AC at E. A line through D drawn parallel to the side AC meets the side BC at F.
Prove that triangles ADE and DBF are congruent.

A

12.2 Recognising similar shapes

Objectives

- You can recognise similar shapes.
- You can find missing sides using facts you know about similar shapes.

Why do this?

Similar shapes allow us to calculate missing dimensions from plans which may be difficult to measure on the real objects.

Get Ready

1. Which of these triangles are congruent?

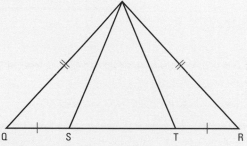

🔍 Key Points

◉ Shapes are **similar** if one shape is an enlargement of the other.
  ◉ The corresponding angles are equal.
  ◉ The corresponding sides are all in the same ratio.

A03

🔍 Example 2

Show that the parallelogram ABCD is not similar to parallelogram EFGH.

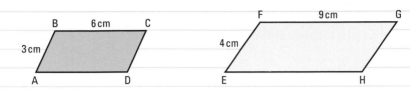

$$\frac{AB}{EF} = \frac{3}{4} = 0.75 \longleftarrow$$

> Work out the ratios of the corresponding sides.

$$\frac{BC}{FG} = \frac{6}{9} = 0.66667$$

The lengths of the corresponding sides are not in the same proportion so the parallelograms are not similar.

⚙️ Exercise 12B

D

1 State which of the pairs of shapes are similar.

a
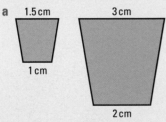

b
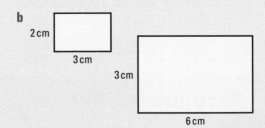

B

2 Show that pentagon ABCDE is similar to pentagon FGHIJ.

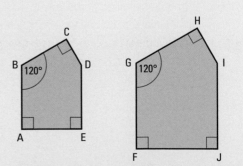

Example 3 These two rectangles are similar. Find the length L of the larger rectangle.

4 cm

2 cm

L cm

3 cm

The widths of these rectangles are in the ratio $2 : 3$.

← Consider the ratio of the widths of the rectangles.

The lengths must be in the same ratio.

← The rectangles are similar so the lengths must be in the same ratio.

$\dfrac{\text{small}}{\text{large}} = \dfrac{2}{3} = \dfrac{4}{L}$

$2L = 12$

$L = 6\text{ cm}$

Results Plus
Examiner's Tip

Make sure you keep corresponding sides together by stating which rectangle they come from.

Exercise 12C

1 A large packet of breakfast cereal has height 35 cm and width 21 cm.
A small packet of cereal is similar to the large packet but has a height of 25 cm.
Find the width of the small packet.

A02 B
A03

2 The diagram shows a design for a metal part.
The sizes of the plan are marked on diagram A.
Diagram B is marked with the actual sizes.
Calculate the value of:
a x
b y.

A02
A03

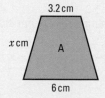

3.2 cm

x cm   A

6 cm

y m

0.72 m   B

0.84 m

3 These cylinders are similar. The height of the smaller cylinder is 5 cm.
Find the height of the larger cylinder.

A

2 cm

6 cm

12.3 Similar triangles

⊙ Objectives

- You can use scale factors to find missing sides in similar triangles.
- You can formally prove that triangles are similar.

◈ Why do this?

Using the fact that triangles are similar can help us to measure lengths and distances which we cannot measure practically.

↑ Get Ready

1. Copy these diagrams and mark the pairs of corresponding angles.

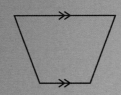

🔵 Key Points

- Two triangles are similar if any of the following is true.
  - The corresponding angles are equal.
  - The corresponding sides are in the same ratio.
  - They have one angle equal and the adjacent sides are in the same ratio.

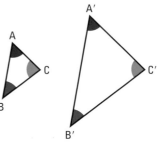

AB and A′B′, AC and A′C′, and BC and B′C′ are corresponding sides.

$$\frac{A'B'}{AB} = \frac{A'C'}{AC} = \frac{B'C'}{BC}$$

A03

🔍 Example 4 ▸ Show that triangle ABC is not similar to triangle DEF.

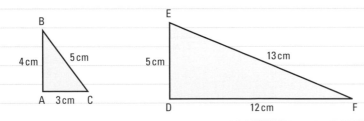

$$\frac{AB}{DE} = \frac{4}{5} = 0.8$$

$$\frac{AC}{DF} = \frac{3}{12} = 0.25$$

ResultsPlus
Examiner's Tip

Make sure you look carefully at the parallel lines in a diagram, as they can give you a lot of information about the angles.

The lengths of the corresponding sides are not in the same proportion so the triangles are not similar.

Example 5 ABC is a triangle.

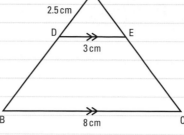

DE is parallel to BC.

a Show that triangle ABC is similar to triangle ADE.

b Find the length of BD.

a $\angle$ADE = $\angle$ABC (corresponding angles).
$\angle$AED = $\angle$ACB (corresponding angles).
$\angle$DAE = $\angle$BAC (common to both triangles).

> Give reasons from what you know about parallel lines.

All angles are equal so triangle ABC is similar to triangle ADE.

b $\dfrac{BC}{DE} = \dfrac{8}{3} = \dfrac{AB}{2.5}$

> The corresponding sides are in the same ratio.

$3 \times AB = 8 \times 2.5$

$3AB = 20$

$AB = 6.67\,cm$

$BD = 6.67 - 2.5$

> BD is only part of the side of the triangle.

$= 4.17\,cm$

Exercise 12D

1 For each pair of similar triangles:
  i name the three pairs of corresponding sides
  ii state which pairs of angles are equal.

a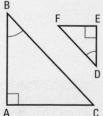

b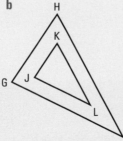

c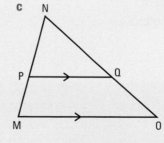

d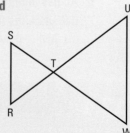

2 Triangle ABC is similar to triangle DEF.

$\angle$ABC = $\angle$DEF

Calculate the length of:

a EF

b FD.

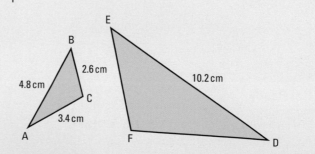

C

A03 A

A* A03

3 The diagram shows triangle ABC which has a line DE drawn across it.

∠ACE = ∠DEB

a Prove that triangle ABC is similar to triangle DBE.
b Calculate the length of AB.
c Calculate the length of AD.
d Calculate the length of EC.

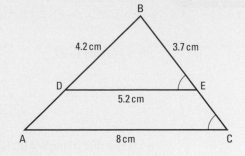

A03

4 The diagram shows triangle ABC which has a line DE drawn across it. ∠CAD = ∠BDE

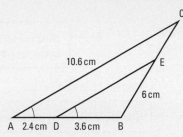

a Prove that triangle ACB is similar to triangle DEB.
b Calculate the length of DE.
c Calculate the length of BC.

A03

5 In the diagram AB is parallel to CD.

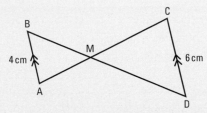

a Prove that triangle ABM is similar to triangle CDM.
b AC has length 20 cm.
Calculate the lengths of:
  i AM ii MC.

12.4 Areas of similar shapes

Objective

- You can solve problems involving the areas of similar shapes.

Why do this?

Designers, architects and surveyors use scale drawings and need to be able to calculate accurate areas from maps, plans and diagrams.

Get Ready

1. Find the square root of **a** 81 **b** 256 **c** 8100
2. Find the squares of **a** 14 **b** 25 **c** 19.6

Key Points

⊙ The diagram shows squares of side 1 cm, 2 cm, 3 cm and 4 cm.

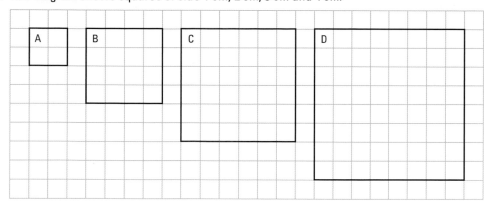

⊙ The squares are all similar.
  ⊙ It takes 4 squares of side 1 cm to fill the square with side 2 cm.
  ⊙ It takes 9 squares of side 1 cm to fill the square with side 3 cm.
  ⊙ It takes 16 squares of side 1 cm to fill the square with side 4 cm.
  ⊙ If the ratio of the corresponding sides is k then the ratio of the areas is k^2.
⊙ k is sometimes called the linear scale factor.
⊙ k^2 is called the area scale factor.

Example 6

The diagram shows a rectangle with side 2 cm and area 2 cm².
A second similar rectangle is drawn with side 6 cm.
Calculate the area of the new rectangle.

$$\text{linear scale factor} = \frac{\text{side of large rectangle}}{\text{side of small rectangle}}$$

> It helps to show which way round you are writing the ratio.

$$= \frac{6}{2}$$
$$= 3$$

area scale factor $= 3^2 = 9$

> Area factor is the square of the linear scale factor.

area of large rectangle = area of small rectangle × area scale factor
$$= 2 \times 1 \times 9 = 18 \text{ cm}^2$$

2 cm

6 cm

Example 7

Shape A is similar to shape B.
The area of A is 150 cm².
Find the area of shape B.

A

← 15 cm →

B

$$\text{linear scale factor} = \frac{67.5}{15} = 4.5$$

$$\text{area scale factor} = 4.5^2 = 20.25$$

67.5 cm

$$\text{area of B} = 20.25 \times 150 = 3037.5 \text{ cm}^2$$

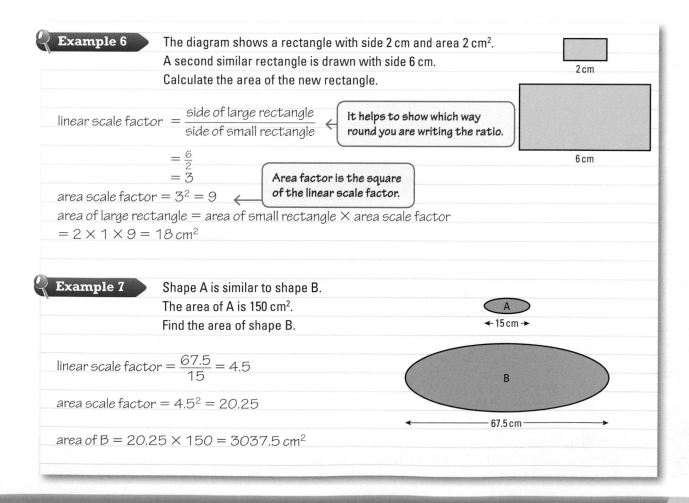

A

Exercise 12E

1 Shapes A and B are similar.
Shape A has area 20 cm² and length 4.5 cm.
Shape B has length 27 cm.
Calculate the area of shape B.

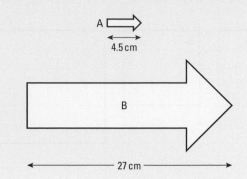

A
4.5 cm
B
27 cm

A02

2 Shapes A and B are similar.
Shape A has area 30 cm² and height 5 cm.
Shape B has height 15 cm.
Calculate the area of shape B.

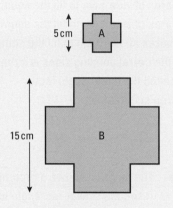

5 cm A

15 cm B

A02

3 Cuboids A and B are similar.
The surface area of cuboid B is 108 cm².
Calculate the surface area of cuboid A.

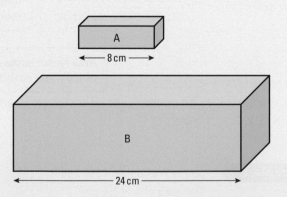

A
8 cm

B
24 cm

A02

4 The diagram shows a small box of chocolates with surface area 500 cm².
The box has a piece of ribbon 50 cm long wrapped around it.
A larger box has a similar piece of ribbon wrapped around it of length 75 cm.
Calculate the surface area of the larger box.

Example 8   Shape C is similar to shape D.
Calculate the length of shape D.

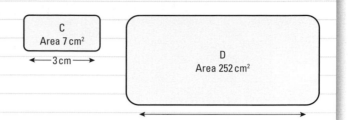

area scale factor = $\frac{252}{7} = 36$ ← | Linear scale factor is the square root of the area scale factor. |

linear scale factor = $\sqrt{36} = 6$

shape D has length $6 \times 3 = 18$ cm

Exercise 12F

1 Triangle A is similar to triangle B.
The area of triangle A is 90 cm².
The area of triangle B is 202.5 cm².
Calculate the value of
a x
b y.

A03 **A**

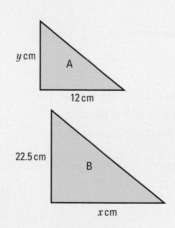

2 Two similar triangles have areas 36 cm² and 64 cm² respectively.

A03 **A***

The base of the smaller triangle is 6 cm.
Find the base of the larger triangle.

12.5 Volumes of similar shapes

◎ Objective

● You can solve problems involving the volumes of similar shapes.

❓ Why do this?

When designing packaging the manufacturer needs to be able to design similar shapes for large, medium and small sizes which will hold specific quantities.

⬆ Get Ready

1. Find the cube roots of **a** 343   **b** 1000
2. Find the cubes of **a** 2.5   **b** 11

🔑 Key Points

● The diagram shows three similar shapes, a cube of side 1 cm, a cube of side 2 cm and a cube of side 3 cm.
 The volume of the cube side 1 cm = $1 \times 1 \times 1 = 1 \, cm^3$
 The volume of the cube side 2 cm = $2 \times 2 \times 2 = 8 \, cm^3$
 The volume of the cube side 3 cm = $3 \times 3 \times 3 = 27 \, cm^3$

 1 cm

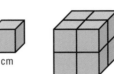 2 cm

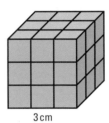 3 cm

● If k is the ratio of the lengths, the ratio of the volumes is k^3.

 When the length is multiplied by k, the volume is multiplied by k^3.

● k^3 is called the volume scale factor.

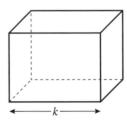

🔍 Example 9

Cuboids P and Q are similar.
The volume of P is 60 cm³.
Calculate the volume of Q.

linear scale factor = $\dfrac{large}{small} = \dfrac{15}{5} = 3$

 15 cm   Q
5 cm   P

volume scale factor = $3^3 = 27$
volume of Q = $27 \times 60 = 1620 \, cm^3$

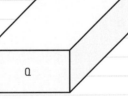

⚙ Exercise 12G

A
A02

1 Prisms A and B are similar.
The volume of A is 10 cm³.
Calculate the volume of prism B.

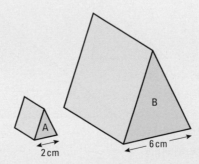
B
A
2 cm
6 cm

2   Cylinders C and D are similar.
The volume of C is 5 cm³.
Calculate the volume of cylinder D.

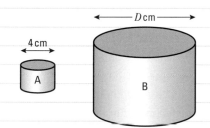

A02   **A**

3   Tetrahedrons E and F are similar.
The volume of E is 12 cm³.
Calculate the volume of tetrahedron F.

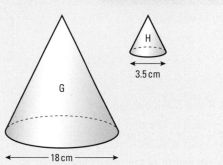

A02

4   Cones G and H are similar.
The volume of G is 306 cm³.
Calculate the volume of cone H.

A02

Example 10   Cylinders A and B are similar.
The diameter of A is 4 cm.
The volume of A is 120 cm³.
The volume of B is 405 cm³.
Work out the diameter D of cylinder B.

$$\text{volume scale factor} = \frac{\text{large}}{\text{small}} = \frac{405}{120} = \frac{27}{8}$$

←   Volume scale factor is the cube of the linear scale factor.

$$\text{linear scale factor} = \sqrt[3]{\frac{27}{8}} = \frac{\sqrt[3]{27}}{\sqrt[3]{8}} = \frac{3}{2} = 1.5$$

$$D = 1.5 \times 4 = 6 \text{ cm}$$

Exercise 12H

1   Sphere K is similar to sphere J.
(All spheres are similar to each other.)
The volume of K is 64 times the volume of J.
Calculate the diameter of J.

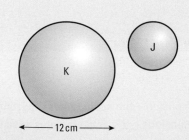

A02
A03   **A**

2 Prisms K and L are similar.
The volume of L is 216 times the volume of K.
Calculate the value of

a x

b y.

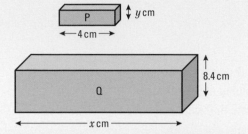

3 Cuboids P and Q are similar.
The volume of cuboid P is 48 cm^3.
The volume of cuboid Q is 16 464 cm^3.
Calculate the value of

a x

b y.

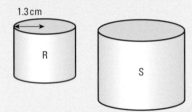

4 Cylinders R and S are similar.
The volume of R is 12π cm^2.
The volume of S is 40.5π cm^2.
Calculate the length of the radius of cylinder S.

5 A bakery sells two sizes of doughnuts.
The small size has a mass of 50 g.
The large size has a mass of 168.75 g.
If the larger doughnut has a diameter of 15 cm,
find the diameter of the small size.

12.6 Lengths, areas and volumes of similar shapes

⊙ Objective

● You understand and use the
relationship between length, area
and volume scale.

⊘ Why do this?

A manufacturer designing a new carton for a drink would
experiment with different heights and widths of the packaging to
find the best shape to hold a certain amount of their product.

⬦ Get Ready

Express these ratios in their simplest form **1.** $14:63$ **2.** $10^2:5^3$ **3.** $2.5:15$ **4.** $4^3:8^2$

◉ Key Points

● A length has 1 dimension – the scale factor is used once.
● An area has 2 dimensions – the scale factor is used twice.
● A volume has 3 dimensions – the scale factor is used three times.

Example 11 ▶ Two similar cylinders have masses of 32 kg and 108 kg.
The area of the label on the small cylinder is 10 cm².
Calculate the area of the label on the large cylinder.

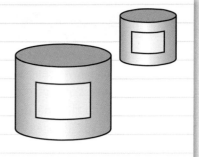

$$\text{volume scale factor} = \frac{\text{large}}{\text{small}} = \frac{108}{32} = 3.375$$

$$\text{linear scale factor} = \sqrt[3]{3.375} = 1.5$$
$$\text{area scale factor} = 1.5^2 = 2.25$$
$$\text{area of the label} = 10 \times 2.25 = 22.5 \text{ cm}^2$$

Exercise 12I

Give your answers to the following questions correct to 3 significant figures.

1 Cones P and Q are similar.
The volume of cone P is 125 times the volume of cone Q.
If the surface area of P is 40 cm²,
calculate the surface area of Q.

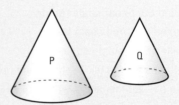

A02
A03

2 Prisms G and H are similar.
The surface area of G is 64 times the surface area of H.
If the volume of G is 2000 cm³,
calculate the volume of H.

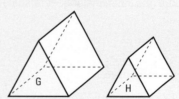

A02
A03

3 A container has a surface area of 5000 cm² and a capacity of 10.6 litres.
Find the surface area of a similar container which has a capacity of 4.8 litres.

A02
A03

4 The volume of a toy is 900 cm³.
A similar larger toy has volume 13 500 cm³.
The surface area of the larger toy is 2700 cm².
Find the surface area of the smaller toy.

A02
A03

5 A detergent manufacturer makes bottles of detergent in three sizes.
The bottles are mathematically similar.
The large bottle holds 5 *l*, the medium bottle holds 3 *l* and
the small bottle holds 1 *l*.
The 1 *l* bottle has a label with an area of 100 cm².
Calculate the area of the labels on
a the large bottle
b the medium bottle.

A02

6 A recycling bin holds 50 *l* of rubbish.
A smaller bin will hold 30 *l* of rubbish.
The surface area of the larger bin is 0.142 m².
Calculate the surface area of the smaller bin.

A02

Chapter review

- Two triangles are **congruent** if they have exactly the same shape and size.
- For two triangles to be congruent one of the following **conditions of congruence** must be true.
  - The three sides of each triangle are equal (SSS).
  - Two sides and the **included angle** are equal (SAS).
  - Two angles and a corresponding side are equal (AAS).
  - Each triangle contains a right angle, and the hypotenuses and another side are equal (RHS).
- Shapes are **similar** if one shape is an enlargement of the other.
  - The corresponding angles are equal.
  - The corresponding sides are all in the same ratio.
- Two triangles are similar if any of the following is true.
  - The corresponding angles are equal.
  - The corresponding sides are in the same ratio.
  - They have one angle equal and the adjacent sides are in the same ratio.
- k is called the linear scale factor.
- k^2 is called the area scale factor.
- k^3 is called the volume scale factor.
- A length has 1 dimension – the scale factor is used once.
- An area has 2 dimensions – the scale factor is used twice.
- A volume has 3 dimensions – the scale factor is used three times.

Review exercise

D

1 Which of these triangles are similar?

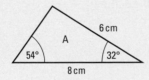
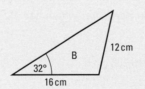
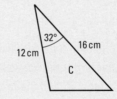

2 Which of these rectangles are similar?

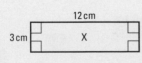
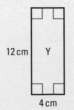
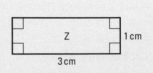

C

A02

3 A car is 4 m long and 1.8 m wide.
A model of the car, similar in all respects, is 5 cm long. How wide is it?

A02

4 A model of a car is 12 cm long and 5.2 cm high.
If the real car is 3.36 m long, how high is it?

5 ABC is an equilateral triangle.
D lies on BC. AD is perpendicular to BC.
a Prove that triangle ADC is congruent to triangle ADB.
b Hence, prove that BD = $\frac{1}{2}$BC.

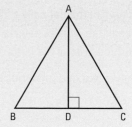

June 2009

6 In the diagram, AB = BC = CD = DA.
Prove that triangle ADB is congruent to triangle CDB.

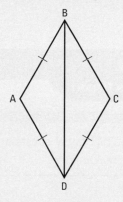

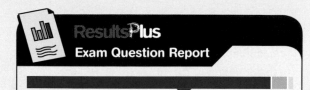

ResultsPlus
Exam Question Report

92% of students answered this sort of question poorly because they did not justify their answers or prove the conditions for congruency.

Nov 2008

7 AB is parallel to DE.
ACE and BCD are straight lines.
AB = 6 cm
AC = 8 cm
CD = 13.5 cm
DE = 9 cm
a Work out the length of CE.
b Work out the length of BC.

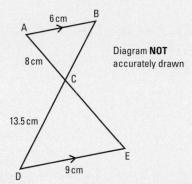

Diagram **NOT** accurately drawn

Nov 2005

8 Parallelogram P is similar to parallelogram Q.

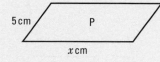

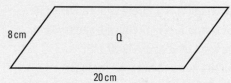

Calculate the value of x.

9 In triangle FGJ, a line IH is drawn parallel to FG.
a Prove that triangle HIJ is similar to triangle GFJ.
b Calculate the length of HI.

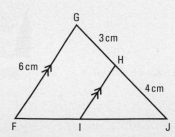

A
AO2

10 The volumes of two mathematically similar solids are in the ratio 27 : 125.
The surface area of the smaller solid is 36 cm^2.
Work out the surface area of the larger solid.

Nov 2007

AO2

11 The diagram shows two quadrilaterals that are mathematically similar.

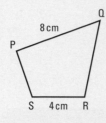
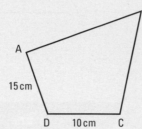

In quadrilateral PQRS, PQ = 8 cm, SR = 4 cm.
In quadrilateral ABCD, AD = 15 cm, DC = 10 cm.
Angle PSR = angle ADC.
Angle SPQ = angle DAB.

a Calculate the length of AB.
b Calculate the length of PS.

ResultsPlus
Exam Question Report

91% of students answered this sort of question well. They showed all of their working.

June 2007

A*

12 BE is parallel to CD.
AB = 9 cm, BC = 3 cm, CD = 7 cm, AE = 6 cm.
a Calculate the length of ED.
b Calculate the length of BE.

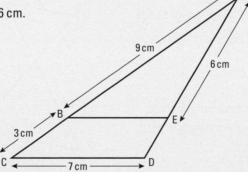

June 2005

AO2

13 Two solid shapes, A and B, are mathematically similar.
The base of shape A is a circle with radius 4 cm.
The base of shape B is a circle with radius 8 cm.
The surface area of shape A is 80 cm^2.
a Work out the surface area of shape B.
The volume of shape B is 600 cm^3.
b Work out the volume of shape A.

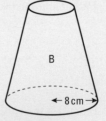

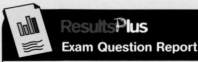
ResultsPlus
Exam Question Report

82% of students answered this question poorly because they used the wrong scale factor.

June 2008

14 Two cones, P and Q, are mathematically similar.
The total surface area of cone P is 24 cm².
The total surface area of cone Q is 96 cm².
The height of cone P is 4 cm.
a Work out the height of cone Q.
The volume of cone P is 12 cm³.
b Work out the volume of cone Q.

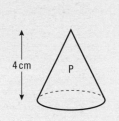
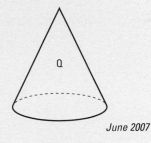

June 2007

15 Two prisms, A and B, are mathematically similar.
The volume of prism A is 12 000 cm³.
The volume of prism B is 49 152 cm³.
The total surface area of prism B is 9728 cm².
Calculate the total surface area of prism A.

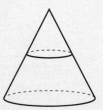

Nov 2006

16 A cone is divided by a cut parallel to the base halfway between the top and the base.
What is the ratio of
a the area of the base of the small cone to the area of the base of the large cone
b the volume of the small cone to the volume of the large cone?

17 **a** Prove that all cubes are similar.
b Two cubes have edges 2 cm and 5 cm.
What is the ratio of the total surface areas of the two cubes?
c 'Two cuboids are similar' – This statement is not always true.
Explain why.

18 A manufacturer makes pots of cream in two sizes.
The small size contains 300 g and has a diameter of 10 cm.
The large size contains 500 g.
The pots are similar.
a Find the diameter of the larger pot.
The front of the large pot has a rectangular label
of area 36 cm².
b Find the area of the label on the smaller pot.

A02
A03
A02
A02
A03
A*

13 CIRCLE GEOMETRY

The O_2 dome is a circle when looked at from above. It has a diameter of 365 m – a metre for every day of the year. The roof structure is incredibly light, weighing less than the air contained within the building. However, it is not strictly a dome as it is not self-supporting.

◎ Objective

In this chapter you will:

- learn theorems about circles and how to prove and apply them.

◁ Before you start

You need to know that:

- the angles in a triangle add up to 180°
- the base angles in an isosceles triangle are equal
- angles on a straight line add up to 180°.

13.1 **Circle theorems**

◉ Objective

◉ You know three theorems about circles and how to prove and apply them.

⍰ Why do this?

You can use circle theorems in engineering as many machines contain circular parts like cogs.

⬆ Get Ready

1. Calculate the size of the angles marked a, b and c.

a  **b**  **c**

⬤ Key Points

◉ **Theorem 1**

The perpendicular from the centre of a circle to a chord bisects the chord (and vice versa: the line drawn from the centre of a circle to the midpoint of a chord is perpendicular to the chord).

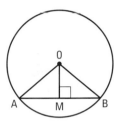

Proof

In triangles OAM and OBM

OA = OB   ← Radii equal.

Angle OMA = angle OMB   ← Both given as 90°.

OM = OM   ← Common side.

So triangle OAM is congruent to triangle OBM (RHS).

So AM = MB.

◉ **Theorem 2**

The angle at the centre of a circle is twice the angle at the circumference, both subtended by the same arc.

In each diagram angle AOB = 2 × angle ACB.

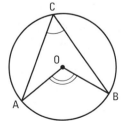 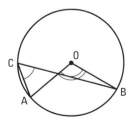 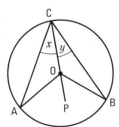

Proof

Draw in the line CO and extend it to P as shown in the diagram above on the right.

Let angle ACO = x and angle BCO = y.

Since triangle CAO and triangle CBO are both
isosceles (radii equal), the angles opposite the ← | Give the reason in words as well as giving the sizes of the angles. |
equal sides are equal.

So angle CAO = x and angle CBO = y.

The exterior angle of a triangle is equal to the sum ← | This rule was learnt in Unit 2 Section 13.3. |
of the two interior opposite angles.

So angle AOP = $2x$ and angle BOP = $2y$.

Angle ACB = $x + y$ ← | We can see this from the diagram. |

Angle AOB = $2x + 2y = 2(x + y)$
= 2(angle ACB)

◉ **Theorem 3**

The angle in a semicircle is a right angle.

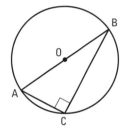

Angle ACB = 90°

Proof

The angle at the centre of the circle is twice the angle ← | We can use the rule proved in theorem 2. |
at the circumference.

So angle AOB = 2 × angle ACB.

But angle AOB = 180° as it is a straight line. ← | AOB is the diameter. |

So angle ACB = $\frac{1}{2}$(180°)
= 90°

| **Example 1** | PQ is a chord of the circle, centre O. |

N is the midpoint of PQ.

Angle QPO = 27°

Work out the size of angle POQ.

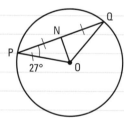

Angle PNO = 90° ← | Line from centre to midpoint of chord is perpendicular to chord (Theorem 1). |

Angle PON = 180° − 90° − 27° ← | The angles in a triangle add up to 180°. |
= 63°

Similarly angle QON = 63° ← | Angle QNO = 90° and triangle PQO is isosceles. |

Angle POQ = angle PON + angle QON

So angle POQ = 126°

Example 2

P, Q and R are points on a circle, centre O.
Angle POQ = 142°.
Work out the size of angle PRQ.
Give a reason for your answer.

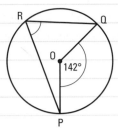

Angle PRQ = 142° ÷ 2 ⟵ | As angle POQ = 2 × angle PRQ (Theorem 2) we can say that angle PRQ = $\frac{1}{2}$ angle POQ.
= 71°

Reason: The angle at the centre of the circle is twice the angle at the circumference, so angle POQ = 2 × angle PRQ. ⟵ Give the reason in words.

Example 3

P, Q and R are points on a circle, centre O.
Angle RQO = 20°

Work out the size of angle PRO.

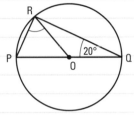

Angle QRO = 20° ⟵ OQ = OR radii, thus triangle ORQ is isosceles.

Angle PRO = angle PRQ − angle QRO ⟵ Angle PRO = 90° as it is in a semicircle (Theorem 3).
= 90° − 20°
= 70°

Exercise 13A

Questions in this chapter are targeted at the grades indicated.

Find the size of each of the angles marked with a letter.
O is the centre of the circle in each case.

1

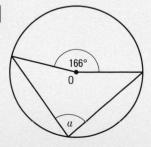

2

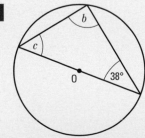

3

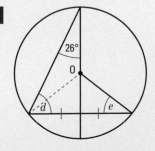

A

A

4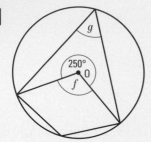

5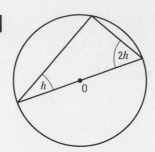

6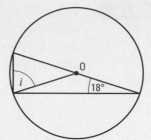

In questions 7–9 give reasons for your answers.

A03

7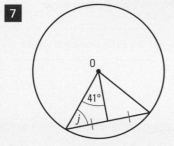

8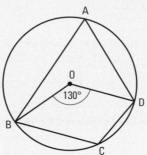

9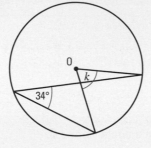

A03

10 A, B, C and D are points on the circle, centre O.

Angle DOB = 130°

a Work out the size of angle DAB.
  Give a reason for your answer.

b Work out the size of reflex angle DOB.
  Give a reason for your answer.

c Work out the size of angle BCD.
  Give a reason for your answer.

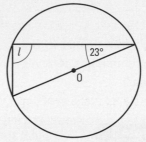

13.2 More circle theorems

⦿ Objectives

● You know three more theorems about circles
  and how to prove and apply them.
● You can use all you have learnt about circles to
  work out angles in more complex problems.

⦿ Why do this?

There are many applications for circle geometry.
Some people believe that the mysterious crop
circles that sometimes appear in our fields are
created using circle geometry.

⦿ Get Ready

1. Calculate the size of the angles marked a, b and c.

a

b

c

Key Points

- **Theorem 4**

  Angles in the same segment are equal.

  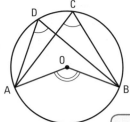

  Angle ACB = angle ADB

  Proof

  Angle AOB = 2 × angle ACB ← | Draw in the angle at the centre and use Theorem 2. |

  The angle at the centre is twice the angle at the circumference.

  Angle AOB = 2 × angle ADB ← | Now do the same for the other angle. |

  The angle at the centre is twice the angle at the circumference.

  So angle ACB = angle ADB. ← | Both are half of angle AOB. |

- **Theorem 5**

  Opposite angles of a **cyclic quadrilateral** add up to 180°.
  (A cyclic quadrilateral is a quadrilateral that has all four
  vertices on the circumference of a circle.)

  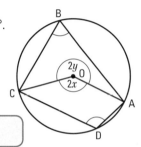

  Angle ABC + angle ADC = 180°

  Proof

  Angle ABC = x ← | First draw in angle AOC. |

  The angle at the centre of the circle is twice the angle at the circumference.

  Angle ADC = y ← | The reflex angle is twice angle ADC. |

  The angle at the centre of the circle is twice the angle at the circumference.

  But $2x + 2y = 2(x + y) = 360°$. ← | The angles at a point add up to 360°. |

  So $x + y = 180°$ ← | Divide each side of the equation by 2. |

- **Theorem 6**

  The angle between a tangent and a chord is equal to the
  angle in the alternate segment.

  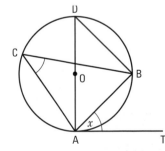

  Angle BAT = angle ACB

  Proof

  Angle DAT = 90° (angle between the tangent and the radius is 90°) ← | Draw the diameter AD. |

  So angle DAB = 90° − x

  Angle DBA = 90° (angle in a semicircle = 90°) ← | Theorem 3 |

  So angle ADB = 180° − 90° − (90° − x)

  = x (angles in a triangle add up to 180°)

  Angle ACB = angle ADB (angles in the same segment) ← | Theorem 4 |

  So angle ACB = x

  So angle BAT = angle ACB. ← | Both equal to x. |

A03

Example 4 P, Q, R and S are points on the circle, centre O.

Angle RSQ = 43°

Work out the size of angle RPQ.

Give a reason for your answer.

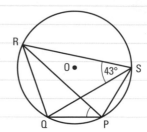

Angle RPQ = 43°

Reason: The angles in the same segment are equal. ← | Write down the reason in words (Theorem 4). |

Example 5 P, Q, R and S are points on the circle, centre O.

Angle ROP = 102°

Work out the size of angle RQP.

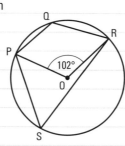

Angle RSP = $\frac{1}{2}$(102°)

= 51° ← | The angle at the centre is twice the angle at the circumference. |

Angle RQP = 180° − 51°

= 129° ← | Opposite angles of cyclic quadrilateral add to 180° (Theorem 5). |

Example 6 PQR is a tangent to the circle, centre O.

S and T are points on the circle.

Angle TQS = 47°

Angle QTS = 76°

Work out the size of angle PQT.

Give reasons for your answer.

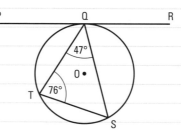

Angle QST = 180° − 47° − 76°

= 57°

The angles in a triangle add up to 180°. ← | Always give the reason. |

Angle PQT = angle QST = 57°

The angle between a tangent and a chord is equal to the angle in the alternate segment.

↑

| Give the second reason too (Theorem 6). |

Exercise 13B

Find the size of each of the angles marked with a letter.

O is the centre of the circle in each case.

1
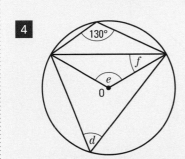
40°

57°

O

a

2
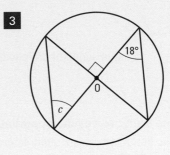
b

O•

82°

3
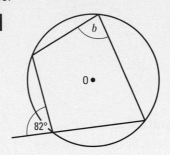
18°

O

c

4
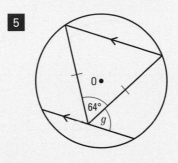
130°

f

e

O•

d

5
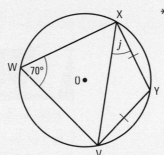
O•

64°

g

6
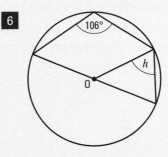
106°

h

O

In questions 7–9 give reasons for your answers.

7
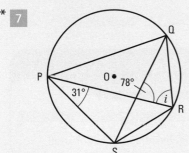
Q

P

O• 78°

31°

i

R

S

8
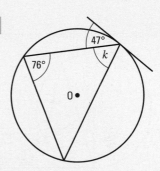
X

j

W 70°

O•

Y

V

9
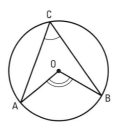
47°

k

76°

O•

A

A03

A03 A★

Chapter review

Circle theorems

1. The perpendicular from the centre of a circle to a chord bisects the chord
   (and vice versa: the line drawn from the centre of a circle to the midpoint of a chord is
   perpendicular to the chord).

2. The angle at the centre of a circle is twice the angle
   at the circumference, both subtended by the same arc.

3. The angle in a semicircle is a right angle.

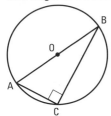

4. Angles in the same segment are equal.

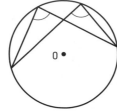

5. Opposite angles of a **cyclic quadrilateral** add up to 180°.
   (A cyclic quadrilateral is a quadrilateral that has all four vertices
   on the circumference of a circle.)

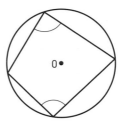

6. The angle between a tangent and a chord is
   equal to the angle in the alternate segment.

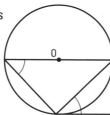

✴ **Review exercise**

A
A03

1 The diagram shows a circle centre O.

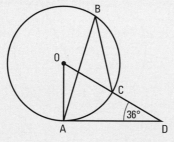

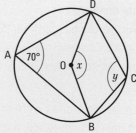

Results Plus
Exam Question Report

82% of students answered this question well.

A, B and C are points on the circumference. DCO is a straight line.
DA is a tangent to the circle. Angle ADO = 36°
Work out the size of angle ABC.

June 2009, adapted

2 In the diagram, A, B, C and D are points on the circumference of a circle, centre O.
Angle BAD = 70°. Angle BOD = x°. Angle BCD = y°.
   a i Work out the value of x.   **ii** Give a reason for your answer.
   b i Work out the value of y.   **ii** Give a reason for your answer.

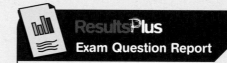

Results Plus
Exam Question Report

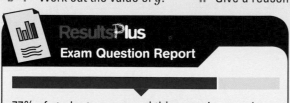

77% of students answered this question poorly as
they had not learnt circle theorems correctly.

June 2008

Find the size of each of the angles marked with a letter.
O is the centre of the circle, where marked.

3

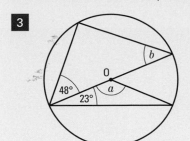

4

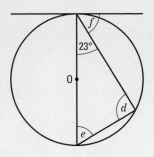

5

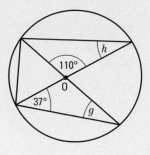

A03

In questions 6–8 give reasons for your answers.

* **6**

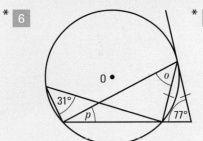

* **7**

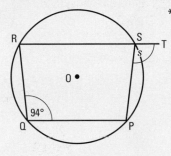

* **8**

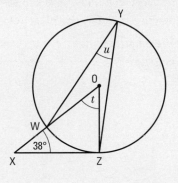

A03 A*

* **9**

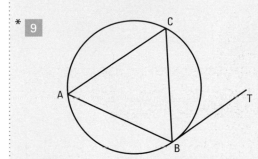

Diagram **NOT** accurately drawn

A, B and C are points on the circumference of the circle.
BT is a tangent to the circle. BC bisects the angle ABT.
Prove that CA = CB.

A03

* **10**

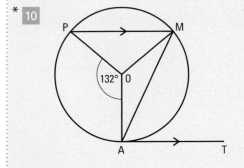

Diagram **NOT** accurately drawn

P, M and A are points on the circumference of a circle, centre O. Angle POA = 132°.
The line PM is parallel to the tangent at A.
Calculate the size of angle OMA.

Give reasons for all your calculations.

A02 A03

The traditional way of measuring the height of mountains used trigonometry. Surveyors measured the angle to the top of the mountain from two different points of known altitude – these are the angles of elevation. They then measured the distance from these points to the mountain. This gave them the side and two angles of the triangle and they could then calculate the third side. The original height of Everest calculated in this way only varied by 8 metres from the current accepted height (8848 m), calculated using modern techniques.

◉ Objectives

In this chapter you will:
- use a ruler and a pair of compasses to draw triangles, given the lengths of the sides
- use a straight edge and a pair of compasses to construct perpendiculars and bisectors
- construct and bisect angles using a pair of compasses
- draw loci and regions
- find the bearing of one point from another
- learn how to draw, use and interpret scale drawings.

◈ Before you start

You need to:
- be able to make accurate drawings of triangles and 2D shapes using a ruler and a protractor
- be able to draw parallel lines using a protractor and ruler
- have some understanding of ratio
- be able to change from one metric unit of length to another.

14.1 Constructing triangles

⊙ Objective

- You can draw a triangle when given the lengths of its sides.

❓ Why do this?

If you were redesigning a garden and wanted a triangular border you would need to make a plan first and draw the triangles accurately.

⬆ Get Ready

1. Use a ruler and protractor to make an accurate drawing of this triangle. Measure AC, BC and angle ACB.

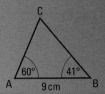

🔑 Key Points

- Two triangles are congruent if they have exactly the same shape and size. One of four conditions must be true for two triangles to be congruent: SSS, SAS, ASA and RHS (see Section 12.1).
- Constructing a triangle using any one of these sets of information therefore creates a unique triangle.
- More than one possible triangle can be created from other sets of information.

🔍 Example 1

Make an accurate drawing of the triangle shown in the sketch.

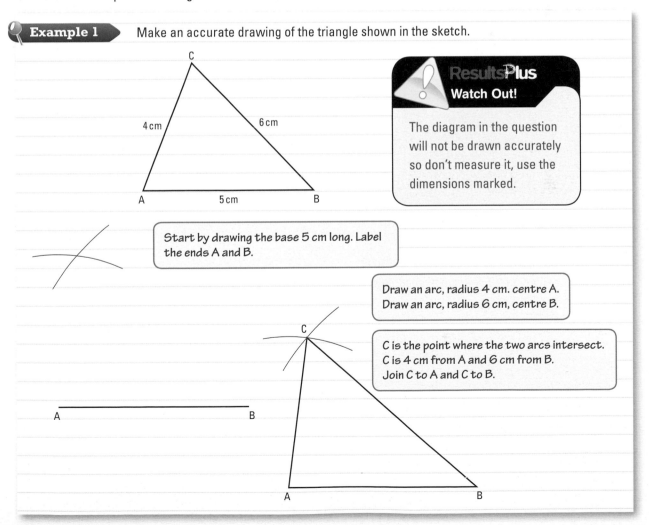

ResultsPlus
Watch Out!

The diagram in the question will not be drawn accurately so don't measure it, use the dimensions marked.

Start by drawing the base 5 cm long. Label the ends A and B.

Draw an arc, radius 4 cm. centre A. Draw an arc, radius 6 cm, centre B.

C is the point where the two arcs intersect. C is 4 cm from A and 6 cm from B. Join C to A and C to B.

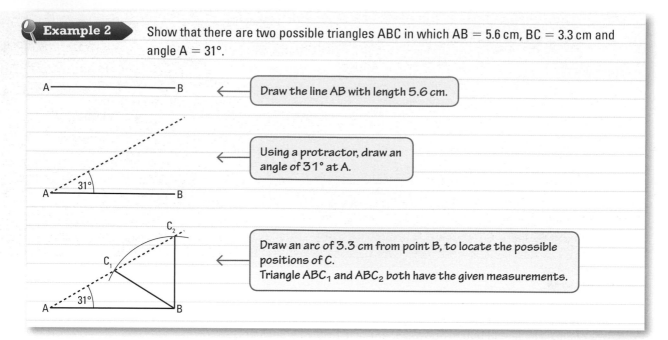

Example 2 ▶ Show that there are two possible triangles ABC in which AB = 5.6 cm, BC = 3.3 cm and angle A = 31°.

← Draw the line AB with length 5.6 cm.

← Using a protractor, draw an angle of 31° at A.

← Draw an arc of 3.3 cm from point B, to locate the possible positions of C.
Triangle ABC₁ and ABC₂ both have the given measurements.

Exercise 14A

Questions in this chapter are targeted at the grades indicated.

D

1 Here is a sketch of triangle XYZ.
Construct triangle XYZ.

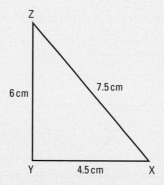

2 Construct an equilateral triangle with sides of length 5 cm.

3 Construct the triangle XYZ with sides XY = 4.2 cm, YZ = 5.8 cm and ZX = 7.5 cm.

4 Here is a sketch of the quadrilateral CDEF.
Make an accurate drawing of quadrilateral CDEF.

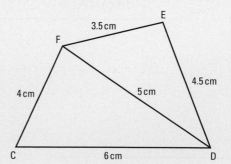

5 The rhombus KLMN has sides of length 5 cm.
The diagonal KM = 6 cm.
Make an accurate drawing of the rhombus KLMN.

6 Explain why it is not possible to construct a triangle with sides of length 4 cm, 3 cm and 8 cm.

14.2 Perpendicular lines

Objective

- You can construct perpendicular lines using a straight edge and compasses.

Why do this?

Many structures involve lines or planes that are perpendicular, for example the walls and floor of a house are perpendicular.

Get Ready

1. Draw a circle with a radius of 4 cm.
2. Mark two points A and B 6 cm apart. Mark the points that are 5 cm from A and 5 cm from B.
3. Draw two straight lines which are perpendicular to each other.

Key Points

- A **bisector** cuts something exactly in half.
- A **perpendicular bisector** is at right angles to the line it is cutting.
- You can use a straight edge and compass in the **construction** of the following:
  - the perpendicular bisector of a **line segment**
  - the perpendicular to a line segment from a point on it
  - the perpendicular to a line segment from a point not on the line.

Example 3

Construct the perpendicular bisector of the line AB.

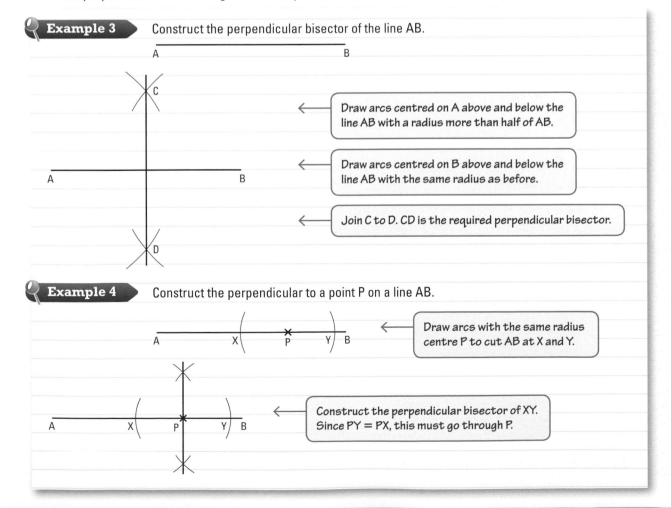

Draw arcs centred on A above and below the line AB with a radius more than half of AB.

Draw arcs centred on B above and below the line AB with the same radius as before.

Join C to D. CD is the required perpendicular bisector.

Example 4

Construct the perpendicular to a point P on a line AB.

Draw arcs with the same radius centre P to cut AB at X and Y.

Construct the perpendicular bisector of XY. Since PY = PX, this must go through P.

Example 5 Construct the perpendicular to a line AB from a point P not on the line.

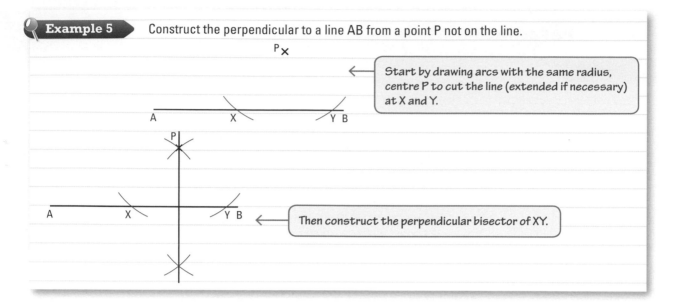

Start by drawing arcs with the same radius, centre P to cut the line (extended if necessary) at X and Y.

Then construct the perpendicular bisector of XY.

Exercise 14B

C

1 Draw line segments of length 10 cm and 8 cm. Using a straight edge and a pair of compasses, construct the perpendicular bisector of each of these line segments.

2 Draw these lines accurately, and then construct the perpendicular from the point P.

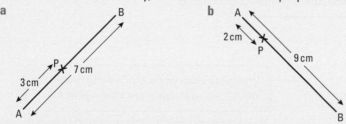

3 Draw a line segment AB, a point above it, P, and a point below it, Q. Construct the perpendicular from P to AB, and from Q to AB.

14.3 Constructing and bisecting angles

◉ Objectives

● You can construct certain angles using compasses.
● You can construct the bisector of an angle using a straight edge and compasses.
● You can construct a regular hexagon inside a circle.

❓ Why do this?

You may need to bisect an angle accurately when cutting a tile to place in an awkward corner.

◈ Get Ready

1. Draw a circle with a radius of 3 cm.
2. Draw an angle of 60°.
3. Use a protractor to bisect an angle of 60°.

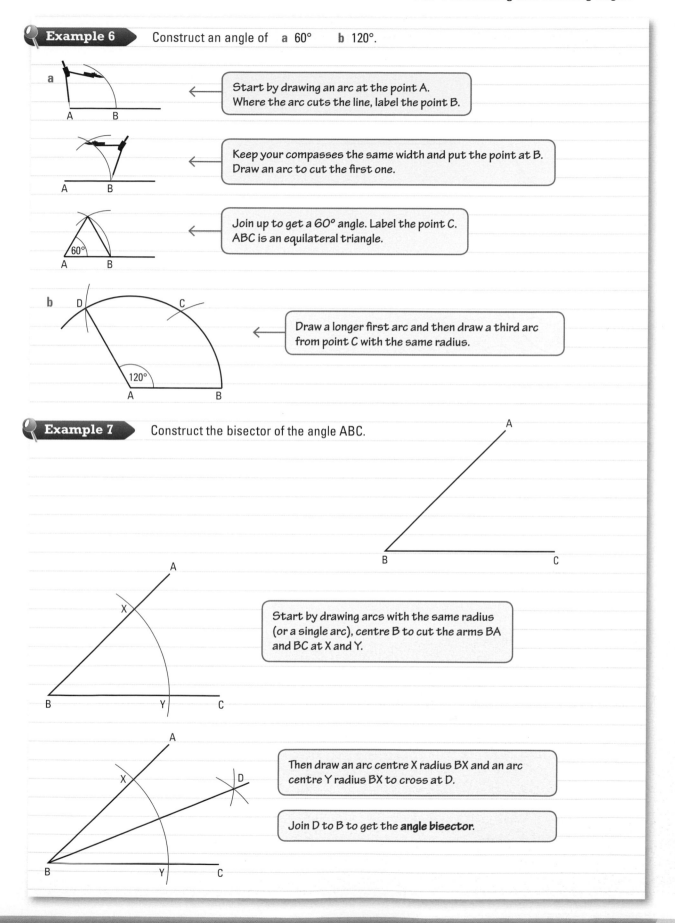

Example 6 Construct an angle of **a** 60° **b** 120°.

a

Start by drawing an arc at the point A.
Where the arc cuts the line, label the point B.

Keep your compasses the same width and put the point at B.
Draw an arc to cut the first one.

Join up to get a 60° angle. Label the point C.
ABC is an equilateral triangle.

b

Draw a longer first arc and then draw a third arc
from point C with the same radius.

Example 7 Construct the bisector of the angle ABC.

Start by drawing arcs with the same radius
(or a single arc), centre B to cut the arms BA
and BC at X and Y.

Then draw an arc centre X radius BX and an arc
centre Y radius BX to cross at D.

Join D to B to get the **angle bisector**.

Example 8

Construct a regular hexagon inside a circle.

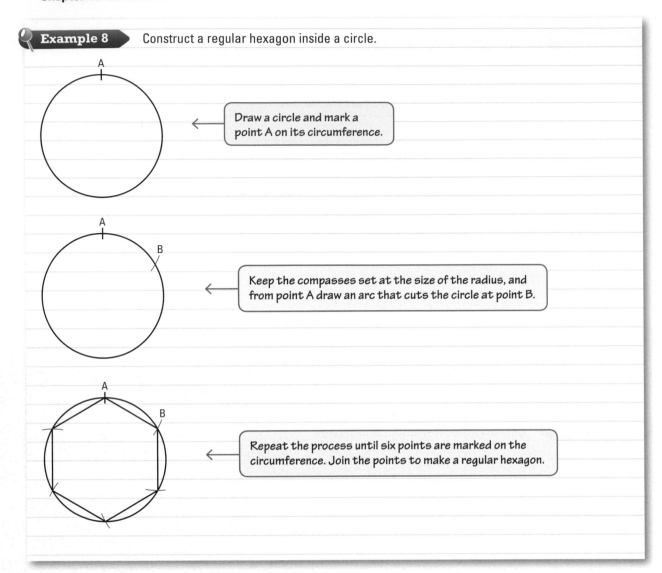

Draw a circle and mark a point A on its circumference.

Keep the compasses set at the size of the radius, and from point A draw an arc that cuts the circle at point B.

Repeat the process until six points are marked on the circumference. Join the points to make a regular hexagon.

Exercise 14C

1   Copy the diagrams and construct the bisector of the angle ABC.

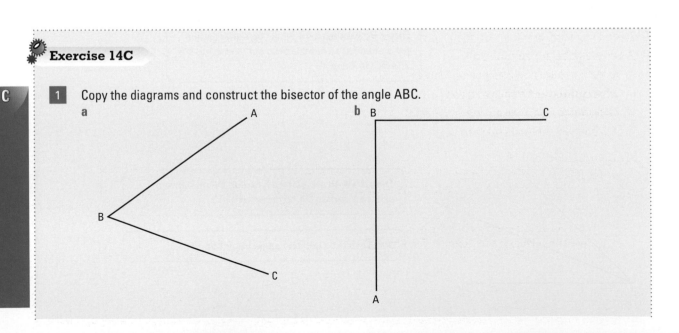

a

b

C

2   Copy the diagrams and construct the bisector of angle Q in the triangle PQR.

a

b

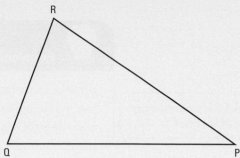

3   Construct each of the following angles.
    a 60°   b 120°   c 90°   d 30°   e 45°

4   Draw a regular hexagon in a circle of radius 4 cm.

5   Draw a regular octagon in a circle of radius 4 cm.

14.4 Loci

◎ Objective

● You can draw the locus
  of a point.

⟐ Why do this?

Scientists studying interference effects of radio waves need to plot paths
that are equidistant from two or more transmitters. They use loci to do this.

⟐ Get Ready

1. Put a cross in your book. Mark some points which are 3 cm from the cross.
2. Put two crosses A and B less than 3 cm apart in your book. Mark points which are 3 cm from each cross.
3. Draw two parallel lines. Mark any points which are the same distance from both lines.

⟐ Key Points

● A **locus** is a line or curve, formed by points that all satisfy a certain condition.
● A locus can be drawn such that:
    ● its distance from a fixed point is constant
    ● it is **equidistant** from two given points
    ● its distance from a given line is constant
    ● it is equidistant from two lines.

⟐ **Example 9**   Show the locus of all points which are at a
                   distance of 3 cm from the fixed point O.

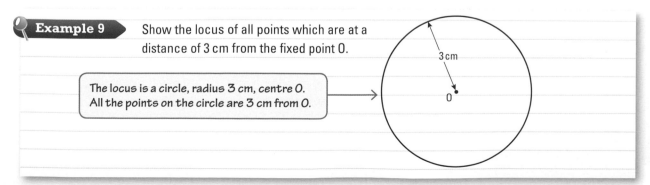

The locus is a circle, radius 3 cm, centre O.
All the points on the circle are 3 cm from O.

Example 10 ▸ Show the locus of all points which are equidistant from the points X and Y.

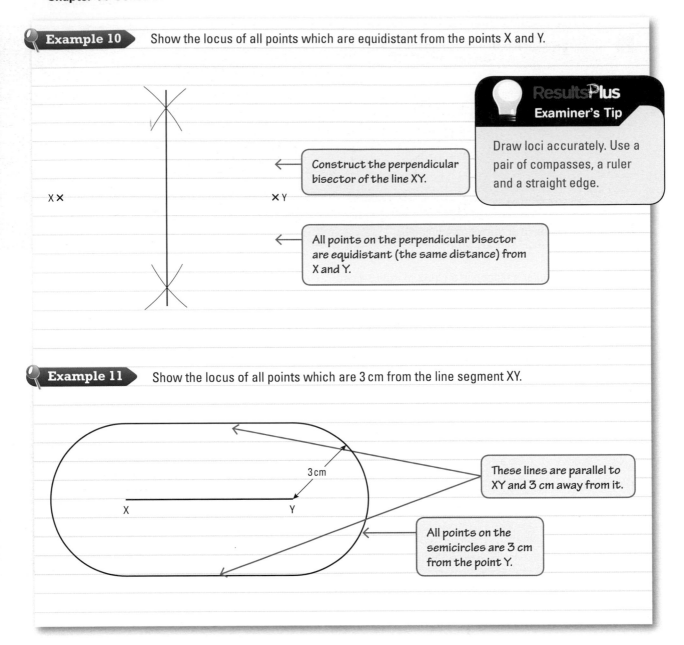

Construct the perpendicular bisector of the line XY.

Results Plus
Examiner's Tip

Draw loci accurately. Use a pair of compasses, a ruler and a straight edge.

All points on the perpendicular bisector are equidistant (the same distance) from X and Y.

Example 11 ▸ Show the locus of all points which are 3 cm from the line segment XY.

3 cm

These lines are parallel to XY and 3 cm away from it.

All points on the semicircles are 3 cm from the point Y.

Exercise 14D

C

1 Mark two points A and B approximately 6 cm apart.
Draw the locus of all points that are equidistant from A and B.

2 Draw the locus of all points which are 3.5 cm from a point P.

3 Draw the locus of a point that moves so that it is always 1.5 cm from a line 5 cm long.

4 Draw two lines PQ and QR, so that the angle PQR is acute. Draw the locus of all points that are equidistant between the two lines PQ and QR.

14.5 Regions

◎ **Objective**

◎ You can draw regions.

⊘ **Why do this?**

If you tether a goat to a point in your garden to eat the grass, you might want to check that the region it can access doesn't include the flowerbed.

🔼 **Get Ready**

1. Put a cross in your book. Mark some points which are less than 3 cm from the cross.
2. Put two crosses A and B in your book. Mark points which are closer to A than to B.
3. Draw two parallel lines. Mark any points which are further from one line than the other.

🔧 **Key Points**

◎ A set of points can lie inside a **region** rather than on a line or curve.

◎ The region of points can be drawn such that:
   ◎ the points are greater than or less than a given distance from a fixed point
   ◎ the points are closer to one given point than to another given point
   ◎ the points are closer to one given line than to another given line.

🔍 **Example 12** ➤ Draw the region of points which are less than 2 cm from the point O.

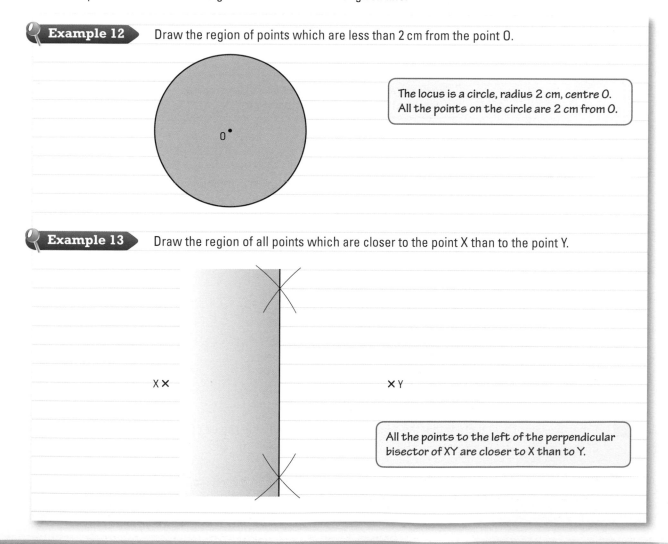

The locus is a circle, radius 2 cm, centre O.
All the points on the circle are 2 cm from O.

🔍 **Example 13** ➤ Draw the region of all points which are closer to the point X than to the point Y.

X ✕

✕ Y

All the points to the left of the perpendicular bisector of XY are closer to X than to Y.

Example 14 ABCD is a square of side 4 cm. Draw the region of points inside the rectangle that are both more than 3 cm from point A and more than 2 cm from the line BC.

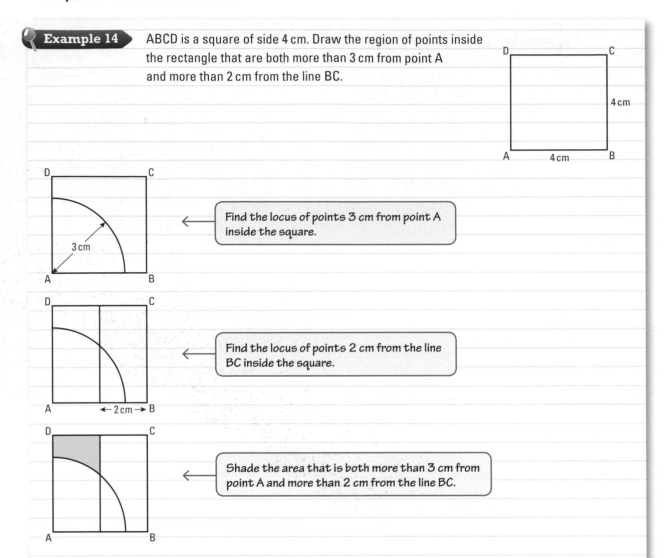

Find the locus of points 3 cm from point A inside the square.

Find the locus of points 2 cm from the line BC inside the square.

Shade the area that is both more than 3 cm from point A and more than 2 cm from the line BC.

Exercise 14E

1. Shade the region of points which are less than 2 cm from a point P.

2. Shade the region of points which are less than 2.6 cm from a line 4 cm long.

3. Mark two points, G and H, roughly 3 cm apart.
   Shade the region of points which are closer to G than to H.

4. Draw two lines DE and EF, so that the angle DEF is acute. Shade the region of points which are closer to EF than to DE.

5. Baby Tommy is placed inside a rectangular playpen measuring 1.4 m by 0.8 m. He can reach 25 cm outside the playpen. Show the region of points Tommy can reach beyond the edge of the playpen.

14.6 Bearings

◎ Objectives

- ◎ You can use bearings for directions.
- ◎ You can find the bearing of one point from another point.
- ◎ You can work out the bearing of point B from point A when you know the bearing of point A from point B.

⟐ Why do this?

When giving directions, it is important to know the exact direction of A from B. This is useful for orienteering.

⟐ Get Ready

Yuen lives in a town. The diagram shows the position of three places in the town in relation to Yuen's home. Give the compass directions from Yuen's home of:

a the cinema
b the park
c the school.

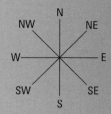

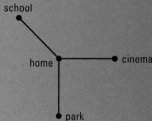

⟐ Key Points

- ◎ **Bearings** are angles measured clockwise from North.
- ◎ Bearings always have three figures.

⟐ Example 15

For each diagram, give the bearing of B from A.

a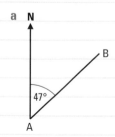

b

Results Plus
Watch Out!

Remember that a bearing has to have three figures and that is why some start with one or two zeros.

a At A, turn 47° clockwise from North to look towards B.
The bearing of B from A is 047°.

b The angle 351° is measured anticlockwise from North.
Clockwise angle = 360° − 351° = 9°
(as there are 360° in a complete turn).
The bearing of B from A is 009°.

Example 16 The bearing of B from A is 106°.
Work out the bearing of A from B.

Bearing of A from B = 106° + 180°
= 286°

ResultsPlus
Examiner's Tip

Remember that bearings are always measured clockwise from the North.

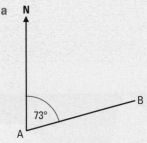

Draw a North line at B.

The angle marked in red is the angle that is the bearing of A from B.

The angle marked in blue is 106° (alternate angles).

Exercise 14F

1 In each of the following, give the bearing of B from A.

a

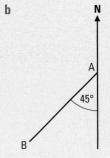

b

c

2 The diagram shows three towns A, B and C.

The bearing of C from A is 038°.
Angle ACB = 116°
CA = CB
Work out the bearing of

a B from A
b A from C
c B from C

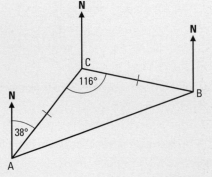
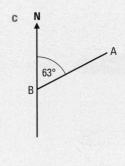

3 The bearing of Norwich from Gloucester is 069°.
Work out the bearing of Gloucester from Norwich.

4 A plane flies from Skegness to Carlisle on a bearing of 132°.
Work out the bearing the plane needs to fly on for the return journey to Skegness.

5 The diagram shows the position of three towns P, Q and R.
Find the bearing of:
a R from Q
b P from Q.

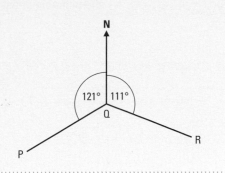

14.7 Scale drawings and maps

C A02 A02 A03

Objectives

- You can read and construct scale drawings.
- You can draw lines and shapes to scale and estimate lengths on scale drawings.
- You can work out lengths using a scale factor.

Why do this?

When a new aeroplane is being designed or an extension to a house is planned, accurate scale drawings have to be made.

Get Ready

1. Convert from cm to km:
   a 5 000 000 cm
   b 250 000 cm.

2. Convert from km to cm:
   a 4 km
   b 0.3 km.

Key Points

- Here is a picture of a scale model of a Saturn rocket. The model has been built to a scale of 1 : 24. This means that every length on the model is shorter than the length on the real rocket, with a length of 1 cm on the model representing a length of 24 cm on the real rocket.
  - The real rocket is an enlargement of the model with a **scale factor** of 24; the model is a smaller version of the real rocket with a scale factor of $\frac{1}{24}$.
- In general, a scale of 1 : n means that:
  - a length on the real object = the length on the **scale diagram** or model × n
  - a length on the scale drawing or model = the length on the real object ÷ n.

Example 17

The Empire State Building is 443 m tall. Bill has a model of the building that is 88.6 cm tall.
a Calculate the scale of the model. Give your answer in the form $1 : n$.
b The pinnacle at the top of Bill's model is 12.4 cm in length. Work out the actual length of the pinnacle at the top of the Empire State Building. Give your answer in metres.

a Height of building = 443 × 100 = 44 300 cm

> Both heights have to be in the same units. Change 443 m to cm by multiplying by 100.

$$\text{Scale factor} = \frac{44\,300}{88.6} = 500$$

> $$\text{Scale factor} = \frac{\text{Height of building}}{\text{Height of model}}$$

Scale of model = 1 : 500

b Length of pinnacle on building = 12.4 × 500
= 6200 cm
Length of pinnacle on building = 6200 ÷ 100
= 62 m

> Length on model = Length on building ÷ 500.
> Length on building = Length on model × 500.

> Change cm to m by dividing by 100.

Example 18

The scale of a map is 1 : 50 000.
a On the map, the distance between two churches is 6 cm. Work out the real distance between the churches. Give your answer in kilometres.
b The real distance between two train stations is 12 km. Work out the distance between the two train stations on the map. Give your answer in centimetres.

Method 1

> A scale of 1 : 50 000 means:
> real distance = map distance × 50 000.

a Real distance between churches
= 6 × 50 000 = 300 000 cm
= 3000 m
= 3 km

> Change cm to m, divide by 100.
> Change m to km, divide by 1000.

b 12 km = 12 × 1000 × 100 = 1 200 000 cm

> Change km to cm by multiplying by 1000 × 100.

Distance between stations on map
= 1 200 000 ÷ 50 000 = 24 cm

> Map distance = real distance ÷ 50 000

Method 2

Map distance of 1 cm represents real distance of 0.5 km.
a 6 cm on the map represents real distance of 6 × 0.5 = 3 km.
Distance between the churches = 3 km.

> 1 : 50 000 means 1 cm : 50 000 cm
> or 1 cm : 500 m
> or 1 cm : 0.5 km

b Real distance of 12 km represents map distance of 12 ÷ 0.5 = 24 cm.
Distance between the stations on map = 24 cm.

Exercise 14G

1 This is an accurate map of a desert island. There is treasure buried on the island at T.

Key to map

P palm trees   R rocks

C cliffs   T treasure

The real distance between the palm trees and the cliffs is 5 km.

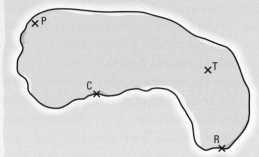

a Find the scale of the map.
Give your answer in the form 1 cm represents n km, giving the value of n.

b Find the real distance of the treasure from:
  i the cliffs **ii** the palm trees **iii** the rocks.

2 On a map of England, 1 cm represents 10 km.

a The distance between Hull and Manchester is 135 km. Work out the distance between Hull and Manchester on the map.

b On the map, the distance between London and York is 31.2 cm. Work out the real distance between London and York.

3 Here is part of a map, not accurately drawn, showing three towns: Alphaville (A), Beecombe (B) and Ceeton (C).

a Using a scale of 1 : 200 000, accurately draw this part of the map.

b Find the real distance, in km, between Beecombe and Ceeton.

c Use the scaled drawing to measure the bearing of Ceeton from Beecombe.

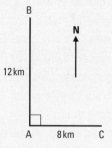

4 This is a sketch of Arfan's bedroom. It is *not* drawn to scale. Draw an accurate scale drawing on cm squared paper of Arfan's bedroom. Use a scale of 1 : 50.

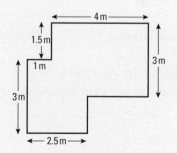

5 A space shuttle has a length of 24 m. A model of the space shuttle has a length of 48 cm.

a Find, in the form 1 : n, the scale of the model.

b The height of the space shuttle is 5 m. Work out the height of the model.

6 The distance between Bristol and Hull is 330 km. On a map, the distance between Bristol and Hull is 6.6 cm.

a Find, as a ratio, the scale of the map.

b The distance between Bristol and London is 183 km. Work out the distance between Bristol and London on the map. Give your answer in centimetres.

Chapter review

- Two triangles are congruent if they have exactly the same shape and size. One of four conditions must be true for two triangles to be congruent: SSS, SAS, ASA and RHS.
- Constructing a triangle using any one of these sets of information therefore creates a unique triangle.
- More than one possible triangle can be created from other sets of information.
- A **bisector** cuts something exactly in half.
- A **perpendicular bisector** is at right angles to the line it is cutting.
- You can use a straight edge and compass in the **construction** of the following:
  - the perpendicular bisector of a **line segment**
  - the perpendicular to a line segment from a point on it
  - the perpendicular to a line segment from a point not on the line.
- A **locus** is a line or curve, formed by points that all satisfy a certain condition.
- A locus can be drawn such that
  - its distance from a fixed point is constant
  - it is **equidistant** from two given points
  - its distance from a given line is constant
  - it is equidistant from two lines.
- A set of points can lie inside a **region** rather than on a line or curve.
- A region of points can be drawn such that:
  - the points are greater than or less than a given distance from a fixed point
  - the points are closer to one given point than to another given point
  - the points are closer to one given line than to another given line.
- **Bearings** are angles measured clockwise from North.
- Bearings always have three figures.
- A scale of $1 : n$ means that:
  - a length on the real object = the length on the **scale diagram** or model $\times n$
  - a length on the scale drawing or model = the length on the real object $\div n$.

Review exercise

1 AB = 8 cm. AC = 6 cm. Angle A = 52°.
Make an accurate drawing of triangle ABC.

Diagram **NOT** accurately drawn

Nov 2008

2 Make an accurate drawing of the quadrilateral ABCD.

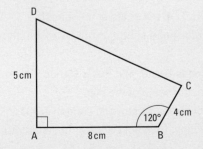

Diagram **NOT** accurately drawn

3 Make an accurate drawing of triangle ABC.

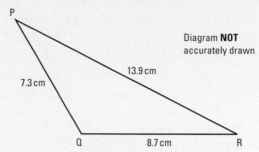

Diagram **NOT** accurately drawn

60°   30°

A   6.5 cm   B

May 2009

4 Use a ruler and compasses to make an accurate drawing of triangle PQR.

Diagram **NOT** accurately drawn

13.9 cm

7.3 cm

Q   8.7 cm   R

5 A model of the Eiffel Tower is made to a scale of 2 millimetres to 1 metre.
The width of the base of the real Eiffel Tower is 125 metres.
  a Work out the width of the base of the model. Give your answer in millimetres.
The height of the model is 648 millimetres.
  b Work out the height of the real Eiffel Tower. Give your answer in metres

June 2008, adapted

6 The diagram shows the position of two boats, P and Q.
The bearing of a boat R from boat P is 060°.
The bearing of boat R from boat Q is 310°.
Draw an accurate diagram to show the position of boat R.
Mark the position of boat R with a cross (×). Label it R.

N

Q

N

P

June 2009

7 **a** Find the bearing of B from A.
  b On a copy of the diagram, draw a line on a bearing of 135° from A.

N

B

A

Nov 2006

8 Beeham is 10 km from Alston.
Corting is 20 km from Beeham.
Deetown is 45 km from Alston.
The diagram below shows the straight road from Alston to Deetown.
This diagram has been drawn accurately using a scale of 1 cm to represent 5 km.

Alston ——————————————————————— Deetown

On a copy of the diagram, mark accurately with crosses (×), the positions of Beeham and Corting.

Nov 2007

D

C

C

9 ABC is a triangle.
Copy the triangle accurately and shade the region
inside the triangle which is **both** less than
4 centimetres from the point B **and** closer to
the line AC than the line AB.

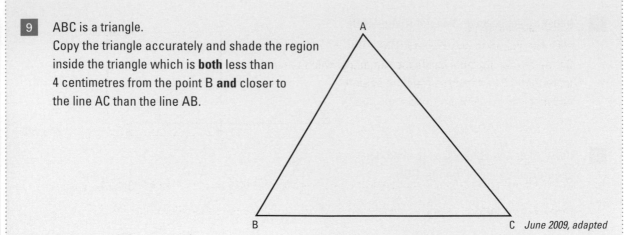

B   C   *June 2009, adapted*

10 On a copy of the diagram, use a ruler and pair of compasses to **construct** an angle of 30° at P.
You **must** show all your construction lines.

P

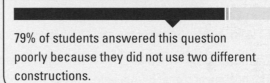

ResultsPlus
Exam Question Report

79% of students answered this question
poorly because they did not use two different
constructions.

Nov 2007, adapted

11   **a**   Mark the points C and D approximately 8 cm apart. Draw the locus of all points that are equidistant
from C and D.

    d   Draw the locus of a point that moves so that it is always 3 cm from a line 4.5 cm long.

12   B is 5 km north of A.
C is 4 km from B.
C is 7 km from A.

    a   Make an accurate scale drawing of triangle ABC.
Use a scale of 1 cm to 1 km.

    b   From your accurate scale drawing, measure the bearing of C from A.

    c   Find the bearing of A from C.

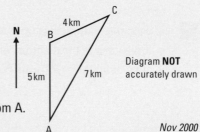

Diagram **NOT**
accurately drawn

Nov 2000

13   On an accurate copy of the diagram use a ruler and pair of compasses to construct the bisector of
angle ABC.
You must show all your construction lines.

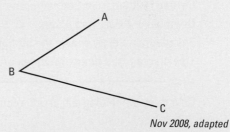

Nov 2008, adapted

14 ABCD is a rectangle.
Make an accurate drawing of ABCD.
Shade the set of points inside the rectangle which are **both**
more than 1.2 centimetres from the point A
and more than 1 centimetre from the line DC.

A _____ B

D _____ C

15 Draw a line segment 7 cm long. Construct the perpendicular bisector of the line segment.

16 Draw a line segment ST and a point above it, M. Construct the perpendicular from M to ST.

17 As a bicycle moves along a flat road, draw the locus of:
a the yellow dot
b the green dot.

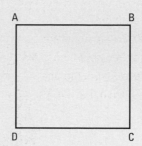

18

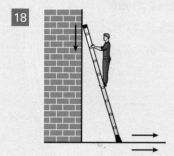

Draw the locus of a man's head as the ladder he is on slips down a wall.

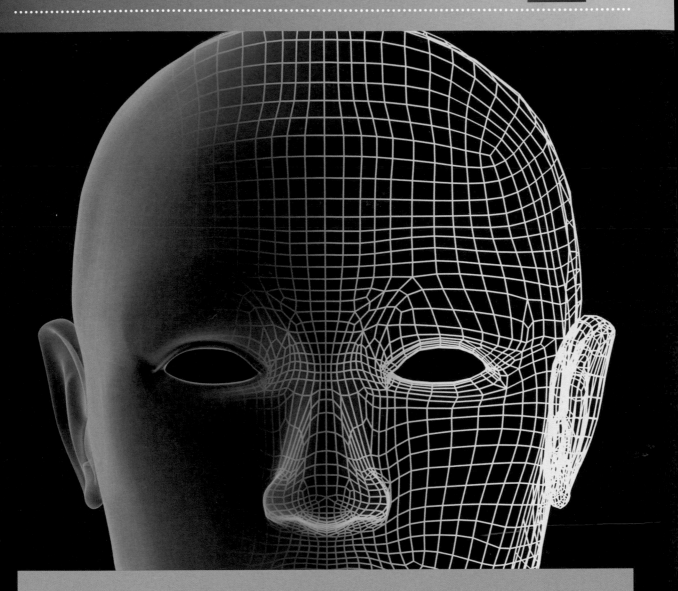

Three-dimensional head models have many uses, including predicting the impact of car crashes. The original computerised head models were made using a mass of 3D points which were then moved around to change the shape of the head and features.

◉ Objectives

In this chapter you will:
- translate, rotate, reflect and enlarge a 2D shape
- describe the translation, rotation, reflection and enlargement of a 2D shape
- combine transformations and describe the single transformation that has the same effect as a combination of transformations.

◈ Before you start

You need to:
- be able to use coordinates in all four quadrants
- understand what it means to move (translate), flip (reflect), turn (rotate) and change the size of a 2D shape
- recognise and be able to draw lines with equations $x = a, y = b, y = \pm x$
- be able to draw scale diagrams.

15.1 Using translations

Objectives

- You know that in a translation all points of a shape move the same distance in the same direction.
- You understand that translations are described by the distance and the direction moved.
- You can use a vector to describe a translation.
- You know that when shape A is mapped to shape B by a translation, shape A and shape B are congruent.

Get Ready

1. Here is a letter square.

U	V	W	X	Y
T	S	R	Q	P
K	L	M	N	O
J	I	H	G	F
A	B	C	D	E

Start at square A, go 2 to the right, then 3 up and stop. Then go 3 down and stop. What mathematical word is this?

Start at square M, go 2 to the right and 2 down and stop. Then go 4 to the left and stop. Then go 3 to the right and 2 up and stop. What mathematical word is this?

Give instructions to get **a** KITE, **b** CIRCLE, **c** ADD.

d Make up some examples of your own.

Key Points

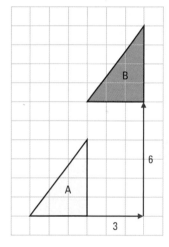

- In the diagram, shape A has been mapped onto shape B by a **translation**.
- All points of shape A move 3 squares to the right and 6 squares up. This can be written as $\begin{pmatrix} 3 \\ 6 \end{pmatrix}$.
- In a translation, all points of the shape move the same distance in the same direction.
- In a translation:
  - the lengths of the sides of the shape do not change
  - the angles of the shape do not change
  - the shape does not turn.
- In a translation, any shape is congruent to its image because the lengths of the sides and angles of the shape are preserved by the translation.

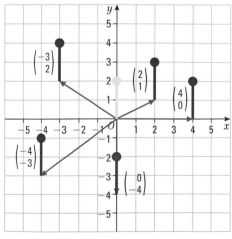

- $\begin{pmatrix} 3 \\ 6 \end{pmatrix}$ is a **vector**. Vectors can be used to describe translations.
  - the top number shows the number of squares moved parallel to the x-axis, to the right or left.
  - the bottom number shows the number of squares moved parallel to the y-axis, up or down.
  - to the right and up are positive.
  - to the left and down are negative.
- Some translations of the yellow shape to the red shape and their **column vectors** are shown on the grid.

Example 1 ▶ Describe the translation that maps triangle P onto triangle Q.

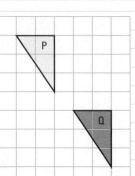

Choose one corner of triangle P.

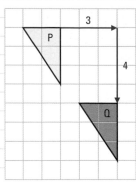

Count the number of squares to the right and the number of squares down from this corner on triangle P to the same corner on triangle Q.

The translation from triangle P to triangle Q is 3 squares to the right and 4 squares down.

This translation can also be written as $\begin{pmatrix} 3 \\ -4 \end{pmatrix}$.

Example 2 ▶
a Describe the transformation that maps shape A onto
   i shape B ii shape C.

b Translate shape A by the vector $\begin{pmatrix} -3 \\ -5 \end{pmatrix}$.

   Label this new shape D.

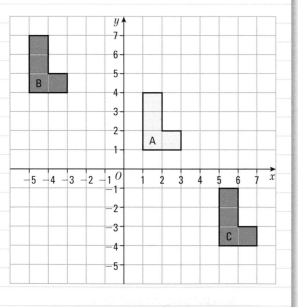

a

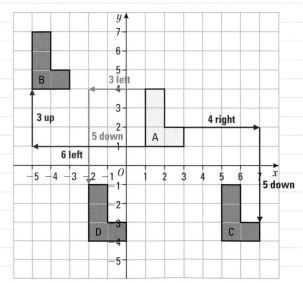

ResultsPlus
Examiner's Tip

The question asks for the transformation, so as well as the vector, you must say it is a translation.

i From A to B is the translation 6 to the left and 3 up, or the translation with vector $\begin{pmatrix} -6 \\ 3 \end{pmatrix}$.

> Count the number of squares moved to the left (negative) and up (positive) from any corner in A to the same corner in B.

$\begin{pmatrix} -3 \\ -5 \end{pmatrix}$ means 3 to the left and 5 down.

ii From A to C is the translation with vector $\begin{pmatrix} 4 \\ -5 \end{pmatrix}$.

> Choose one corner of shape A.
> Count from this corner 3 squares to the left and then count 5 squares down to find where this corner has moved to.
> The new shape is the same as shape A.
> Draw the new shape and label it D.

b D is marked on the diagram.

Exercise 15A

> Questions in this chapter are targeted at the grades indicated.

1 Describe, with a vector, the translation that maps triangle A onto:
  a triangle B
  b triangle C
  c triangle D
  d triangle E
  e triangle F
  f triangle G.

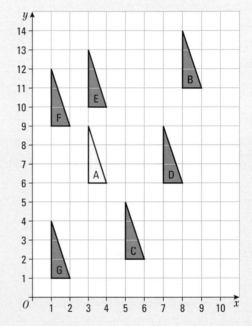

2 On a copy of the diagram translate triangle A:
  a 5 to the right and 4 up.
  Label your new triangle B.
  b 4 to the right and 6 down.
  Label your new triangle C.
  c 7 to the left. Label your new triangle D.
  d by the vector $\begin{pmatrix} 3 \\ 2 \end{pmatrix}$.
  Label your new triangle E.
  e by the vector $\begin{pmatrix} -6 \\ -4 \end{pmatrix}$.
  Label your new triangle F.

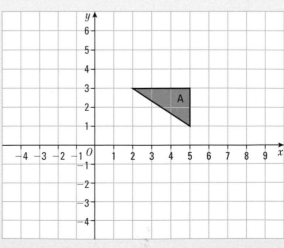

C

C

3 The coordinates of point A of this kite are $(-2, 1)$.
The kite is translated so that the point A is mapped
onto the point $(3, 4)$.

   a On a copy of the diagram draw the image of the
kite after this translation.

   b Describe this translation with a vector.

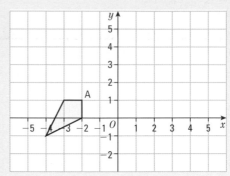

4 Draw the following translations on a copy of the diagram.

   a Translate kite A by the vector $\begin{pmatrix} 4 \\ 7 \end{pmatrix}$.
Label this new kite B.

   b Translate kite B by the vector $\begin{pmatrix} -6 \\ -3 \end{pmatrix}$.
Label this new kite C.

   c Describe, with a vector, the translation that
maps kite A onto kite C.

   d Describe, with a vector, the translation that
maps kite C onto kite A.

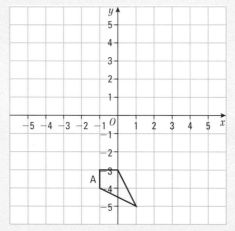

15.2 Transforming shapes using reflections

◎ Objectives

- You know that in a reflection the image is as far behind the mirror line as
the object is in front of the mirror line.
- You understand that reflections are described by the mirror line.
- You can find an equation of a mirror line.
- You know that when shape A is mapped to shape B by a reflection,
shape A and shape B are congruent.

◈ Why do this?

Many interesting patterns can
be produced using reflections.
The patterns in a kaleidoscope
are caused by light being
reflected many times.

◈ Get Ready

1. Write down the equation of each of the lines A, B, C and D.

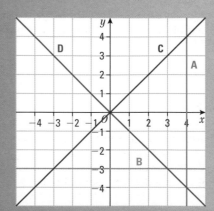

Key Points

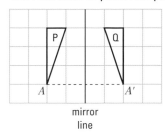

- When you look in a mirror, you see your **reflection**.

  The diagram below shows triangle P reflected in the mirror line to triangle Q.

  The reflection of point A is point A′ so A and A′ are corresponding points.

  Point A′ is the same distance behind the mirror line as point A is in front.

  The line joining points A and A′ is perpendicular to the mirror line.

  Triangle Q is the reflection of triangle P in the mirror line. Each corner of Q is the reflection in the mirror line of the corresponding corner of P.

  Triangle Q is as far behind the mirror line as triangle P is in front.

  In mathematics all mirror lines are two-way mirrors so triangle P is also the reflection of triangle Q in the mirror line.

- To describe a reflection, give the mirror line.

- In a reflection:
  - the lengths of the sides of the shape do not change
  - the angles of the shape do not change
  - the reflection of a shape (the **image**) is as far behind the mirror line as the shape is in front.

- In a reflection, any shape is congruent to its image because the lengths of the sides and angles of the shape are preserved by the reflection.

- The mirror line is the line of symmetry.

Example 3

Reflect trapezium T in the mirror line.
Label the new trapezium U.

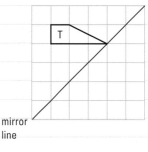

Method 1

> Reflect each corner of T in the mirror line so that its reflection is the same distance behind the mirror line as the corner is in front.
> Notice that:
> the line joining each corner to its image is perpendicular to the mirror line
> the image of the corner which is on the mirror line is also on the mirror line.

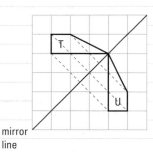

Method 2

> Put the edge of a sheet of tracing paper on the mirror line and make a tracing of the trapezium.
> Turn the tracing paper over and put the edge of the tracing paper back on the mirror line.
> Mark the images of the corners with a pencil or compass point.
> Method 2 is particularly useful when the shape is not a polygon or not drawn on a grid.

Example 4

Triangle T is a reflection of triangle S.
Draw the mirror line of the reflection.

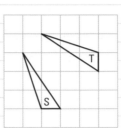

Join each corner of triangle S to its image on triangle T.
The mirror line passes through the mid-points (marked
with crosses) of these lines.

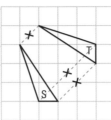

Draw the mirror line by joining the crosses.

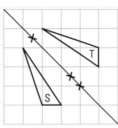

Example 5

Describe fully the transformation which
maps triangle P onto triangle Q.

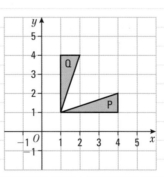

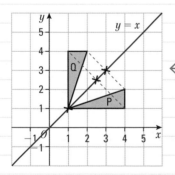

The transformation is a
reflection as triangle P has
been 'flipped over' to triangle Q.
Notice that the point on the
mirror line does not move.

ResultsPlus
Examiner's Tip

Make sure that you can recognise the lines
with equations $x = a$, $y = b$, $y = \pm x$

The transformation is a reflection in the line with equation $y = x$.

Exercise 15B

1. Make a copy of the diagram and complete the following reflections.
   a Reflect triangle P in the line $x = 1$.
     Label this new triangle Q.
   b Reflect triangle P in the line $y = 2$.
     Label this new triangle R.
   c Describe the reflection that maps triangle Q onto triangle T.

   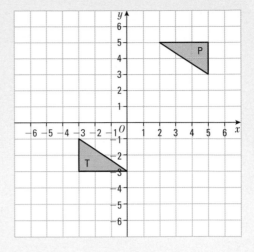

2. On a copy of the diagram, complete the following reflections.
   a Reflect triangle A in the line $y = x$.
     Label this new triangle B.
   b Reflect triangle A in the line $y = -x$.
     Label this new triangle C.
   c Describe fully the transformation that maps triangle B onto triangle A.

   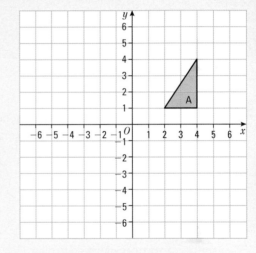

3. a Give the equation of the mirror line of the reflection that maps:
     i shape P onto shape Q
     ii shape P onto shape R.
   b Describe fully the transformation that maps shape Q onto shape P.

   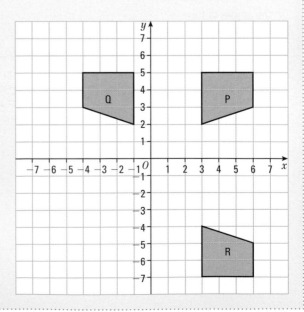

15.3 Transforming shapes using rotations

◉ Objectives

- ● You know that in a rotation all points of a shape move around circles with the same centre.
- ● You understand that rotations are described by a centre and an angle of turn.
- ● You can find a centre of rotation.
- ● You know that when shape A is mapped to shape B by a rotation, shape A and shape B are congruent.

❓ Why do this?

Many everyday objects turn or rotate, for example, cycle wheels and the hands of a clock. It is often necessary to describe the rotation.

◈ Get Ready

1. Here is a clock face with only one hand. The hand is pointing to 12.
   a The hand is turned 90° anticlockwise. What number is the hand pointing to now?
   b Describe as fully as you can how the hand can turn to point to:
   i 3   **ii** 6   **iii** 5.
2. Imagine that the hand is pointing to 5. Describe as fully as you can how the hand can turn to point to:
   a 8   **b** 2   **c** 11   **d** 12.

🔑 Key Points

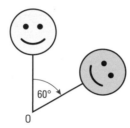

- ◉ To rotate means to turn. This face on a stick has rotated 60° clockwise about the point O. The size of the face has not changed.

💡 ResultsPlus
Examiner's Tip

A common mistake when describing a rotation is to call it a turn instead of a rotation and forgetting to say where the centre of rotation is.

- ◉ To describe a **rotation** you need to give:
  - ◉ the angle of turn
  - ◉ the direction of turn (clockwise or anticlockwise)
  - ◉ the point the shape turns about (the **centre of rotation**).
- ◉ In a rotation:
  - ◉ the lengths of the sides of the shape do not change
  - ◉ the angles of the shape do not change
  - ◉ the shape turns
  - ◉ the centre of rotation does not move.
- ◉ In a rotation, any shape is congruent to its image because the lengths of the sides and angles of the shape are preserved by the rotation.

Example 6
Rotate the triangle a quarter turn clockwise about the point A.

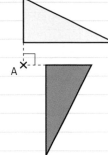

Tracing paper can be used to rotate the shape. Trace the triangle and mark the point A.
Fix the point A with a pencil or a compass point so that the point A does not move. Turn the tracing paper about A, clockwise through a quarter turn (90°).
Now the position of the image of the triangle can be seen.
Notice that each line of the triangle has turned through a quarter turn clockwise.

Example 7
Describe the transformation that maps triangle A onto triangle B.

Triangle A is mapped onto triangle B by a rotation of 180° (a half turn) about the point (5, 5).

Tracing paper can be used to check that the transformation is a rotation of 180° with the centre of rotation the point (5, 5).

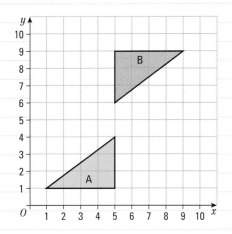

Exercise 15C

1 On a copy of the diagram, complete the following rotations.

a Rotate trapezium A a half turn about the origin O. Label the new trapezium B.

b Rotate trapezium A a quarter turn clockwise about the origin O. Label the new trapezium C.

c Rotate trapezium A a quarter turn anticlockwise about the origin O. Label the new trapezium D.

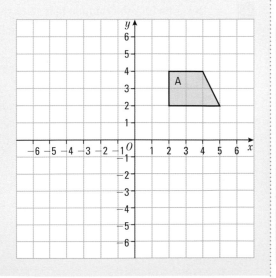

D

C

2 Make three copies of this diagram showing trapezium P.

  a On copy 1 of the diagram, rotate trapezium P 180°
     about the point (2, 0). Label the new trapezium Q.

  b On copy 2 of the diagram, rotate trapezium P 90°
     clockwise about the point (−2, 2). Label the new
     trapezium R.

  c On copy 3 of the diagram, rotate trapezium P 90°
     anticlockwise about the point (−1, −1). Label the
     new trapezium S.

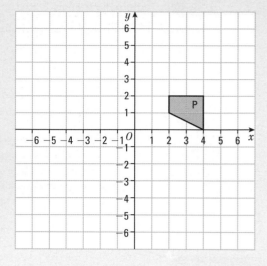

A03

3 **a** Describe fully the rotation that maps shape A
     onto: **i** shape B **ii** shape C **iii** shape D.

  b Describe fully the rotation that maps shape B
     onto shape A.

  c Describe fully the rotation that maps shape B
     onto shape D.

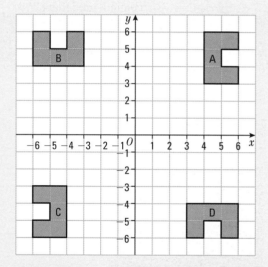

A03

4 **a** Describe fully the rotation that maps triangle A onto:
     i triangle B **ii** triangle C **iii** triangle D **iv** triangle E
     v triangle F.

  b Describe the transformation that maps triangle B onto
     triangle E.

  c Describe the transformation that maps:
     i triangle D onto triangle B **ii** triangle F onto triangle E.

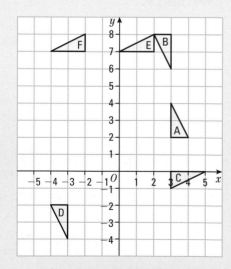

15.4 Enlargements and scale factors

Objectives

- You can enlarge a shape given the scale factor.
- You know that enlargements preserve angles but change lengths.
- You understand that enlargements are described by a centre and a scale factor.
- You can find the centre of an enlargement.
- You can use positive, negative and fractional scale factors.

Why do this?

If you have holiday photos blown up for a poster, you are making an enlarged version of the original photo.

Get Ready

1. Plot the following points on graph paper and join them up.
   a (0, 1)   **b** (1, 1)   **c** (1, 0)   **d** (0, 0)
2. Then plot the following points on the same graph and join them up.
   a (0, 2)   **b** (2, 2)   **c** (2, 0)   **d** (0, 0)
3. What can you say about these two shapes?

Scale factors

Key Points

- Here is a photograph of a shark.

  Here is an **enlargement** of the photograph.

  The sharks in the two photographs are the same but each length in the enlargement is 2 times the corresponding length in the original photograph.

  For example, the length of the shark's fin in the enlargement is 2 times the length of the fin in the original photograph.

- The scale factor of an enlargement is the number of times by which each original length has been multiplied.
  So the larger photograph is an enlargement with scale factor 2 of the smaller photograph as

  $$\text{scale factor} = \frac{\text{length of side in image}}{\text{length of corresponding side in object}}$$

- The scale factor can be found from the ratio of the lengths of two corresponding sides; in this case the ratio is 1 : 2.

- An enlargement changes the size of an object but not the shape of the object.
- Notice that each angle in the original photograph has the same size as the corresponding angle in the enlargement.
- So in an enlargement:
  - the lengths of the sides of the shape change
  - the angles of the shape do not change.

Example 8 Triangle B is an enlargement of triangle A.

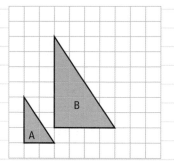

a Work out the scale factor of the enlargement that maps triangle A onto triangle B.

b Work out the scale factor of the enlargement that maps triangle B onto triangle A.

a Scale factor of the enlargement that maps triangle A onto triangle B $= \frac{4}{2}$.
Scale factor $= 2$

> The base of triangle A is 2 squares. The base of triangle B is 4 squares. Notice that pairs of corresponding sides are parallel.

b Scale factor of the enlargement that maps triangle B onto triangle A $= \frac{2}{4}$.
Scale factor $= \frac{1}{2}$

> This answer means that the length of each side of triangle A is $\frac{1}{2}$ the length of the corresponding side of triangle B.

ResultsPlus
Examiner's Tip

The transformation is still called an enlargement when the scale factor is a positive fraction less than 1, so that the image is smaller than the object.

Exercise 15D

1 Here is a right-angled triangle.
The triangle is enlarged with a scale factor of 4.
a Work out the length of each side of the enlarged triangle.
b Compare the perimeter of the enlarged triangle with the perimeter of the original triangle.

2 Copy the shape on squared paper and draw:
a an enlargement of shape A with scale factor 3.
Label this enlargement shape B.
b an enlargement of shape A with scale factor $\frac{1}{2}$.
Label this enlargement shape C.
c Shape B is an enlargement of shape C.
Work out the scale factor of the enlargement.

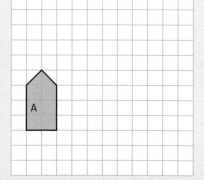

3 Rectangle P has a base of 4 cm and a height of 2 cm.

Rectangle Q is an enlargement of rectangle P with a scale factor of 2.

Rectangle R is an enlargement of rectangle P with a scale factor of 3.

a On squared paper, draw rectangles P, Q and R.

b Find the perimeter of: **i** rectangle P **ii** rectangle Q **iii** rectangle R.

c Find the area of: **i** rectangle P **ii** rectangle Q **iii** rectangle R.

d Work out the value of: **i** $\dfrac{\text{perimeter of Q}}{\text{perimeter of P}}$ **ii** $\dfrac{\text{perimeter of R}}{\text{perimeter of P}}$.

Write down anything that you notice about these values.

e Work out the value of: **i** $\dfrac{\text{area of Q}}{\text{area of P}}$ **ii** $\dfrac{\text{area of R}}{\text{area of P}}$.

Write down anything that you notice about these values.

f Rectangle S is an enlargement of rectangle P with a scale factor of 8.

What is the perimeter of rectangle S?

Centre of enlargement

 Key Points

◉ In the diagram, triangle P has been enlarged by a scale factor of 2 to give triangle Q.

The corner A of triangle P is mapped onto the corner A′ of triangle Q. A line has been drawn joining A and A′.

Lines have also been drawn joining the other pairs of corresponding points of triangles P and Q.

The lines meet at a point C called the **centre of enlargement**.

C to A is 2 squares across and 3 squares up.

C to A′ is 4 squares across and 6 squares up.

So $\dfrac{CA'}{CA} = 2$, the scale factor of the enlargement.

◉ To describe an enlargement you need to give:

 ◉ the scale factor

 ◉ the centre of enlargement.

◉ In general, when shape P is mapped onto shape Q by an enlargement with centre C and scale factor k, $CA' = k \times CA$ for any point A of shape P and the corresponding point A′ of shape Q.

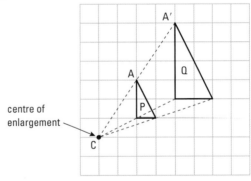

centre of enlargement

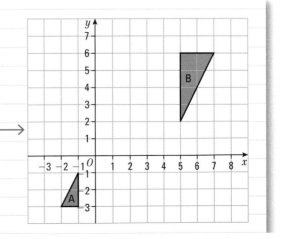

Example 9 Describe fully the transformation which maps triangle A onto triangle B.

> The lengths of the sides of triangle B are twice those of triangle A.
> This means that the transformation is an enlargement.
> To find the centre of enlargement, join each corner (vertex) of triangle A to the corresponding vertex of triangle B.
> The centre of enlargement C is the point where these lines cross.

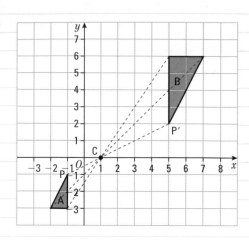

Notice that point C is between the object A and the image B. From C to P' is twice the distance from C to P but in the opposite direction.
The scale factor of the enlargement is −2.

ResultsPlus
Watch Out!

When a shape is enlarged by a negative scale factor, the image is on the opposite side of the centre of enlargement to the object.

The transformation is an enlargement with scale factor −2, centre (1, 0).

Example 10

a Enlarge triangle PQR by a scale factor of $-\frac{1}{3}$ with centre of enlargement C (3, 5).

b Describe fully the transformation that maps triangle P'Q'R' onto triangle PQR.

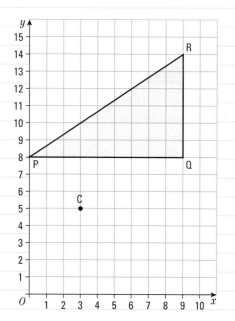

a

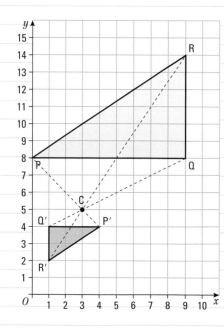

From C to P is 3 squares to the left and 3 squares up. So from C to P' is $-\frac{1}{3} \times 3 = -1$ square to the left, or 1 square to the right, and $-\frac{1}{3} \times 3 = -1$ square up, or 1 square down.

In the same way, from C to Q' is 2 squares to the left and 1 square down, from C to R' is 2 squares to the left and 3 squares down

b The transformation that maps triangle P′Q′R′ onto triangle PQR is an enlargement with scale factor −3, centre (3, 5).

The lengths of the sides of triangle PQR are 3 times those of triangle P′Q′R′ and the centre of enlargement is between the two triangles.

Results Plus
Examiner's Tip

The word 'enlargement' is used even when the new shape is smaller than the original shape.

Results Plus
Examiner's Tip

In an enlargement, corresponding sides in the object and the image are parallel.

Exercise 15E

1 Copy the shape on squared paper and draw the enlargement of the shape with the given scale factor and centre of enlargement marked with a dot (•).

a Scale factor 3.

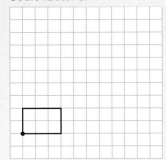

b **i** Scale factor 3.
  ii Scale factor 2.
  iii Scale factor $\frac{1}{2}$.

   Draw all three enlargements on the same diagram.

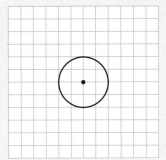

2 On a copy of the diagram complete the following enlargements.
  a Enlarge triangle A with a scale factor of −2, centre (0, 0). Label this new triangle B.
  b Enlarge triangle A with a scale factor of −$\frac{1}{3}$, centre (1, 6). Label this new triangle C.
  c Find the scale factor of the enlargement that maps triangle C onto triangle B.

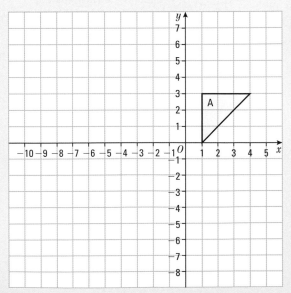

C

A03

C

3   a On a copy of the diagram, enlarge shape P with a scale factor
       of −1, centre (1, 2). Label this new shape Q.
    The mapping of shape P onto shape Q is also a rotation.

A03

    b Describe fully the rotation that maps shape P onto shape Q.

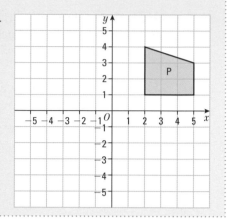

15.5 Combinations of transformations

Objectives

- You can transform a shape using combined translations,
  rotations, reflections or enlargements.
- You can find a single transformation which has the same
  effect as a combination of transformations.

Why do this?

Many designs for wallpaper and fabric are
based on combinations of transformations.

Key Points

- A combination of transformations is when shape P is transformed to shape Q and then shape Q is transformed to
  shape R. It may be possible to find a single transformation which maps shape P onto shape R.
  For example, a reflection in the y-axis has the same effect as a reflection in the x-axis followed by a rotation of
  180° about the origin.

Example 11

a Reflect triangle P in the x-axis.
   Label the new triangle Q.

b Rotate triangle Q 180° about the origin O.
   Label the new triangle R.

c Describe fully the single transformation
   which maps triangle P onto triangle R.

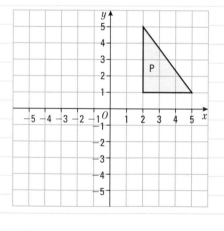

a, b

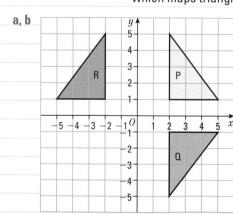

c The single transformation which maps triangle P
   onto triangle R is a reflection in the y-axis.

Example 12

a Enlarge triangle P with scale factor 3 and centre of enlargement (2, 1).
Label the new triangle Q.

b Enlarge triangle Q with scale factor $\frac{1}{3}$ and centre of enlargement (8, 10).
Label the new triangle R.

c Describe fully the single transformation which maps triangle P onto triangle R.

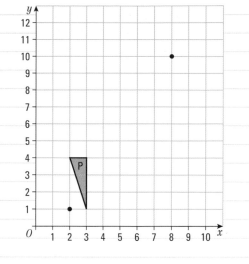

a, b

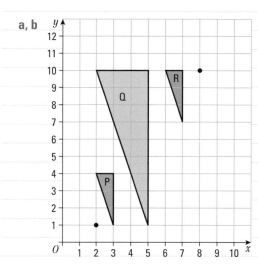

c From P to R is 4 to the right and 6 up.
The single transformation which maps triangle P onto triangle R is the translation with vector $\begin{pmatrix} 4 \\ 6 \end{pmatrix}$.

Exercise 15F

For each question, make a copy of the diagram.

1 Complete the following translations.

a Translate flag F by the vector $\begin{pmatrix} 3 \\ 8 \end{pmatrix}$.
Label the new flag G.

b Translate flag G by the vector $\begin{pmatrix} 6 \\ -4 \end{pmatrix}$.
Label the new flag H.

c Describe fully the single transformation which maps flag F onto flag H.

d Describe fully the single transformation which maps flag H onto flag F.

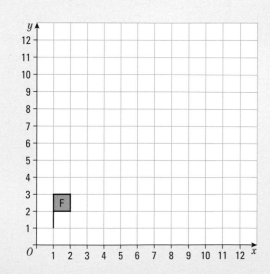

A03

A03

C

B

2 Complete the following transformations.

a Rotate triangle T 180° about (2, 1).
Label the new triangle U.

b Translate triangle U by the vector $\begin{pmatrix} 4 \\ 4 \end{pmatrix}$.
Label the new triangle V.

A03

c Describe fully the single transformation which maps triangle T onto triangle V.

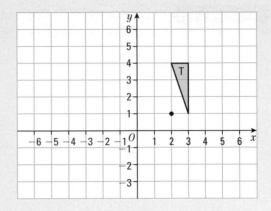

3 Complete the following transformations.

a Rotate triangle T 90° clockwise about the origin O.
Label the new triangle U.

b Reflect triangle U in the line $y = -x$. Label the new triangle V.

A03

c Describe fully the single transformation which has the same effect as a rotation of 90° clockwise about the origin O followed by a reflection in the line $y = -x$.

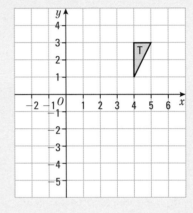

A02
A03

* 4 Use your copy of the graph paper to find and describe fully the single transformation which has the same effect as a translation with vector $\begin{pmatrix} 4 \\ 0 \end{pmatrix}$ followed by a reflection in the line $x = 7$.

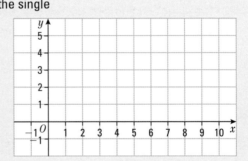

A02
A03

* 5

Use your copy of the graph paper to find and describe fully the single transformation which has the same effect as a rotation of 180° about (0, 0) followed by a reflection in the y-axis.

Chapter review

- In a **translation**, all points of the shape move the same distance in the same direction.
- In a translation
  - the lengths of the sides of the shape do not change
  - the angles of the shape do not change
  - the shape does not turn.
- In a translation, any shape is congruent to its image because the lengths of the sides and angles of the shape are preserved by the translation.
- **Vectors** can be used to describe translations. In a **column vector**:
  - the top number shows the number of squares moved parallel to the x-axis, to the right or left.
  - the bottom number shows the number of squares moved parallel to the y-axis, up or down.
  - to the right and up are positive.
  - to the left and down are negative.
- In a **reflection**:
  - the lengths of the sides of the shape do not change
  - the angles of the shape do not change
  - the **image** is as far behind the mirror line as the shape is in front.
- To describe a reflection, give the mirror line. The mirror line is the line of symmetry.
- In a reflection, any shape is congruent to its image because the lengths of the sides and angles of the shape are preserved by the reflection.
- To describe a **rotation**, give:
  - the angle of turn
  - the direction of turn (clockwise or anticlockwise)
  - the point the shape turns about (the **centre of rotation**).
- In a rotation:
  - the lengths of the sides of the shape do not change
  - the angles of the shape do not change
  - the shape turns
  - the centre of rotation does not move.
- In a rotation, any shape is congruent to its image because the lengths of the sides and angles of the shape are preserved by the rotation.
- In an **enlargement**:
  - the lengths of the sides of the shape change
  - the angles of the shape do not change.
- To describe an enlargement, give:
  - the scale factor
  - the centre of enlargement.
- If each vertex of shape P is joined to the corresponding vertex of shape Q, the joining lines intersect at the **centre of enlargement**.
- In general, when shape P is mapped onto shape Q by an enlargement with centre C and scale factor k, $CA' = k \times CA$ for any point A of shape P and the corresponding point A' of shape Q.
- A combination of transformations is when shape P is transformed to shape Q and then shape Q is transformed to shape R. It may be possible to find a single transformation which maps shape P onto shape R.

Review exercise

1 On a copy of the grid, draw an enlargement of the shaded shape with a scale factor of 3.

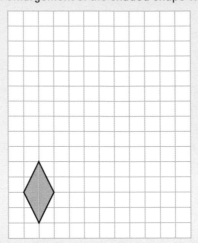

Nov 2006

2 **a** On a copy of the grid, rotate the shaded shape 90° clockwise about the point O.

b Describe fully the single transformation that will map shape P onto shape Q.

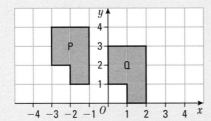

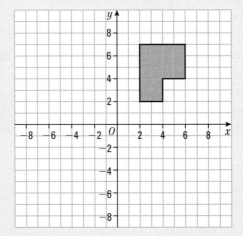

May 2009

3

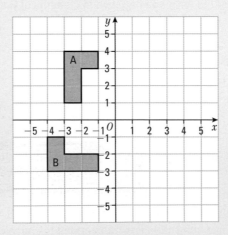

a On a copy of the grid, reflect shape A in the y-axis.

b Describe fully the **single** transformation which takes shape A to shape B.

Nov 2008

4

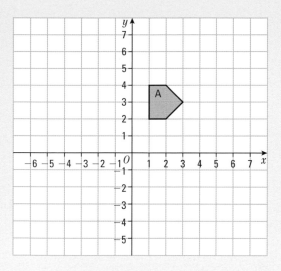

On a copy of the grid:

a reflect shape A in the y-axis. Label your new shape B.

b translate shape A by 3 squares right and 2 squares down. Label your new shape C.

Nov 2007

5 You have been asked to design a bathroom tile with reflective symmetry.

Draw a design in the top left 4 by 4 corner.

Then reflect your design in the vertical and horizontal lines to create the full pattern.

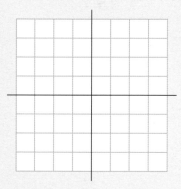

6 Describe fully the single transformation that will map shape P onto shape Q.

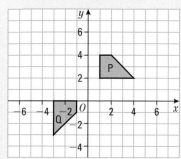

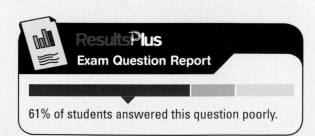

ResultsPlus

Exam Question Report

61% of students answered this question poorly.

A03 C

Nov 2007

C

7

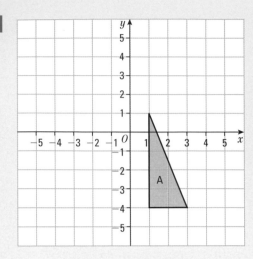

On a copy of the grid, enlarge triangle A by scale factor $-\frac{1}{2}$, centre $(-1, -2)$.
Label your triangle B.

Nov 2005

8 **a** Describe fully the single transformation that maps triangle A onto triangle B.
b On a copy of the grid, rotate triangle A 90° anticlockwise about the point $(-1, 1)$. Label your new triangle C.

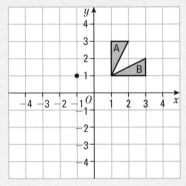

Nov 2006

9

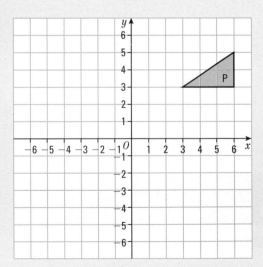
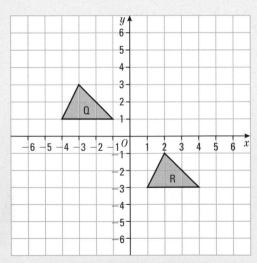

a On a copy of the grid, reflect triangle P in the line $x = 2$.

b Describe fully the **single** transformation that takes triangle Q to triangle R.

Nov 2006

10

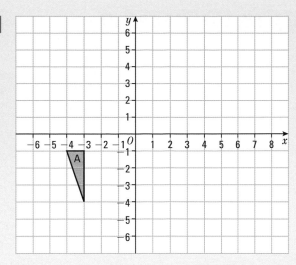

On a copy of the grid, enlarge triangle A by scale factor −2, centre (0, −1).

* **11**   Triangle A is reflected in the x-axis to give triangle B.
Triangle B is reflected in the line $x = 1$ to give triangle C.
Describe the **single** transformation that takes
triangle A to triangle C.

A03   B

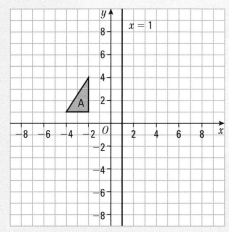

June 2008

* **12**   Triangle A is reflected in the y-axis to give triangle B.
Triangle B is then reflected in the x-axis to give triangle C.
Describe the single transformation that takes triangle A
to triangle C.

A03

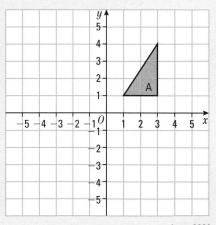

June 2006

16 PYTHAGORAS' THEOREM AND TRIGONOMETRY 1

Pythagoras lived in Greece in about 500 BC, and though one of the most famous results in mathematics was named after him, it is thought possible that the Egyptians used the result of the theorem to build their pyramids many years before.

◉ Objectives

In this chapter you will:
- use Pythagoras' Theorem to find the length of sides in right-angled triangles
- find the length of a line segment
- use trigonometry to find the sizes of angles and the lengths of sides in right-angled triangles.

◈ Before you start

You need to:
- know how to find the square and square root of a number
- be able to recognise a right-angled triangle.

16.1 Pythagoras' Theorem

◎ Objectives

◎ You can use Pythagoras' Theorem to find the length of the hypotenuse in a right-angled triangle.
◎ You can use Pythagoras' Theorem to find the length of a shorter side in a right-angled triangle.

⬥ Why do this?

The size of a TV is given as the length of the diagonal across the screen. You can use Pythagoras' Theorem to work out the length and width of the screen.

◈ Get Ready

1. Find the value of **a** 12^2 **b** 7.4^2
2. Work out **a** $9^2 + 40^2$ **b** $13.8^2 + 9.3^2$
3. Work out **a** $17^2 - 8^2$ **b** $16.4^2 - 12.9^2$
4. Work out **a** $\sqrt{24^2 + 7^2}$ **b** $\sqrt{15^2 - 9^2}$
5. Work out **a** $\sqrt{3.8^2 + 5.2^2}$ **b** $\sqrt{13.1^2 - 9.6^2}$

Give your answers correct to 3 significant figures.

◔ Key Points

◉ The right-angled triangle in this diagram has sides of length 3 cm, 4 cm and 5 cm. Squares have been drawn on each side of the triangle and each square has been divided up into squares of side 1 cm.
The area of the square on the side of length 3 cm is 9 cm².
The area of the square on the side of length 4 cm is 16 cm².
The area of the square on the side of length 5 cm (the **hypotenuse**) is 25 cm².
Notice that $25 = 9 + 16$, that is, $5^2 = 3^2 + 4^2$.

In other words, 5^2 (the area of the square on the hypotenuse) is equal to the sum of 3^2 and 4^2 (the areas of the squares on the other two sides added together).
This is an example of **Pythagoras' Theorem**. It is only true for right-angled triangles.

◉ Here is a right-angled triangle, ABC.
The angle at C is the right angle.
The side, AB, opposite the right angle is called the hypotenuse.
It is the longest side in the triangle.

◉ Pythagoras' Theorem states that:
In a right-angled triangle, the square of the hypotenuse is equal to the sum of the squares of the other two sides.

$c^2 = a^2 + b^2$
or
$AB^2 = BC^2 + CA^2$
Where AB^2 means the length of the side AB squared.

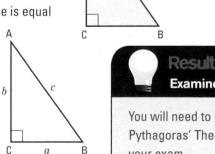

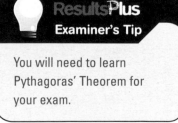

ResultsPlus

Examiner's Tip

You will need to learn Pythagoras' Theorem for your exam.

◉ You can also use Pythagoras' Theorem to work out the length of one of the shorter sides in a right-angled triangle when you know the lengths of the other two sides.

Example 1 Work out the length of the hypotenuse in this triangle.

15 cm   c cm

8 cm

$c^2 = a^2 + b^2$ ← Write down Pythagoras' Theorem.

$c^2 = 8^2 + 15^2$ ← Put in the known values.

$c^2 = 64 + 225 = 289$ ← Work out the value of c^2.

$c = \sqrt{289} = 17$ ← Square root to find the value of c.

Length of hypotenuse = 17 cm.

Example 2 In triangle XYZ, angle X = 90°, XY = 8.6 cm and
XZ = 13.9 cm. Work out the length of YZ.
Give your answer correct to 3 significant figures.

X

8.6 cm

Y

13.9 cm

Z

$YZ^2 = XY^2 + XZ^2$ ← YZ is the hypotenuse as it is opposite the right angle.

$YZ^2 = 8.6^2 + 13.9^2$ ← Put in the values.

$YZ^2 = 73.96 + 193.21$

$YZ^2 = 267.17$ ← Work out the value of YZ^2.

$YZ = \sqrt{267.17} = 16.34...$ ← Square root.

$YZ = 16.3$ cm (to 3 s.f.)

ResultsPlus
Examiner's Tip

When you are showing your
calculations on your exam paper,
write down at least four figures of
the calculator display.

Exercise 16A   Questions in this chapter are targeted at the grades indicated.

C

1   Work out the length of each hypotenuse marked with letters in these triangles.
Where appropriate, give each answer correct to 3 significant figures.

a

12 cm a

5 cm

b   6 cm

8 cm b

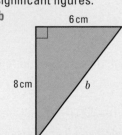

C

c

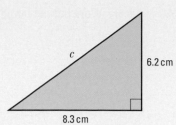

6.2 cm

8.3 cm

d

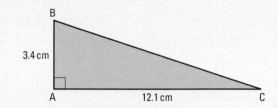

d

10.6 cm

4.8 cm

2 **a** In triangle ABC, angle A = 90°, AB = 3.4 cm and
AC = 12.1 cm. Work out the length of BC.
Give your answer correct to 3 significant figures.

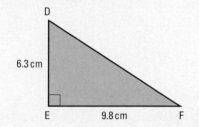

B

3.4 cm

A   12.1 cm   C

b In triangle DEF, angle E = 90°, DE = 6.3 cm and
EF = 9.8 cm. Work out the length of DF.
Give your answer correct to 3 significant figures.

D

6.3 cm

E   9.8 cm   F

c In triangle PQR, angle R = 90°, PR = 5.9 cm and QR = 13.1 cm.
Work out the length of PQ. Give your answer correct to 3 significant figures.

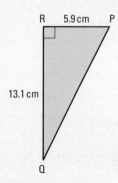

R   5.9 cm   P

13.1 cm

Q

d

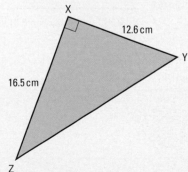

X

12.6 cm

Y

16.5 cm

Z

In triangle XYZ, angle X = 90°, XY = 12.6 cm and XZ = 16.5 cm.
Work out the length of YZ. Give your answer correct to 3 significant figures.

Example 3

In triangle ABC, angle A = 90°, BC = 17.4 cm and AC = 5.8 cm. Work out the length of AB. Give your answer correct to 3 significant figures.

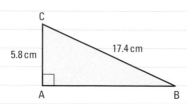

$BC^2 = AC^2 + AB^2$ ← Angle A is the right angle so the hypotenuse is BC.

$17.4^2 = 5.8^2 + AB^2$ ← Write down Pythagoras' Theorem and put in the values.

$302.76 = 33.64 + AB^2$

$302.76 - 33.64 = AB^2$ ← Work out the value of AB^2 by subtracting 33.64 from both sides.
$269.12 = AB^2$

$AB = \sqrt{269.12} = 16.4...$

ResultsPlus
Examiner's Tip

Check that the hypotenuse is the longest side of the triangle.

$AB = 16.4$ cm (to 3 s.f.) ← Give the length of AB correct to 3 s.f.

Exercise 16B

C

1 Work out the lengths of the sides marked with letters in these triangles.
Where appropriate give each answer correct to 3 significant figures.

a

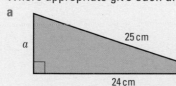
25 cm
a
24 cm

b

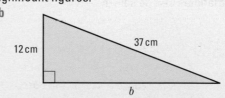
12 cm
37 cm
b

c

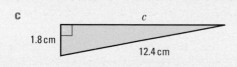
c
1.8 cm
12.4 cm

d

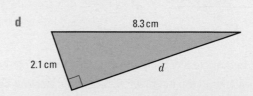
8.3 cm
2.1 cm
d

2 **a** In triangle ABC, angle A = 90°, AB = 5.9 cm and BC = 16.3 cm. Work out the length of AC.
Give your answer correct to 3 significant figures.

B
5.9 cm
16.3 cm
A
C

b In triangle PQR, angle R = 90°, PQ = 11.2 cm and QR = 9.6 cm.
Work out the length of RP. Give your answer correct to 3 significant figures.

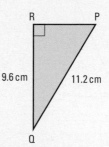
R   P
9.6 cm   11.2 cm
Q

c In triangle DEF, angle E = 90°, DF = 10.1 cm and EF = 7.8 cm.
  i Draw a sketch of the right-angled triangle DEF and label sides DF and EF with their lengths.
  ii Work out the length of DE. Give your answer correct to 3 significant figures.

16.2 Applying Pythagoras' Theorem

Objective

⦿ You can solve problems using Pythagoras' Theorem.

Why do this?

If you travel 30 kilometres North and then 20 kilometres East, you can use Pythagoras' Theorem to work out the direct distance 'as the crow flies'.

Get Ready

1. Draw a sketch of an A4 page with sides of lengths 21.2 cm and 29.7 cm. Round the lengths to the nearest centimetre and work out an estimate of the length of the diagonal of the page.

Key Points

⦿ Pythagoras' Theorem can be used to solve problems.
⦿ Isosceles triangles can be split into two right-angled triangles. You can then use Pythagoras' Theorem.

Example 4 A boat travels due North for 5.7 km. The boat then turns and travels due East for 7.2 km.
Work out the distance between the boat's finishing point and its starting point.
Give your answer in km correct to 3 significant figures.

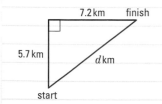
7.2 km   finish
5.7 km
d km
start

$d^2 = 5.7^2 + 7.2^2$
$d^2 = 32.49 + 51.84$
$d^2 = 84.33$
$d = \sqrt{84.33} = 9.183\ldots$
Distance = 9.18 km (to 3 s.f.)

Draw a sketch of the boat's journey. A reminder about bearings can be found in Section 14.6.

The triangle is right-angled and the distance between the start and the finish is the hypotenuse, so you can use Pythagoras' Theorem.

Example 5 The diagram shows an isosceles triangle ABC.
The midpoint of BC is the point M. In the triangle,
AB = AC = 8 cm and BC = 6 cm.
Work out the height, AM, of the triangle.
Give your answer correct to 3 significant figures

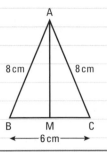

By Pythagoras
$AB^2 = AM^2 + BM^2$
$8^2 = h^2 + 3^2$
$64 = h^2 + 9$
$64 - 9 = h^2$
$55 = h^2$
$h = \sqrt{55} = 7.416...$
$h = 7.42$
Height of triangle =
7.42 cm (to 3 s.f.)

> Pythagoras' Theorem cannot be used in triangle ABC as it is not a right-angled triangle.
> As M is the midpoint of the base of the isosceles triangle, the line AM is the line of symmetry of triangle ABC.
> So AM is perpendicular to the base and angle AMB = 90°.
> BM = 3 cm as M is the midpoint of BC.

> Draw a sketch of triangle ABM.

> Triangle ABM is right-angled with hypotenuse AB. The height, AM, of the triangle is marked h cm on the sketch.

Exercise 16C

C

1 Find the lengths of the sides marked with letters in each of these triangles.
Give each answer correct to 3 significant figures.

a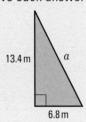

b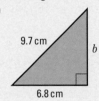

c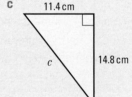

A02
A03

2 The diagram shows a ladder leaning against a vertical wall.
The foot of the ladder is on horizontal ground, 3.6 m from the wall. The length of the ladder is 5 m.
Work out how far up the wall the ladder reaches.
Give your answer correct to 3 significant figures.

A03

3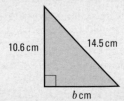

Work out the area of the triangle.
Give your answer correct to 3 significant figures.

4

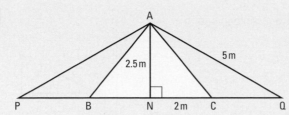

The diagram represents the end view of a tent, triangle ABC; two guy-ropes, AP and AQ; and a vertical tent pole, AN. The tent is on horizontal ground so that PBNCQ is a straight horizontal line. Triangles ABC and APQ are both isosceles triangles. BN = NC = 2 m, AN = 2.5 m and AP = AQ = 5 m

A02
A03 B

a Work out the length of the side AC of the tent. Give your answer correct to 3 significant figures.

b Work the length of **i** NQ, **ii** CQ.   Give your answers correct to 3 significant figures.

There is a tent peg at P and a tent peg at Q.

c Work out the distance between the two tent pegs at P and Q.

Give your answer correct to 3 significant figures.

5 Here are the lengths of sides of six triangles.

A03

Triangle 1 5 cm, 12 cm and 13 cm	Triangle 2 9 cm, 40 cm and 41 cm
Triangle 3 10 cm, 17 cm and 18 cm	Triangle 4 20 cm, 21 cm and 29 cm
Triangle 5 8 cm, 17 cm and 20 cm	Triangle 6 33 cm, 56 cm and 65 cm

Which of these triangles are right-angled triangles?

6 The diagram shows two right-angled triangles.
Work out the length of the side marked a.
Give your answer correct to 3 significant figures.

A02
A03

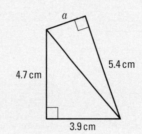

7 The size of a computer monitor or TV screen is determined by the length of the diagonal across the screen.

A02
A03

a A computer monitor has a screen that is a rectangle with dimensions 13.2 inches by 10.8 inches. Work out the size of this monitor by working out, correct to the nearest inch, the length of the diagonal of the rectangle.

b A widescreen TV has a 37 inch screen which is a rectangle with sides of lengths in the ratio 16 : 9. Work out the height and width of the TV screen.

16.3 **Finding the length of a line segment**

◉ Objective

◉ You can calculate the length of a line segment.

◈ Get Ready

1. Calculate   **a** $\sqrt{10^2 + 24^2}$ **b** $\sqrt{18^2 + 7^2}$ **c** $\sqrt{4.8^2 + 9.1^2}$

🔍 Key Points

- By creating a right-angled triangle, Pythagoras' Theorem can be used to find the length between two points on a line.

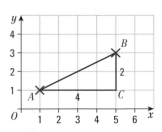

- The length of the line segment AB between
  $A(x_1, y_1)$ and $B(x_2, y_2)$ is $\sqrt{(x_2 - x_1)^2 + (y_2 - y_1)^2}$

🔍 Example 6

Find the length of the line joining   **a** $A(3, 2)$ and $B(15, 7)$ **b** $P(-9, 4)$ and $Q(7, -5)$

a

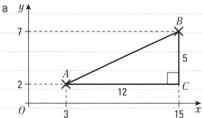

$AC = 15 - 3 = 12$
$BC = 7 - 2 = 5$

> Draw a sketch showing A and B and complete the right-angled triangle ABC.

$AB^2 = 12^2 + 5^2$
$AB^2 = 144 + 25 = 169$
$AB = \sqrt{169} = 13$

> Use Pythagoras' Theorem to find the length of AB.

b

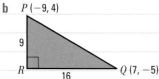

$QR = 7 - -9 = 7 + 9 = 16$
$PR = 4 - -5 = 4 + 5 = 9$

> Draw a sketch showing P and Q and complete the right-angled triangle PQR.

$PQ^2 = 16^2 + 9^2$
$PQ^2 = 256 + 81 = 337$
$PQ = \sqrt{337} = 18.4$ (to 3 s.f.)

> Use Pythagoras' Theorem to find the length of PQ.

⚙ Exercise 16D

A02

1 Work out the length of the line joining each of these pairs of points.
  a (3, 1) and (11, 7)
  b (2, 5) and (12, 29)
  c (−6, 9) and (8, 13)
  d (−4, −6) and (6, 12)
  e (9, −15) and (−11, 6)
  f (0, −5) and (9, −11)

A02
A03

2 The point A has coordinates (5, 2), the point B has coordinates (8, 6) and the point C has coordinates (1, 5).
  a Work out the length of   **i** AB   **ii** BC   **iii** AC
  b What does your answer to part a tell you about triangle ABC?

3 A circle has centre point O (4, 2). The point A (9, 14) lies on the circle.

a Work out the radius of the circle.

b Determine by calculation which of the following points also lie on the circle.

   i B (16, 7)   **ii** C (−1, −10)   **iii** D (7, 16)   **iv** E (4, 15)

A02
A03 C

16.4 Trigonometry in right-angled triangles

Objective

○ You can use trigonometry to find the sizes of angles in right-angled triangles.

Why do this?

Trigonometry is used in science and engineering. It is used to plan the movements of the robotic arm on the International Space Station.

Get Ready

Look at these right-angled triangles. For each one, name the side that is

a the hypotenuse **b** the side opposite **c** the side adjacent to the angle marked $x°$.

1.

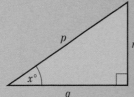

2.

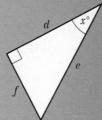

3.

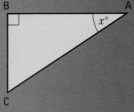

Key Points

○ The hypotenuse (hyp) of a right-angled triangle is the longest side of the triangle and is opposite the right angle. The other two sides of the triangle are named adjacent and opposite. The side opposite an angle is called the opposite side (opp).

The side next to this angle is called the adjacent side (adj).

○ Here is a right-angled triangle with its hypotenuse of length 1.

The length of the opposite side (opp) in this triangle is known accurately and is called the **sine** of 70° and is written sin 70°.

Its value can be found on any scientific calculator. Not all calculators are the same but the key sequence to find sin 70° applies to many calculators.

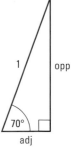

○ Make sure that the angle mode of your calculator is degrees, usually shown by 'D' on the calculator screen.

Press ⌈sin⌉   Key in ⌈7⌉ ⌈0⌉   Press ⌈=⌉

The number ⌈ 0.93969262 ⌉ should appear on your calculator screen.

So, correct to four decimal places, sin 70° = 0.9397

- The length of the adjacent side (adj) is called the **cosine** of 70° and is written cos 70°. Using a similar sequence to the one above, but using the $\boxed{\cos}$ key, correct to four decimal places, cos 70° = 0.3420

- The terms sine and cosine are called trigonometric ratios, or trig ratios.

- There is another trig ratio called the **tangent** of 70° and written tan 70°. As above, but using the $\boxed{\tan}$ key, correct to four decimal places, tan 70° = 2.7475

- You can find the sine, cosine and tangent of any angle.

- Here are three right-angled triangles.

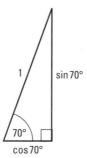  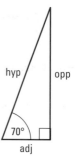

The second triangle is an enlargement of the first triangle with a scale factor of 3.
This means that $a = 3 \times \sin 70°$ or $3\sin 70°$ and $b = 3 \times \cos 70°$ or $3\cos 70°$
The third triangle is an enlargement of the first triangle with a scale factor of hyp.
This means that opp = hyp $\times \sin 70°$ and adj = hyp $\times \cos 70°$

These results can also be written as $\sin 70° = \dfrac{\text{opp}}{\text{hyp}}$ and $\cos 70° = \dfrac{\text{adj}}{\text{hyp}}$.

- Results like these are true for all right-angled triangles so that

$$\sin x° = \frac{\textbf{opp}}{\textbf{hyp}} \qquad \cos x° = \frac{\textbf{adj}}{\textbf{hyp}}$$

When the opposite side and the adjacent side are involved

$$\tan x° = \frac{\textbf{opp}}{\textbf{adj}}$$

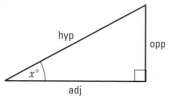

SOHCAHTOA might help you remember these results.

Sin Opp Hyp Cos Adj Hyp Tan Opp Adj

Example 7   Use a calculator to write down, correct to 4 decimal places, the value of cos 74.6°.

> Make sure that your calculator is in degree mode.

cos 74.6° = 0.265 556 117 ←   Use the $\boxed{\cos}$ key and key in 74.6.

cos 74.6° = 0.2656 correct to 4 decimal places

   cosine   tangent   SOHCAHTOA

Example 8 Find the value of x when $\tan x° = 2.7$.
Give your answer correct to 1 decimal place.

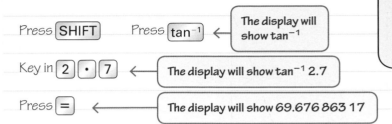

When the value of sine, cosine or tangent is given, a calculator can be used to find the size of the angle. To do this, use the SHIFT or INV key on your calculator.

ResultsPlus
Watch Out!

Not all calculators are the same. It is important that you know how to work your calculator. Make sure that your calculator is in degree mode.

Press SHIFT   Press tan⁻¹ ← The display will show tan⁻¹

Key in 2 · 7 ← The display will show tan⁻¹ 2.7

Press = ← The display will show 69.676 863 17

$x = 69.7$ correct to 1 decimal place.

Exercise 16E

1 Use a calculator to find the value of
    a $\sin 20°$ **b** $\sin 72.6°$ **c** $\cos 60°$ **d** $\cos 18.9°$
    e $\tan 45°$ **f** $\tan 86.4°$ **g** $\cos 137.8°$ **h** $\tan 4°$
    i $\sin 127.2°$ **j** $\sin 14.7°$ **k** $\tan 159.5°$ **l** $\cos 87.3°$
Give each answer correct to four decimal places, where necessary.

2 Use a calculator to find the value of x when
    a $\cos x° = 0.6$ **b** $\sin x° = 0.43$ **c** $\cos x° = 0.5$
    d $\tan x° = 0.96$ **e** $\sin x° = 0.8516$ **f** $\tan x° = 2.03$
    g $\sin x° = 0.047$ **h** $\tan x° = \sqrt{3}$ **i** $\cos x° = \dfrac{\sqrt{2}}{2}$
Give each answer correct to 1 decimal place where necessary.

Example 9 Write down the trigonometric ratio needed to calculate:
    a the size of the angle marked $x°$ **b** the length of the side marked p.

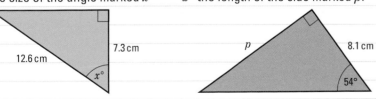

a The given sides are the hypotenuse and the side adjacent to the angle $x°$ so cosine is needed.
b The hypotenuse is not involved, p is opposite the given angle and 8.1 cm is adjacent so tangent is needed.

Exercise 16F

1 Write down which trigonometric ratio is needed to calculate either the length of the side marked p or the size of the angle marked x in each of these triangles. You do not have to calculate anything.

a
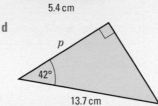

b

c

d

Example 10 Work out the size of each of the marked angles.

Give each answer correct to one decimal place.

a

b

c
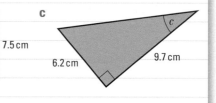

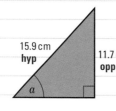

a $\sin a = \dfrac{11.7}{15.9} = 0.7358\ldots$ ←

$a = 47.379\ldots°$

$a = 47.4°$ correct to 1 d.p.

> 15.9 cm is the hypotenuse.
> 11.7 cm is opposite angle a.
> $\sin = \dfrac{\text{opp}}{\text{hyp}}$
> Use your calculator to find:
> $\sin^{-1} 0.7358\ldots$ which is $47.379\ldots°$.

b $\cos b = \dfrac{7.5}{16.1} = 0.4658\ldots$ ←

$b = 62.235\ldots°$

$b = 62.2°$ correct to 1 d.p.

> 16.1 cm is the hypotenuse.
> 7.5 cm is adjacent to angle b.
> $\cos = \dfrac{\text{adj}}{\text{hyp}}$
> Use your calculator to find:
> $\cos^{-1} 0.4658\ldots$ which is $62.235\ldots°$.

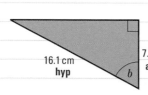

c $\tan c = \dfrac{6.2}{9.7} = 0.6391\ldots$ ←

$c = 32.585\ldots°$

$c = 32.6°$ correct to 1 d.p.

> 6.2 cm is opposite angle c.
> 9.7 cm is adjacent to angle c.
> $\tan = \dfrac{\text{opp}}{\text{adj}}$
> Use your calculator to find:
> $\tan^{-1} 0.6391\ldots$ which is $32.585\ldots°$.

Exercise 16G

1. Work out the size of each of the lettered angles. Give each answer correct to one decimal place.

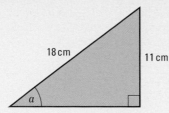

a

18 cm   11 cm

a

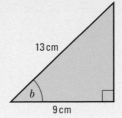

b

13 cm

b

9 cm

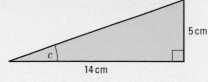

c

5 cm

c   14 cm

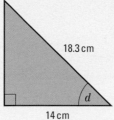

d

18.3 cm

d

14 cm

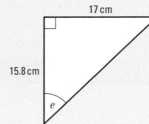

e

17 cm

15.8 cm

e

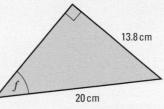

f

13.8 cm

f   20 cm

2. Triangle ABC is right-angled at B.

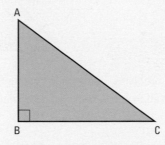

A

B   C

a AB = 8.9 cm and BC = 12.1 cm. Calculate the size of angle ACB.
Give your answer correct to 0.1°.

b BC = 15.5 cm and AC = 24.7 cm. Calculate the size of angle BAC.
Give your answer correct to 0.1°.

c AB = 6.3 cm and AC = 11.8 cm. Calculate the size of angle ACB.
Give your answer correct to 0.1°.

3. In triangle ACD, the point B lies on AD so that
CB and AD are perpendicular.

a Using triangle ABC, calculate the size of angle ACB.
Give your answer correct to one decimal place.

b Using triangle BCD, calculate the size of angle BCD.
Give your answer correct to one decimal place.

c Hence calculate the size of angle ACD.
Give your answer to the nearest degree.

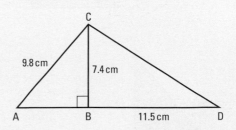

C

9.8 cm   7.4 cm

A   B   11.5 cm   D

16.5 Working out lengths of sides using trigonometry

⊕ Get Ready

1. Draw sketch diagrams to show these situations. (You do not need to do any working out.)
   a A ship is 20 km from lighthouse X on a bearing of 055°.
   b A ladder 8 m in length leans against a wall, with the foot of the ladder 3 m from the wall.
2. For question **1b**, which trig ratio would you use to find the angle the ladder makes with the ground?

🔍 Key Points

- The results used in the last section can be written as
  $$opp = hyp \times \sin x°$$
  $$adj = hyp \times \cos x°$$
  $$opp = adj \times \tan x°$$

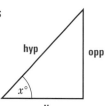

- Trigonometry can be used to solve problems. Sometimes Pythagoras' Theorem is needed as well. Some questions involve bearings and angles of elevation and depression (see Section 14.6 in Unit 3 and Section 13.4 in Unit 2).

Example 11   Work out the length of each of the lettered sides.
Give each answer correct to 3 significant figures.

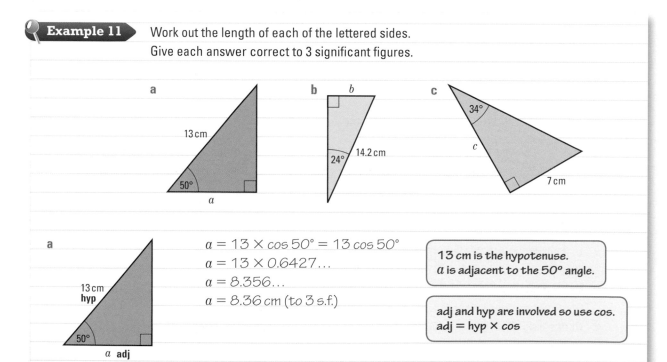

$a = 13 \times \cos 50° = 13 \cos 50°$

$a = 13 \times 0.6427...$

$a = 8.356...$

$a = 8.36$ cm (to 3 s.f.)

> 13 cm is the hypotenuse.
> a is adjacent to the 50° angle.

> adj and hyp are involved so use cos.
> adj = hyp × cos

b

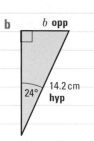

$b = 14.2 \times \sin 24° = 14.2 \sin 24°$
$b = 14.2 \times 0.4067…$
$b = 5.7756…$
$b = 5.78 \, cm \, (to \, 3 \, s.f.)$

> 14.2 cm is the hypotenuse.
> b is opposite to the 24° angle.

> opp and hyp are involved so use sin.
> opp = hyp × sin

c

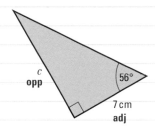

$c = 7 \times \tan 56° = 7 \tan 56°$
$c = 7 \times 1.4825…$
$c = 10.3779…$

$c = 10.4 \, cm \, (to \, 3 \, s.f.)$

> The hypotenuse is not given and opposite and adjacent are involved so use tan.
> opp = adj × tan

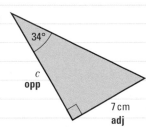

> When using tan to find a length, it is easiest to find the opposite side. Relative to the angle of 34°, 7 cm is the opposite side and c is the adjacent side.
> The third angle in the triangle is $(180 − 90 − 34)° = 56°$
> Relative to the angle of 56°, c is the opposite side and 7 cm is the adjacent side.

🔍 **Example 12** The diagram shows a lighthouse 30 m in height, standing on horizontal ground.

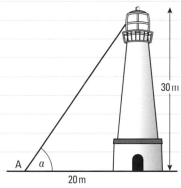

a Work out the angle of elevation of the top of the lighthouse from point A on the ground.
b Work out the angle of depression of point A from the top of the lighthouse.

a $\tan a = \dfrac{30}{20}$ ← $\tan a = \dfrac{opp}{adj}$

$\tan a = 1.5$
$a = 56.3° \, (to \, 3 \, s.f.)$

b Angle of depression $= 56.3° \, (to \, 3 \, s.f.)$ ←

> The angles of elevation and depression are the same; alternate angles.

Exercise 16H

B

1   Work out the length of each lettered side.
Give each answer correct to three significant figures.

a

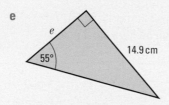

16 cm

a

20°

b

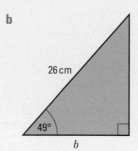

26 cm

49°

b

c

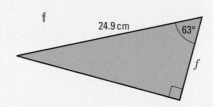

c

37°

15.4 cm

d

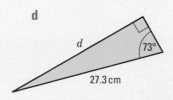

d

73°

27.3 cm

e

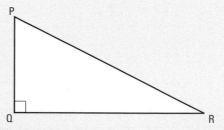

e

14.9 cm

55°

f

24.9 cm

63°

f

2   Triangle PQR is right-angled at Q.

In each part, calculate the length of QR.
Give each answer correct to three significant figures.

a   PQ = 7.3 cm, angle QPR = 68°

b   PR = 17.2 m, angle QRP = 39°

c   PR = 12.6 cm, angle QPR = 59°

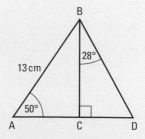

P

Q   R

A03

3   In triangle ABD, the point C lies on AD so that BC and AD are perpendicular.

a   Using triangle ABC, work out the length of **i** BC **ii** AC.
Give each answer correct to three significant figures.

b   Using triangle BCD, work out the length of CD,
correct to three significant figures.

c   Hence calculate the length of AD, correct to three significant figures.

d   Calculate the area of triangle ABD.
Give your answer correct to the nearest cm².

B

28°

13 cm

50°

A   C   D

Example 13 Two towns, Aytown and Beeville, are 40 km apart.

The bearing of Beeville from Aytown is 067°.

a Calculate how far East and how far North Beeville is from Aytown.

Give your answers to 3 significant figures.

Ceeham is 60 km East of Beeville.

b Calculate the distance between Aytown and Ceeham.

Give your answer to the nearest km.

c Calculate the bearing of Ceeham from Aytown.

Give your answer to the nearest degree.

a

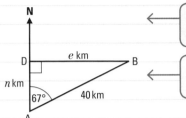

> First draw a diagram showing the positions of Aytown (A) and Beeville (B).
> Draw a line from B 'West' to meet the 'North' line from A at D.

> In the right-angled triangle ABD, the length of AD gives how far B is
> North of A (n km). The length of DB gives how far B is East of A (e km).

> In triangle ABD, the 40 km is the hypotenuse and e km
> is opposite the 67° angle.
> opp = hyp × sin

$e = 40\sin 67° = 36.82\ldots$

Distance East = 36.8 km (to 3 s.f.)

> n km is adjacent to the 67° angle.
> adj = hyp × cos

$n = 40\cos 67° = 15.629\ldots$

Distance North = 15.6 km (to 3 s.f.)

> Having worked out the value of e, the value of n could be
> found using Pythagoras' Theorem.

b

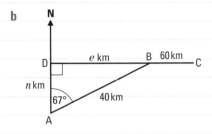

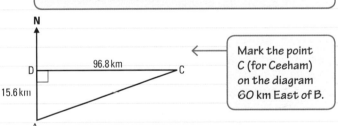

> Mark the point
> C (for Ceeham)
> on the diagram
> 60 km East of B.

$AC^2 = AD^2 + DC^2$

$\quad = 15.6^2 + 96.8^2$

$AC^2 = 9613.6$

$AC = 98.04\ldots$ km

> Ceeham is 15.6 km North of Aytown.
> Ceeham is 60 + 36.8 = 96.8 km East of Aytown.
> Draw triangle ADC.

Distance between Aytown and
Ceeham is 98 km (to nearest km).

> The distance between Aytown and Ceeham is the length of AC.
> Find the length of AC using Pythagoras' Theorem.

c $\tan DAC = \dfrac{96.8}{15.6} = 6.205\ldots$

$DAC = 80.8\ldots°$

$\quad = 81°$ (to nearest degree)

> To find the bearing of C from A, calculate the size of angle DAC.
> Any of the trigonometric ratios can be used as all three sides
> of triangle ACD are known.
> Using $\tan = \dfrac{\text{opp}}{\text{adj}}$

Bearing of Ceeham from Aytown is 081° (to nearest degree).

> The bearing should be given
> as a three-figure bearing.

Exercise 16I

Where necessary give lengths correct to 3 significant figures and angles correct to 1 decimal place.

1 The diagram shows the plans for the sails of a boat.

   a Work out the length of the side marked

     i *a*   **ii** *b*   **iii** *c*

   b Work out the size of the angle marked *d*.

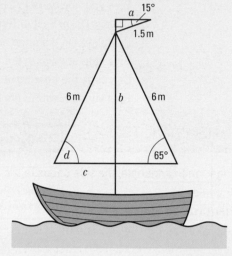

2 The diagram shows a vertical building standing on horizontal ground.

The points A, B and C are in a straight line on the ground.

The point T is at the top of the building so that TC is vertical.

The angle of elevation of T from A is 40°, as shown in the diagram.

   a Work out the height, TC, of the building.

   b Work out the size of the angle of elevation of T from B.

   c Work out the size of angle ATB.

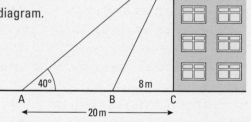

3 The points P and Q are marked on a horizontal field. The distance from P to Q is 100 m. The bearing of Q from P is 062°. Work out how far:

   a Q is North of P

   b Q is East of P.

4 The diagram shows a circle centre O.

The line ABC is the tangent to the circle at B so that the size of angle OBA is 90°.

   a Work out the radius of the circle.

   b Work out the size of angle OCB.

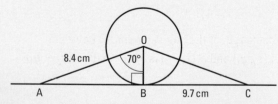

5 A, B and C are three buoys marking the course of a yacht race.

   a Calculate how far B is: **i** North of A **ii** East of A.

   b Calculate how far C is: **i** North of B **ii** East of B.

   c Hence calculate how far C is: **i** North of A **ii** East of A.

   d Calculate the distance and bearing of C from A.

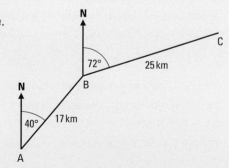

6 The diagram shows an isosceles trapezium.

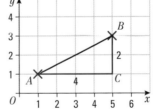

a Work out the distance, h cm, between the two parallel sides of the trapezium.

The length of the shorter parallel side of the trapezium is 5.8 cm, as shown in the diagram.

b Work out the length of the longer parallel side of the trapezium.

c Calculate the area of the trapezium. Give your answer to the nearest cm².

A02
A03
A

Chapter review

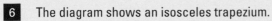

- In a right-angled triangle, the side opposite the right angle is called the **hypotenuse**. It is the longest side in the triangle.
- **Pythagoras' Theorem** states that:
  In a right-angled triangle, the square of the hypotenuse is equal to the sum of the squares of the other two sides.
  That is $c^2 = a^2 + b^2$
  or
  $AB^2 = BC^2 + CA^2$
  where AB^2 means the length of the side AB squared.
- You can also use Pythagoras' Theorem to work out the length of one of the shorter sides in a right-angled triangle when you know the lengths of the other two sides.
- Pythagoras' Theorem can be used to solve problems.
- Isosceles triangles can be split into two right-angled triangles. You can then use Pythagoras' Theorem.
- By creating a right-angled triangle, Pythagoras' Theorem can be used to find the length between two points on a line.

- The length of the line segment AB between $A\,(x_1, y_1)$ and $B\,(x_2, y_2)$ is $\sqrt{(x_2 - x_1)^2 + (y_2 - y_1)^2}$

- The three sides of a triangle are named hypotenuse, adjacent and opposite. The side opposite an angle is called the opposite side (opp). The side next to this angle is called the adjacent side (adj).
- The terms **sine** (sin), **cosine** (cos) and **tangent** (tan) are called trigonometric ratios, or trig ratios.

  $\sin x° = \dfrac{\text{opp}}{\text{hyp}}$ $\cos x° = \dfrac{\text{adj}}{\text{hyp}}$ $\tan x° = \dfrac{\text{opp}}{\text{adj}}$

  SOHCAHTOA might help you remember these results.
- These trig rations can also be written as
  opp = hyp × sin $x°$
  adj = hyp × cos $x°$
  opp = adj × tan $x°$

Review exercise

1 AC = 12 cm. Angle ABC = 90°. Angle ACB = 32°.
Calculate the length of AB.
Give your answer correct to 3 significant figures.

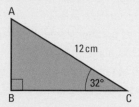

June 2007

2 ABC is a right-angled triangle.
AC = 6 cm.
BC = 9 cm.
Work out the length of AB.
Give your answer correct to 3 significant figures.

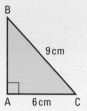

June 2009

3 The diagram shows three cities.
Norwich is 168 km due east of Leicester.
York is 157 km due north of Leicester.
Calculate the distance between Norwich and York.
Give your answer correct to the nearest kilometre.

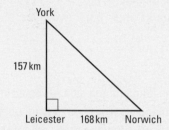

Nov 2006

4 In triangle ABC
Angle ABC = 90°
BC = 8 cm
AC = 21 cm
Work out the length of AB.
Give your answer correct to 3 significant figures.

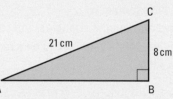

March 2007

5 PQR is a right-angled triangle. PR = 12 cm. QR = 4.5 cm. Angle PRQ = 90°.
Work out the value of x.
Give your answer correct to one decimal place.

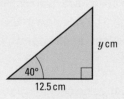

Nov 2007

6 Here is a right-angled triangle.   Here is another right-angled triangle.

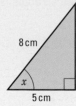

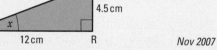

a Calculate the size of the angle marked x.
Give your answer correct to 1 decimal place.

b Calculate the value of y.
Give your answer correct to 1 decimal place.

June 2009

7 The diagram shows a vertical tower DC on horizontal ground ABC.
ABC is a straight line.
The angle of elevation of D from A is 28°.
The angle of elevation of D from B is 54°.
AB = 25 m
Calculate the height of the tower.
Give your answer to 3 significant figures.

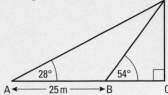

June 2006

8 Paul flies his helicopter from Ashwell to Birton.
He flies due west from Ashwell for 4.8 km. He then flies due south for 7.4 km to Birton.
Calculate the bearing from Birton to Ashwell.

9

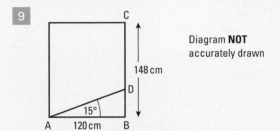

Diagram **NOT** accurately drawn

AB and BC are two sides of a rectangle.
AB = 120 cm and BC = 148 cm.
D is a point on BC.
Angle BAD = 15°.
Work out the length of CD.
Give your answer correct to the nearest centimetre.

10 The diagram shows a field 80 metres by 60 metres.
Alan runs around the field from point A to C (via point B) at 5 m/s.
Bhavana sets off at the same time, but runs directly across the diagonal of the field (shown by the dotted line) from A to C at 3 m/s.
  a Who will reach point C first?
  b How long will it be before the second person arrives?

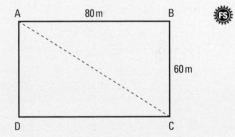

11 **a** Calculate the length of the side of the largest square that fits inside a 10 cm diameter circle.
  b Work out the length of the side of the smallest square that surrounds a 10 cm diameter circle.

12 Hamish wants to refelt the roof of his lean-to shed.
Felt is sold in 5 m rolls that are 1 m wide. They cost £12 each.
  a How many rolls will he need to buy?
The felt is stuck on with an adhesive which costs £6.99 for a 2.5 litre tin.
It will cover 6 m².
  b How much will Hamish have to pay for the materials to do the job?

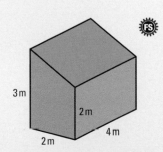

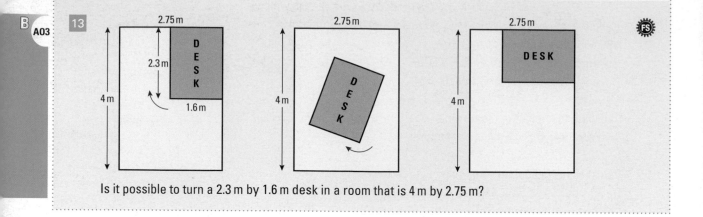

Is it possible to turn a 2.3 m by 1.6 m desk in a room that is 4 m by 2.75 m?

17 PYTHAGORAS' THEOREM AND TRIGONOMETRY 2

GPS works by sending and receiving signals to up to four satellites in orbit around the Earth. The paths of the signals create triangles, and the GPS uses trigonometry to work out your distance from each point. The satellites are 150 miles away from Earth but can still work out your position to within 15 metres.

Objectives

In this chapter you will:

- use Pythagoras' Theorem and trigonometry in three dimensions
- find the size of an angle between a line and a plane
- draw, sketch and recognise graphs of the trigonometric functions $y = \sin x$ and $y = \cos x$
- work out the area of a triangle using $\frac{1}{2}ab\sin C$
- use the sine rule and cosine rule to solve problems.

Before you start

- You should know the trigonometric ratios and how to apply Pythagoras' Theorem to right-angled triangles.

17.1 Pythagoras' Theorem and trigonometry in three dimensions

◉ Objective

● You can use Pythagoras' Theorem and trigonometry to solve problems in three dimensions.

❓ Why do this?

You could calculate the diagonal distance across a room. For example, in adventure training centres, zip lines are often attached between diagonally opposite corners of a room.

⬆ Get Ready

1. Here is a cuboid.
   Name as many right-angled triangles in this cuboid as you can.
   You should try to find at least six triangles.

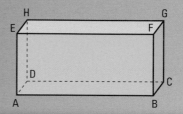

🌐 Key Points

● Problems on cuboids and other three-dimensional shapes involve identifying right-angled triangles and using Pythagoras' Theorem and trigonometry. It is important to draw the relevant triangles separately.

● The length of the longest diagonal of a cuboid with dimensions a, b, c is
$d = \sqrt{a^2 + b^2 + c^2}$

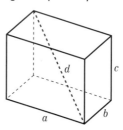

🔍 Example 1

ABCDEFGH is a cuboid, with length 8 cm, width 6 cm and height 9 cm.

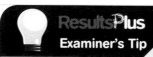

ResultsPlus
Examiner's Tip

Remember that you need to know Pythagoras' Theorem and the trigonometric results.

a Calculate the length of
  i AC **ii** AG. Give your answers correct to 3 significant figures.

b Calculate the size of angle FAB.
  Give your answer correct to the nearest degree.

a i

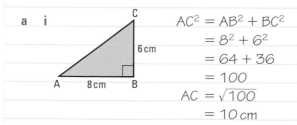

$AC^2 = AB^2 + BC^2$
$= 8^2 + 6^2$
$= 64 + 36$
$= 100$
$AC = \sqrt{100}$
$= 10 \text{ cm}$

> Triangle ABC is a right-angled triangle with AC as one side and the lengths of the other two sides known. Draw a sketch of triangle ABC. Use Pythagoras' Theorem for triangle ABC.

ii

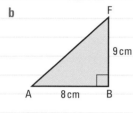

> Triangle ACG is a right-angled triangle with AG as one side and the lengths of the other two sides known. Use Pythagoras' Theorem for triangle ACG.

$$AG^2 = AC^2 + CG^2$$
$$= 10^2 + 9^2$$
$$= 100 + 81$$
$$= 181$$
$$AG = \sqrt{181} = 13.4536$$
$$AG = 13.5 \text{ cm (to 3 s.f.)}$$

Results Plus
Examiner's Tip

Write down at least four figures of the calculator display.

b

$$\tan \text{ angle FAB} = \frac{9}{8} = 1.125$$

> For angle FAB
> 9 cm is the opposite side
> 8 cm is the adjacent side
> $$\tan = \frac{\text{opp}}{\text{adj}}$$

$$\text{angle FAB} = 48.366\ldots°$$

$$\text{angle FAB} = 48° \text{ (to the nearest degree)}$$

Exercise 17A

> Questions in this chapter are targeted at the grades indicated.

Where necessary give lengths correct to 3 significant figures and angles correct to one decimal place.

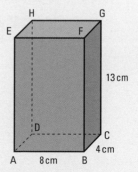

1 ABCDEFGH is a cuboid of length 8 cm, width 4 cm and height 13 cm.
  a Calculate the length of:
   i AC **ii** GB **iii** FA **iv** GA.
  b Calculate the size of:
   i angle FAB **ii** angle GBC.

A

2 For the cuboid ABCDEFG, show that
$$AG^2 = AB^2 + BC^2 + CG^2.$$

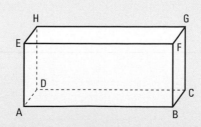

A03

A

3 A box is in the shape of a cuboid. The length of the box is 12 cm, the width of the box is 6 cm and the height of the box is 4 cm. The length of a needle is 15 cm. The needle cannot be broken. Can the needle fit inside the box?

AO2 AO3

4 The diagram shows a cylinder and a stick.
The cylinder has a base but no top.
The cylinder is standing on a horizontal table.
The radius of the cylinder is 8 cm.
The height of the cylinder is 12 cm.
The length of the stick is 25 cm.
The stick rests in the cylinder as shown so that as much of the stick is inside the cylinder as possible.
Work out the length of the stick that is not inside the cylinder.

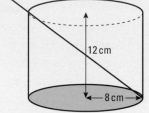

A*

5 The diagram shows a square-based pyramid.
The lengths of sides of the square base, ABCD, are 10 cm and the base is on a horizontal plane.
The centre of the base is the point M and the vertex of the pyramid is O, so that OM is vertical.
The point E is the midpoint of the side AB.
OA = OB = OC = OD = 15 cm

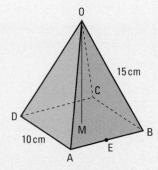

 a Calculate the length of **i** AC **ii** AM.
 b Calculate the length of OM.
 c Calculate the size of angle OAM.
 d Hence find the size of angle AOC.
 e Calculate the length of OE.
 f Calculate the size of angle OAB.

AO3

17.2 Angle between a line and a plane

◎ Objective

● You can find the size of the angle between a line and a plane.

⊘ Why do this?

You can use knowledge of angles between lines and planes to set up correctly the guy-lines for a tent.

⬡ Get Ready

1. Look at these three diagrams. Each diagram shows a thin pole stuck into the ground.

It looks as if these are three diagrams of three different poles. Explain how these diagrams could be of the same pole.

Key Points

- The angle between the pole and the ground seems to depend on how you look at the pole. So it is necessary to define what is meant by the angle between a line and a plane.
  Imagine a light shining directly above AB onto the **plane**.
  AN is the shadow of AB on the plane.
  AN is called the **projection** of AB on the plane.
  A line drawn from the point B perpendicular to the plane will meet the line AN and form a right angle with this line.
- Angle BAN is the angle between the line AB and the plane.

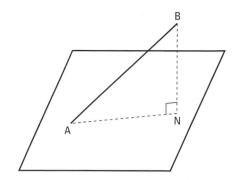

Example 2 The diagram shows a pyramid.
The base, ABCD, is a horizontal rectangle in which AB = 12 cm and AD = 9 cm. The vertex, O, is vertically above the midpoint of the base and OB = 18 cm. Calculate the size of the angle that OB makes with the horizontal plane. Give your answer to one decimal place.

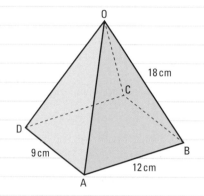

The base ABCD of the pyramid is horizontal so the angle that OB makes with the horizontal plane is the angle that OB makes with the base, ABCD.
Let M, directly below O, be the midpoint of the base and join O to M and M to B.
As OM is perpendicular to the base of the pyramid, the angle OBM is the angle between OB and the base and is the required angle.

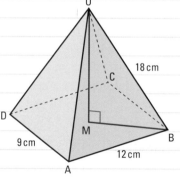

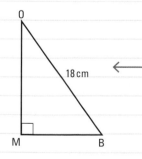

Draw the right-angled triangle OBM.
To find the size of angle OBM, find the length of either MB or OM.
The length of MB is $\frac{1}{2}$DB.

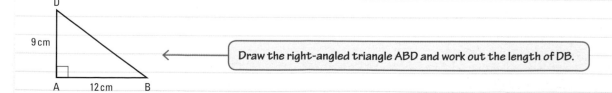

Draw the right-angled triangle ABD and work out the length of DB.

$DB^2 = 9^2 + 12^2 = 81 + 144 = 225$ ← Use Pythagoras' Theorem to calculate the length of DB.

$DB = \sqrt{225} = 15$

$MB = \frac{1}{2}DB = 7.5\ cm$

$cos\ (angle\ OBM) = \frac{7.5}{18}$ ← For angle OBM, 18 cm is the hypotenuse and 7.5 cm is the adjacent side.

$angle\ OBM = 65.37568...°$ ← $cos = \dfrac{adj}{hyp}$

The angle between OB and the horizontal plane is 65.4° (to 1 d.p.).

Exercise 17B

Where necessary give lengths correct to 3 significant figures and angles correct to one decimal place.

1 The diagram shows a pyramid.
The base, ABCD, is a horizontal rectangle in which AB = 15 cm
and AD = 8 cm. The vertex, O, is vertically above the centre of the
base and OA = 24 cm. Calculate the size of the angle that OA
makes with the horizontal plane.

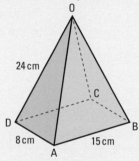

2 ABCDEFGH is a cuboid with a rectangular base in which AB = 12 cm and BC = 5 cm.
The height, AE, of the cuboid is 15 cm.

a Calculate the size of the angle:

  i between FA and ABCD

  ii between GA and ABCD

  iii between BE and ADHE.

b Write down the size of the angle between HE and ABFE.

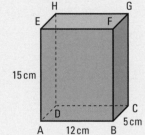

3 The diagram shows a learners' ski slope, ABCD, of length, AB, 500 m.
Triangles BAF and CDE are congruent right-angled
triangles and ABCD, AFED and BCEF are rectangles.
The rectangle BCEF is horizontal and the rectangle AFED
is vertical.
The angle between AB and the horizontal is 20° and the
angle between AC and the horizontal is 10°. Calculate:

a the length FB

b the height of A above F

c the distance AC

d the width, BC, of the ski slope.

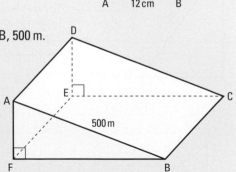

4 ABCD is a horizontal rectangular lawn in a garden and TC is a vertical pole. Ropes run from the top of the pole, T, to the corners A, B and D of the lawn.

a Calculate the length of the rope TA.

b Calculate the size of the angle made with the lawn by:
 i the rope TB **ii** the rope TD **iii** the rope TA.

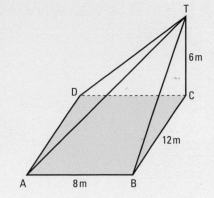

5 Diagram 1 shows a square-based pyramid, OABCD. Each side of the square is of length 60 cm and OA = OB = OC = OD = 50 cm.

Diagram 1

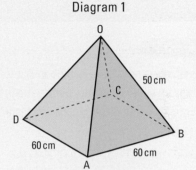

Diagram 2 shows a cube, ABCDEFGH, in which each edge is of length 60 cm.

A solid is made by placing the pyramid on top of the cube so that the base, ABCD, of the pyramid is on the top, ABCD, of the cube. The solid is placed on a horizontal table with the face EFGH on the table.

a Calculate the height of the vertex O above the table.

b Calculate the size of the angle between OE and the horizontal.

Diagram 2

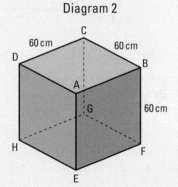

6 The diagram shows a solid cube of side 12 cm. The points P, Q and R are the midpoints of the edges on which they lie. The pyramid OPQR is removed from the cube.

a Taking OPQ as the base of the pyramid, draw a sketch of the pyramid, marking the size of angles POQ, QOR and ROP and the lengths of sides OP, OQ and OR.

b Find the size of the angle between:
 i RP and the plane OPQ
 ii RQ and the plane OPQ.

c Work out the volume of the solid remaining when the pyramid is removed from the cube.

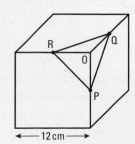

17.3 Trigonometric ratios for any angle

◎ Objective

● You can draw, sketch and recognise graphs of the trigonometric functions $y = \sin x$ and $y = \cos x$.

❓ Why do this?

The sine and cosine wave patterns can be seen in light waves, sound waves and ocean waves.

⬆ Get Ready

1. P, Q, R and S are points on a circle. The coordinates of P are (u, v). What are the coordinates of:
   a Q   **b** R   **c** S?

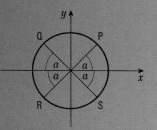

2. Using your calculator, find:
   a sin 30° and sin 150°.
   b cos 50° and cos 130°.
   c What do you notice?

🔑 Key Points

● Values of sine and cosine can be found for any angle.

● The diagram shows a circle, centre the origin O and radius 1 unit. A line, OP, of length 1 unit fixed at O, rotates in an anticlockwise direction about O, starting from the x-axis. The diagram shows OP when it has rotated through 40°.

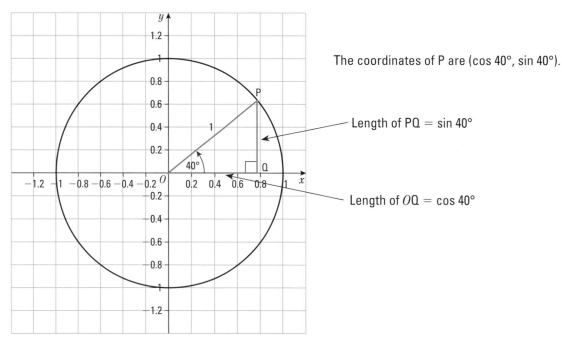

The coordinates of P are $(\cos 40°, \sin 40°)$.

Length of PQ $= \sin 40°$

Length of $OQ = \cos 40°$

● Trigonometry can be used to determine the lengths of side PQ and side OQ in the right-angled triangle OPQ, and the coordinates of point P.

● In general, when OP rotates through the angle $\theta°$, the position of P on the circle, radius $= 1$ is given by $x = \cos \theta°$, $y = \sin \theta°$. The coordinates of P are $(\cos \theta°, \sin \theta°)$.

● A rotation of 400° is one complete revolution of 360° plus a further rotation of 40°. The position of P is the same in the previous diagram so $(\cos 400°, \sin 400°)$ is the same point as $(\cos 40°, \sin 40°)$, therefore $\cos 400° = \cos 40°$ and $\sin 400° = \sin 40°$.

● A rotation through $-40°$ means the line OP rotates through 40° in a clockwise direction.

◉ For $\theta° = 136$, $\theta° = 225$, $\theta° = 304$ and $\theta° = -40$ the position of P is shown on the diagram.

◉ **In this quadrant:**
  ◉ sin x is positive
  ◉ cos x is negative

◉ In this quadrant:
  ◉ sin x is positive
  ◉ cos x is positive

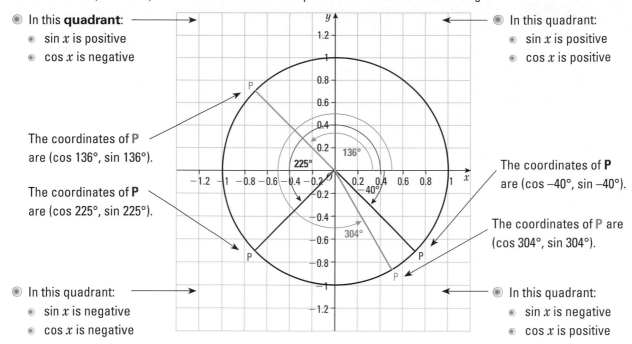

The coordinates of **P** are (cos 136°, sin 136°).

The coordinates of **P** are (cos 225°, sin 225°).

The coordinates of **P** are (cos −40°, sin −40°).

The coordinates of **P** are (cos 304°, sin 304°).

◉ In this quadrant:
  ◉ sin x is negative
  ◉ cos x is negative

◉ In this quadrant:
  ◉ sin x is negative
  ◉ cos x is positive

◉ The sine and cosine of any angle can be found using your calculator. Using these values the graphs of $y = \sin \theta°$ and $y = \cos \theta°$ can be drawn.

◉ **Graph of $y = \sin \theta°$**

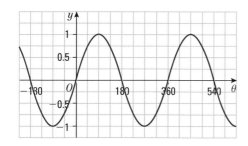

◉ The graph of $y = \sin \theta°$:
  ◉ cuts the θ-axis at …, -180, 0, 180, 360, 540, …
  ◉ repeats itself every 360°, that is, it has a **period** of 360°
  ◉ has a maximum value of 1 at $\theta° = $ …, 90, 450, …
  ◉ has a minimum value of -1 at $\theta° = $ …, -90, 270, …

◉ **Graph of $y = \cos \theta°$**

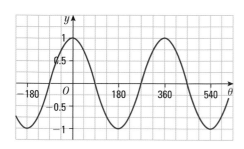

◉ The graph of $y = \cos \theta°$:
  ◉ cuts the θ-axis at …, -90, 90, 270, 450, …
  ◉ repeats itself every 360°, that is, it has a period of 360°
  ◉ has a maximum value of 1 at $\theta° = $ …, 0, 360, …
  ◉ has a minimum value of -1 at $\theta° = $ …, -180, 180, 540, …

◉ The graph of $y = \sin \theta°$ and the graph of $y = \cos \theta°$ are horizontal translations of each other.

Example 3 — For values of θ in the interval -180 to 360 solve the equation

    a $\sin \theta° = 0.7$

    b $5\cos \theta° = 2$.

    Give each answer correct to one decimal place.

a $\sin \theta° = 0.7$

   $\theta = 44.4$

> Use a calculator to find one value of θ.

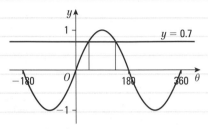

> To find the other solutions draw a sketch of $y = \sin \theta$ for $\theta°$ from -180 to 360.

> The sketch shows that there are two values of θ in the interval -180 to 360 for which $\sin \theta° = 0.7$.

$\theta = 44.4, 180 - 44.4$

$\theta = 44.4, 135.6$

> One solution is $\theta = 44.4$ and by symmetry the other solution is $\theta = 180 - 44.4$.

b $5\cos \theta° = 2$

  $\cos \theta° = \dfrac{2}{5} = 0.4$

  $\theta = 66.4$

> Divide each side of the equation by 5.

> Use a calculator to find one value of θ.

> To find the other solutions draw a sketch of $y = \cos \theta°$ for θ from -180 to 360.

> The sketch shows that there are three values of θ in the interval -180 to 360 for which $\cos \theta° = 0.4$.

$\theta = 66.4, -66.4, 360 + -66.4$

$\theta = 66.4, -66.4, 293.6$

> One solution is $\theta = 66.4$ and by symmetry another solution is $\theta = -66.4$.
> Using the period of the graph the other solution is $\theta = 360 + -66.4$.

Exercise 17C

A

A*

1 For $-360 \leqslant \theta \leqslant 360$, sketch the graph of

    a $y = \sin \theta°$ **b** $y = \cos \theta°$.

2 Find all values of θ in the interval 0 to 360 for which

    a $\sin \theta° = 0.5$ **b** $\cos \theta° = 0.1$.

3 **a** Show that one solution of the equation $3\sin \theta° = 1$ is 19.5, correct to 1 decimal place.

    b Hence solve the equation $3\sin \theta° = 1$ for values of θ in the interval 0 to 720.

4 **a** Show that one solution of the equation $10\cos \theta° = -3$ is 107.5 correct to 1 decimal place.

    b Hence find all values of θ in the interval -360 to 360 for which $10\cos \theta° = -3$.

17.4 Finding the area of a triangle using $\frac{1}{2}ab\sin C$

◉ Objectives

◉ You can work out the area of a triangle using $\frac{1}{2}ab\sin C$.

◉ You can work out the area of a segment of a circle.

⦾ Why do this?

You could use the formula $\frac{1}{2}ab\sin C$ to work out the area of a triangular lake if you could measure the lengths of the edges but not the distance across.

⬙ Get Ready

1. Work out the area of these triangles.

a

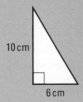

10 cm, 6 cm

b

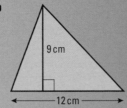

9 cm, 12 cm

c

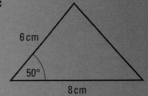

6 cm, 50°, 8 cm

◖ Key Points

◉ The vertices of a triangle are labelled with capital letters.
The triangle shown is triangle ABC.
The sides opposite the angles are labelled so that a is the length of the side opposite angle A, b is the length of the side opposite angle B and c is the length of the side opposite angle C.

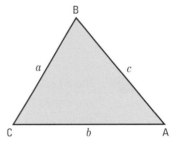

ResultsPlus
Examiner's Tip

The formula $\frac{1}{2}ab\sin C$ will be given on the formula sheet.

◉ Area of a triangle $= \frac{1}{2}$ base $\times$ height

Area of triangle ABC $= \frac{1}{2}bh$

In the right-angled triangle BCN, $h = a\sin C$

So area of triangle ABC $= \frac{1}{2}b \times a\sin C$

that is

area of triangle ABC $= \frac{1}{2}ab\sin C$

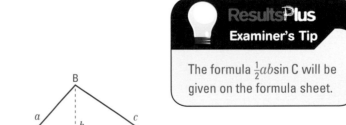

◉ The angle C is the angle between the sides of length a and b and is called the included angle.
The formula for the area of a triangle means that

area of a triangle $= \frac{1}{2}$ product of two sides $\times$ sine of the included angle

For triangle ABC, there are two more formulae for the area.

Area of triangle ABC $= \frac{1}{2}ab\sin C = \frac{1}{2}bc\sin A = \frac{1}{2}ac\sin B$

These formulae give the area of a triangle, whether the included angle is acute or obtuse.

Example 4   Calculate the area of each of the triangles correct to 3 significant figures.

a

B, 6.8 cm, 66°, C, 7.6 cm, A

b

5.1 m, 108°, 6.2 m

a Area $= \frac{1}{2} \times 6.8 \times 7.6 \times \sin 66$

   Area $= 23.606\ldots$

   Area $= 23.6 \text{ cm}^2$ (3 s.f.)

> Substitute $a = 6.8, b = 7.6, C = 66°$ into area $= \frac{1}{2}ab\sin C$.
> Give the area correct to 3 significant figures and give the units.

b Area $= \frac{1}{2} \times 5.1 \times 6.2 \times \sin 108$

   Area $= 15.036\ldots$

   Area $= 15.0 \text{ m}^2$ (3 s.f.)

> Substitute into area of a triangle
> $= \frac{1}{2}$ product of two sides $\times$ sine of the included angle.

Example 5   The diagram shows a circle of radius 6 cm
and centre O.
AB is a chord of the circle and angle AOB $= 56°$.
Work out the area of the shaded segment.
Give your answer correct to 3 significant figures.

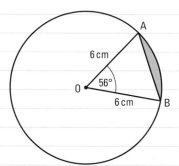

Area of segment $=$ area of sector OAB $-$ area of triangle OAB

> See Section 10.2 for the
> area of a sector.

Area of sector OAB $= \frac{56}{360} \times \pi \times 6^2 = 17.5929\ldots \text{ cm}^2$

> Write down at least four figures
> from your calculator display.

Area of triangle OAB $= \frac{1}{2} \times 6 \times 6 \times \sin 56 = 14.9226\ldots \text{ cm}^2$

> Use area of triangle $= \frac{1}{2}ab\sin C$.

Area of segment $= 17.5929\ldots - 14.9226\ldots = 2.6703$

Area of segment $= 2.67 \text{ cm}^2$ (3 s.f.)

> Give your answer correct to 3 significant figures.

Exercise 17D

Give lengths and areas correct to three significant figures and angles correct to one decimal place.

1 Work out the area of each of these triangles.

a
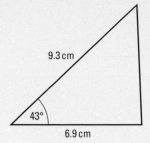
9.3 cm
43°
6.9 cm

b
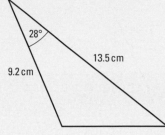
28°
13.5 cm
9.2 cm

c
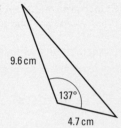
9.6 cm
137°
4.7 cm

d
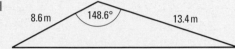
8.6 m   148.6°   13.4 m

2 The area of triangle ABC is 60.7 m².
Work out the length of BC.

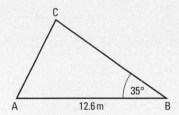
C
A   12.6 m   35°   B

3 The area of triangle ABC is 15 cm².
Angle A is acute.
Work out the size of angle A.

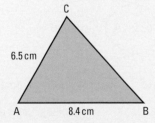
C
6.5 cm
A   8.4 cm   B

4 a Triangle ABC is such that $a = 6$ cm, $b = 9$ cm and angle C = 25°.
   Work out the area of triangle ABC.
  b Triangle PQR is such that $p = 6$ cm, $q = 9$ cm and angle R = 155°.
   Work out the area of triangle PQR.
  c What do you notice about your answers? Why do you think this is true?

5 The diagram shows a regular octagon, with centre O.
  a Work out the size of angle AOB.
   OA = OB = 6 cm.
  b Work out the area of triangle AOB.
  c Hence work out the area of the octagon.

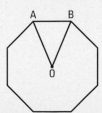
A   B
O

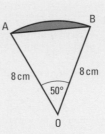

6 The diagram shows a sector, OAB, of a circle, centre O.
The radius of the circle is 8 cm and the size of angle AOB is 50°.
Work out the area of the segment of the circle shown shaded
in the diagram.

17.5 The sine rule

⊙ Objective

● You can use the sine rule to work out the length of
a side in a triangle.

⊕ Why do this?

You can use the sine rule to find a length that isn't
easily accessible to measure, for example, the
height of a mountain, using the angle of elevation
to the top of the mountain from two points a
certain distance apart.

⊕ Get Ready

1. Work out the value of   **a** $\dfrac{7.9 \times \sin 67°}{8.4}$ **b** $\dfrac{14.8 \times \sin 58°}{\sin 67}$

🌐 Key Points

● The last section showed that

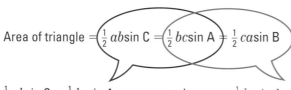

Area of triangle $= \frac{1}{2}ab\sin C = \frac{1}{2}bc\sin A = \frac{1}{2}ca\sin B$

$\frac{1}{2}ab\sin C = \frac{1}{2}bc\sin A$ and $\frac{1}{2}bc\sin A = \frac{1}{2}ca\sin B$

cancelling $\frac{1}{2}$ and b from both sides   cancelling $\frac{1}{2}$ and c from both sides

$a\sin C = c\sin A$ and $b\sin A = a\sin B$

or

$\dfrac{a}{\sin A} = \dfrac{c}{\sin C}$ and $\dfrac{b}{\sin B} = \dfrac{a}{\sin A}$

These results are combined to get the **sine rule**
which can be used in any triangle.

$\dfrac{a}{\sin A} = \dfrac{b}{\sin B} = \dfrac{c}{\sin C}$

ResultsPlus
Examiner's Tip

The sine rule will be given on
the formula sheet.

To use the sine rule to find a length in a triangle, it is necessary to know any two angles and a side (ASA).

Example 6

Find the length of the side marked a in the triangle.
Give your answer correct to three significant figures.

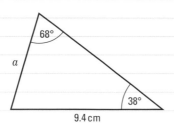

Results Plus

Watch Out!

You can only use Pythagoras' Theorem in a right-angled triangle.

$$\frac{a}{\sin 38°} = \frac{9.4 \text{ cm}}{\sin 68°}$$

Substitute A = 38°, b = 9.4 cm, B = 68° into $\frac{a}{\sin A} = \frac{b}{\sin B}$

Results Plus

Examiner's Tip

Check that your answer is sensible: the greater length is always opposite the greater angle.

$$a = \frac{9.4 \times \sin 38°}{\sin 68°}$$

Multiply both sides by sin 38°.

$$a = 6.2417\ldots$$

$$a = 6.24 \text{ cm (3 s.f.)}$$

Example 7

Find the length of the side marked x in the triangle.
Give your answer correct to three significant figures.

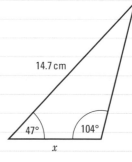

Missing angle = 180° − (47° + 104°) = 29°

The angle opposite x must be known before the sine rule can be used. Use the angle sum of a triangle.

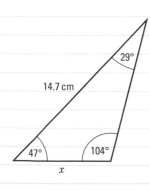

$$\frac{x}{\sin 29°} = \frac{14.7}{\sin 104°}$$

Write down the sine rule with x opposite 29° and 14.7 cm opposite 104°.

$$x = \frac{14.7 \times \sin 29°}{\sin 104°}$$

Multiply both sides by sin 29°.

$$x = 7.3448\ldots$$

$$x = 7.34 \text{ cm (3 s.f.)}$$

Exercise 17E

Give lengths correct to three significant figures.

A | A02

1 Find the lengths of the sides marked with letters in these triangles.

a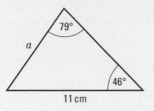

b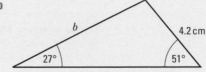

c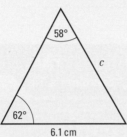

d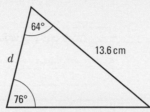

e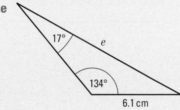

f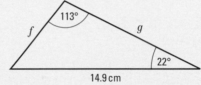

17.6 Using the sine rule to calculate an angle

◎ Objective

● You can use the sine rule to work out the size of an angle in a triangle.

? Why do this?

Astronomers sometimes use the sine rule to calculate unknown angles between stars in space.

◈ Get Ready

1. Find the size of the acute angle x when $\sin x = \dfrac{14 \times 0.7}{17}$.

Key Points

● To use the sine rule to find an angle in a triangle, it is necessary to know two sides and the non-included angle (SSA).
● When the sine rule is used to calculate an angle, it is a good idea to turn each fraction upside down (the reciprocal).
This gives $\dfrac{\sin A}{a} = \dfrac{\sin B}{b} = \dfrac{\sin C}{c}$

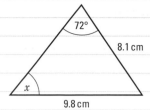

Example 8 — Find the size of the acute angle x in the triangle.

Give your answer correct to one decimal place.

$$\frac{\sin x}{8.1\text{ cm}} = \frac{\sin 72°}{9.8\text{ cm}}$$

← Write down the sine rule with x opposite 8.1 cm and 72° opposite 9.8 cm.

$$\sin x = \frac{8.1 \times \sin 72°}{9.8}$$

← Multiply both sides by 8.1.

$$\sin x = 0.7860\ldots$$ ← Work out the value of $\sin x$.

$$x = 51.820\ldots°$$

$$x = 51.8° \text{ (1 d.p.)}$$

Exercise 17F

Give lengths and areas correct to three significant figures and angles correct to one decimal place.

1 Calculate the size of each of the acute angles marked with a letter.

a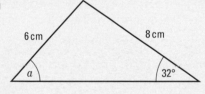

b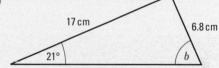

c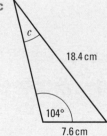

d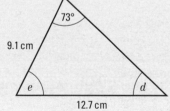

2 The diagram shows quadrilateral ABCD and its diagonal AC.
a In triangle ABC, work out the length of AC.
b In triangle ACD, work out the size of angle DAC.
c Work out the size of angle BCD.

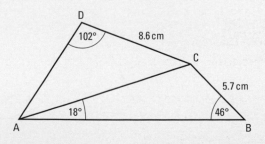

3 In triangle ABC, BC = 8.6 cm, angle BAC = 52° and angle ABC =63°.

   a Calculate the length of AC.

   b Calculate the length of AB.

   c Calculate the area of triangle ABC.

4 In triangle PQR all the angles are acute. PR = 7.8 cm and PQ = 8.4 cm. Angle PQR = 58°.

   a Work out the size of angle PRQ.

   b Work out the length of QR.

5 The diagram shows the position of a port (P), a lighthouse (L) and a buoy (B). The lighthouse is due East of the buoy.

The lighthouse is on a bearing of 035° from the port and the buoy is on a bearing of 312° from the port.

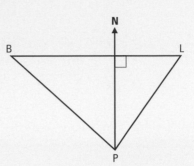

   a Work out the size of: **i** angle PBL **ii** angle PLB.

The lighthouse is 8 km from the port.

   b Work out the distance PB.

   c Work out the distance BL.

   d Work out the shortest distance from the port (P) to the line BL.

17.7 The cosine rule

◎ Objective

● You can use the cosine rule to work out the length of a side in a triangle.

⍰ Why do this?

You can use the cosine rule to find a length that isn't easily accessible to measure. For example, mapmakers can calculate the distance between two towns separated by a lake using the distance to each town from a fixed point and the angle between these two lengths.

⊕ Get Ready

1. Work out the value of **a** $5 + 3 \times 2$ **b** $9 + 8 - 2 \times 7$ **c** $5^2 + 6^2 - 2 \times 5 \times 6 \times \cos 50°$

2. Work out the positive value of x when $x^2 = 7^2 + 4^2 - 2 \times 7 \times 4 \times 0.9$

🔍 Key Points

◉ The diagram shows triangle ABC.

The line BN is perpendicular to AC and meets the line AC at N so that AN = x and NC = $(b - x)$. The length of BN is h.

In triangle ANB,
Pythagoras' Theorem gives
$c^2 = x^2 + h^2$ (1)

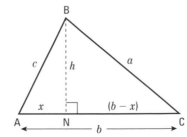

In triangle BNC,
Pythagoras' Theorem gives
$a^2 = (b - x)^2 + h^2$
$a^2 = b^2 - 2bx + x^2 + h^2$
Using (1), substitute c^2 for
$x^2 + h^2$
$a^2 = b^2 - 2bx + c^2$
$= b^2 + c^2 - 2bx$ (2)

In the right-angled triangle ANB, $x = c \cos A$
Substituting this into (2)
$a^2 = b^2 + c^2 - 2bc\cos A$
This result is known as the **cosine rule** and can be
used in any triangle.

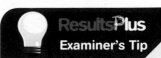

Results Plus
Examiner's Tip

The cosine rule will be given
on the formula sheet.

● Similarly $b^2 = a^2 + c^2 - 2ac\cos B$ and $c^2 = a^2 + b^2 - 2ab\cos C$
● To use the cosine rule to find a length in a triangle, it is necessary
to know the other two sides and the angle between these sides (SAS).

Example 9 Find the length of the side marked with a letter in each triangle.
Give your answers correct to three significant figures.

a

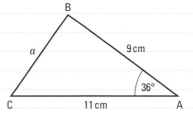

b

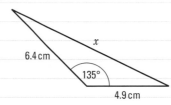

a $a^2 = 11^2 + 9^2 - 2 \times 11 \times 9 \times \cos 36°$
$a^2 = 121 + 81 - 160.1853\ldots$
$a^2 = 41.8146$
$a = \sqrt{41.8146}$ ← Take the square root.
$a = 6.4664$
$a = 6.47$ cm

Substitute $b = 11$ cm, $c = 9$ cm, A = 36° into
$a^2 = b^2 + c^2 - 2bc\cos A$.
Evaluate each term separately.

b $x^2 = 6.4^2 + 4.9^2 - 2 \times 6.4 \times 4.9 \times \cos 135°$
$x^2 = 40.96 + 24.01 - 62.72 \times (-0.70710\ldots)$
$x^2 = 64.97 + 44.3497\ldots$
$x^2 = 109.3197\ldots$
$x = \sqrt{109.3197\ldots}$ ← Take the square root.
$x = 10.4556\ldots$
$x = 10.5$ cm

Substitute the two given lengths and
the included angle into the cosine rule.

The cosine of an obtuse angle is
negative so $\cos 135° < 0$.

Exercise 17G

Give lengths correct to three significant figures.

1 Calculate the length of the sides marked with letters in these triangles.

a

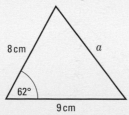

b

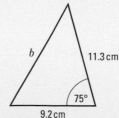

A
A02

A

c

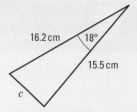

d

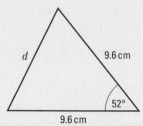

e

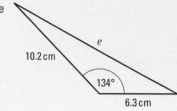

f

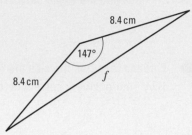

A02 **2** In triangle XYZ, XY = 20.3 cm, XZ = 14.5 cm and angle YXZ = 38°.
Calculate the length of YZ.

17.8 Using the cosine rule to calculate an angle

◎ Objective

● You can use the cosine rule to work out the size of an angle in a triangle.

⦿ Why do this?

The cosine rule can be used in sailing to calculate bearings to a destination, given different wind speeds and the angle at which it strikes the boat.

⬦ Get Ready

1. Find the size of angle A when:

a $\cos A = \dfrac{12}{19}$ **b** $\cos A = \dfrac{7}{23}$ **c** $\cos A = \dfrac{11^2 + 6^2 - 9^2}{2 \times 11 \times 6}$

🔍 Key Points

◉ To use the cosine rule to find the size of an angle in a triangle, it is necessary to know all three sides (SSS).

◉ To find an angle using the cosine rule, rearrange $a^2 = b^2 + c^2 - 2bc\cos A$

$2bc\cos A = b^2 + c^2 - a^2$

$\cos A = \dfrac{b^2 + c^2 - a^2}{2bc}$

Similarly $\cos B = \dfrac{a^2 + c^2 - b^2}{2ac}$ and $\cos C = \dfrac{a^2 + b^2 - c^2}{2ab}$

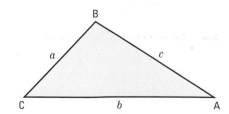

Example 10 Find the size of: **a** angle BAC **b** angle x.

Give your answers correct to one decimal place.

a

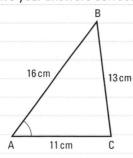

b

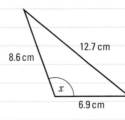

a $\cos A = \dfrac{11^2 + 16^2 - 13^2}{2 \times 11 \times 16} = \dfrac{208}{352}$

Substitute $b = 11$ cm, $c = 16$ cm, $a = 13$ cm into $\cos A = \dfrac{b^2 + c^2 - a^2}{2bc}$.

$\cos A = 0.590909\ldots$

$A = 53.77\ldots$

$A = 53.8°$ (1 d.p.)

b $\cos x = \dfrac{8.6^2 + 6.9^2 - 12.7^2}{2 \times 8.6 \times 6.9}$

Substitute the three lengths into the cosine rule, noting that 12.7 cm is opposite the angle to be found.

$\cos x = -\dfrac{39.72}{118.68}$

$\cos x = -0.33468\ldots$

The value of $\cos x$ is negative so x is an obtuse angle.

$x = 109.553\ldots$

$x = 109.6°$ (1 d.p.)

Exercise 17H

Give lengths and areas correct to three significant figures and angles correct to one decimal place.

1 Calculate the size of each of the angles marked with a letter in these triangles.

a

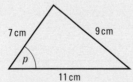

b

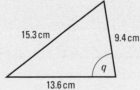

c

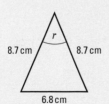

d

2 AB is a chord of a circle with centre O.
The radius of the circle is 7 cm and the
length of the chord is 11 cm.
Calculate the size of angle AOB.

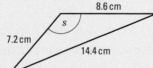

A

A02

A02

A **A02 A03**

3 The region ABC is marked on a school field.
The point B is 70 m from A on a bearing of 064°.
The point C is 90 m from A on a bearing of 132°.
 a Work out the size of angle BAC.
 b Work out the length of BC.

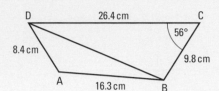

A★

4 The diagram shows the quadrilateral ABCD.
 a Work out the length of DB.
 b Work out the size of angle DAB.

A03
 c Work out the area of quadrilateral ABCD.

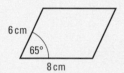

A02 A03

5 Chris ran 4 km on a bearing of 036° from P to Q. He then ran in a straight line from Q to R, where R is 7 km due East of P. Chris then ran in a straight line from R to P. Calculate the total distance that Chris ran.

A03

6 The diagram shows a parallelogram.
Work out the length of each diagonal of the parallelogram.

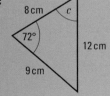

17.9 Using trigonometry to solve problems

⊙ Objectives

- ● You can identify whether to use the sine rule or the cosine rule when solving 2D and 3D problems.
- ● You can solve problems involving non right-angled triangles.

⌘ Why do this?

Some engineers use the sine and cosine rules to find unknown lengths and angles when they create maps of large features on the Earth's surface, such as a mountain range or an ocean floor.

⬥ Get Ready

1. Look at these triangles. To find the length of the lettered side or the size of the lettered angle, which one of the sine rule or the cosine rule should be used?

a

b

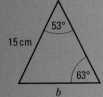

c

8 cm, *c*, 72°, 12 cm, 9 cm

◈ Key Points

- ◉ Use the sine rule when a problem involves two sides and two angles.
- ◉ Use the cosine rule when a problem involves three sides and one angle.
- ◉ The formulae for the sine rule and the cosine rule are given on the formula sheet.

Example 11

The area of triangle ABC is 12 cm².
AB = 3.8 cm and angle ABC = 70°.
a Find the length of: **i** BC **ii** AC.
b Find the size of angle BAC.

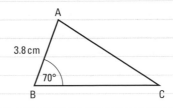

a **i** $\frac{1}{2} \times BC \times 3.8 \times \sin 70° = 12$ ← Substitute $c = 3.8$ cm, $B = 70°$ into area $= \frac{1}{2}ac\sin B$.

$BC = \dfrac{2 \times 12}{3.8 \times \sin 70°}$

$BC = 6.721...$
$BC = 6.72$ cm (3 s.f.)

ii $b^2 = 6.721...^2 + 3.8^2 - 2 \times 6.721... \times 3.8 \times \cos 70°$
$= 59.613... - 17.470...$
$= 42.142...$
$b = 6.491...$
$b = 6.49$ cm (3 s.f.)

Substitute $a = 6.721...$ cm, $c = 3.8$ cm and $B = 70°$ into $b^2 = a^2 + c^2 - 2ac\cos B$.

ResultsPlus
Examiner's Tip

Remember to use uncorrected values of your answers for subsequent calculations.

b $\dfrac{\sin A}{6.721...} = \dfrac{\sin 70°}{6.491...}$ ←

Substitute $a = 6.721...$ cm, $b = 6.491$ cm... and $B = 70°$ into $\dfrac{\sin A}{a} = \dfrac{\sin B}{b}$.

$\sin A = \dfrac{6.721... \times \sin 70°}{6.491...}$

$\sin A = 0.9728...$
$A = 76.62...°$
$A = 76.7°$ (1 d.p.)

Exercise 17I

Where necessary, give lengths and areas correct to three significant figures and angles correct to one decimal place, unless the question states otherwise.

1 A triangle has sides of lengths 9 cm, 10 cm and 11 cm.
　a Calculate the size of each angle of the triangle.
　b Calculate the area of the triangle.

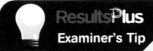

2 In the diagram, ABC is a straight line.
　a Calculate the length of BD.
　b Calculate the size of angle DAB.
　c Calculate the length of AC.

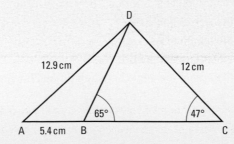

A*

A03

3 The area of triangle ABC is 15 cm².

AB = 4.6 cm and angle BAC = 63°.

a Work out the length of AC.

b Work out the length of BC.

c Work out the size of angle ABC.

A03

4 ABCD is a kite, with diagonal DB.

a Calculate the length of DB.

b Calculate the size of angle BDC.

c Calculate the value of x.

d Calculate the length of AC.

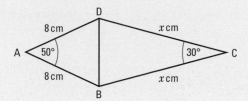

A02
A03

5 James walked 9 km due south from point A to point B.

He then changed direction and walked 5 km to point C.

James was then 6 km from his starting point A.

a Work out the bearing of point C from point B.

Give your answer correct to the nearest degree.

b Work out the bearing of point C from point A.

Give your answer correct to the nearest degree.

A03

6 The diagram shows a pyramid. The base of the pyramid, ABCD, is a rectangle in which AB = 15 cm and AD = 8 cm.

The vertex of the pyramid is O where

OA = OB = OC = OD = 20 cm.

Work out the size of angle DOB,

correct to the nearest degree.

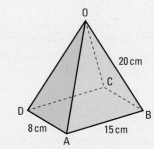

A02
A03

7 The diagram shows a vertical pole, PQ, standing on a hill.

The hill is at an angle of 8° to the horizontal.

The point R is 20 m downhill from Q and the line PR is at 12° to the hill.

a Calculate the size of angle RPQ.

b Calculate the length, PQ, of the pole.

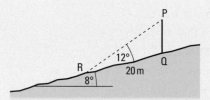

A03

8 A, B and C are points on horizontal ground so that AB = 30 m, BC = 24 m and angle CAB = 50°.

AP and BQ are vertical posts, where AP = BP = 10 m.

a Work out the size of angle ACB.

b Work out the length of AC.

c Work out the size of angle PCQ.

d Work out the size of the angle between QC and the ground.

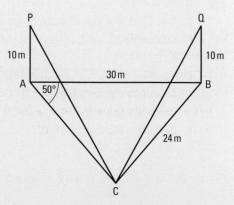

9 The diagram shows a port P and two buoys A and B.
The buoy A is at a distance of 15 km and on a bearing of 020° from P.
The buoy B is at a distance of 20 km and on a bearing of 310° from P.
Calculate the distance between A and B.
Give your answer in km correct to 3 significant figures.

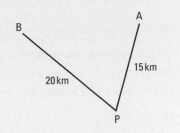

Chapter review

◉ Problems on cuboids and other 3D shapes involve identifying right-angled triangles and using Pythagoras' Theorem and trigonometry. It is important to draw these triangles separately.

◉ AN is called the **projection** of AB on the **plane**.

◉ Angle BAN is the angle between the line AB and the plane.

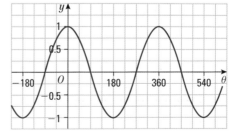

◉ The diagram shows for each **quadrant** whether the sine and cosine of angles in that quadrant are positive or negative.

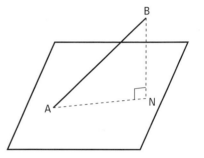

Graph of $y = \sin \theta°$

Notice that the graph:
◉ cuts the θ-axis at ..., -180, 0, 180, 360, 540, ...
◉ repeats itself every 360°, that is, it has a **period** of 360°
◉ has a maximum value of 1 at $\theta° = ...$, 90, 450, ...
◉ has a minimum value of -1 at $\theta° = ...$, -90, 270, ...

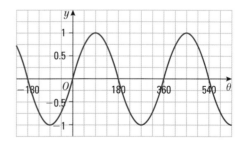

Graph of $y = \cos \theta°$

Notice that the graph:
◉ cuts the θ-axis at ..., -90, 90, 270, 450, ...
◉ repeats itself every 360°, that is, it has a period of 360°
◉ has a maximum value of 1 at $\theta° = ...$, 0, 360, ...
◉ has a minimum value of -1 at $\theta° = ...$, -180, 180, 540, ...

◉ Area of triangle ABC $= \frac{1}{2}ab\sin C = \frac{1}{2}bc\sin A = \frac{1}{2}ac\sin B.$

- The **sine rule** can be used in any triangle: $\dfrac{a}{\sin A} = \dfrac{b}{\sin B} = \dfrac{c}{\sin C}$

  which can also be written as $\dfrac{\sin A}{a} = \dfrac{\sin B}{b} = \dfrac{\sin C}{c}$

- The **cosine rule** can be used in any triangle: $a^2 = b^2 + c^2 - 2bc\cos A$

  Similarly $b^2 = a^2 + c^2 - 2ac\cos B$ and $c^2 = a^2 + b^2 - 2ab\cos C$

  which can also be written as

  $$\cos A = \frac{b^2 + c^2 - a^2}{2bc} \qquad \cos B = \frac{a^2 + c^2 - b^2}{2ac} \qquad \cos C = \frac{a^2 + b^2 - c^2}{2ab}$$

- Use the sine rule when a problem involves two sides and two angles.

- Use the cosine rule when a problem involves three sides and one angle.

Review exercise

A

1 A cuboid has length 3 cm, width 4 cm and height 12 cm.
Work out the length of PQ.

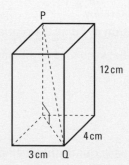

Nov 2007

2 ABC is a triangle.

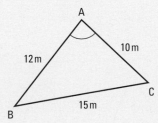

AB = 12 m.
AC = 10 m.
BC = 15 m.
Calculate the size of angle BAC.
Give your answer correct to one decimal place.

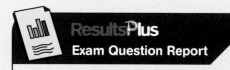

Exam Question Report

80% of students answered this question poorly because they did not take care substituting values into the equation.

June 2008

3 ABC is a triangle.
AC = 8 cm.
BC = 9 cm.
Angle ACB = 40°.
Calculate the length of AB.
Give your answer correct to 3 significant figures.

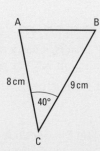

June 2007

4 The diagram represents a cuboid ABCDEFGH.
AB = 5 cm.
BC = 7 cm.
AE = 3 cm.

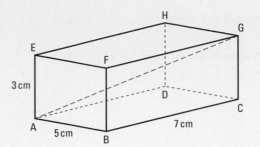

a Calculate the length of AG.
Give your answer to 3 significant figures.

b Calculate the size of the angle between AG
and the face ABCD.
Give your answer correct to 1 decimal place.

Nov 2004

5 AB = 3.2 cm.
BC = 8.4 cm.
The area of triangle ABC is 10 cm².
Calculate the perimeter of triangle ABC.
Give your answer correct to three significant figures.

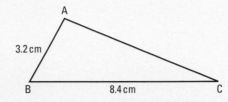

June 2004

6 The diagram shows a tetrahedron. AD is perpendicular to both AB and AC.
AB = 10 cm. AC = 8 cm. AD = 5 cm. Angle BAC = 90°.

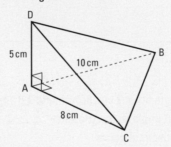

Calculate the size of angle BDC.
Give your answer correct to 1 decimal place.

Nov 2007

7 The diagram shows a sketch of the curve $y = \sin x°$ for $0 \leqslant x \leqslant 360$.

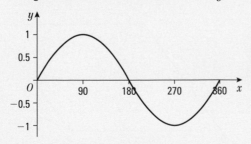

ResultsPlus
Exam Question Report

92% of students answered this question poorly
because they did not use the information given in
the question.

The exact value of $\sin 60° = \dfrac{\sqrt{3}}{2}$.

Write down the exact value of: **a** $\sin 120°$ **b** $\sin 240°$.

Nov 2008, adapted

A*

8 Here is a graph of the curve $y = \cos x°$
for $0 \leqslant x \leqslant 360$.
Use the graph to solve $\cos x° = 0.75$
for $0 \leqslant x \leqslant 360$.

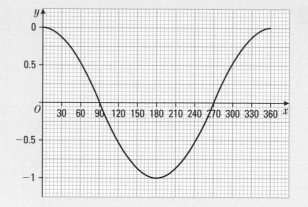

June 2007

A02
A03

9 The diagram shows an equilateral triangle ABC with sides of length 6 cm.
P is the midpoint of AB.
Q is the midpoint of AC.
APQ is a sector of a circle, centre A.
Calculate the area of the shaded region.
Give your answer correct to 3 significant figures.

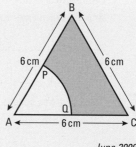

June 2009

A02
A03

10 The lengths of the sides of a triangle are 4.2 cm, 5.3 cm and 7.6 cm.
  a Calculate the size of the largest angle of the triangle.
    Give your answer correct to 1 decimal place.
  b Calculate the area of the triangle.
    Give your answer correct to 3 significant figures.

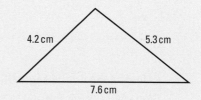

Nov 2006

A02
A03

11 The diagram represents a prism.
AEFD is a rectangle.
ABCD is a square.
EB and FC are perpendicular to plane ABCD.
AB = 60 cm.
AD = 60 cm.
Angle ABE = 90°.
Angle BAE = 30°.
Calculate the size of the angle that the line DE makes with the plane ABCD.
Give your answer correct to 1 decimal place.

June 2004

12 The diagram shows a sector OABC of a circle with centre O.
OA = OC = 10.4 cm.
Angle AOC = 120°.

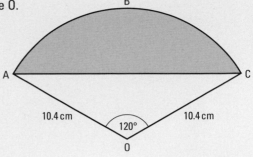

a Calculate the length of the arc ABC of the sector.
Give your answer correct to 3 significant figures.

b Calculate the area of the shaded segment ABC.
Give your answer correct to 3 significant figures.

June 2006

13

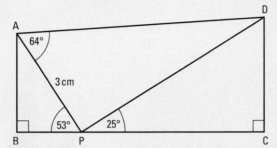

Diagram **NOT**
accurately drawn

BPC is a straight line. Angles ABP = angle DCP = 90°.
Calculate the length of PD. Give your answer correct to 3 significant figures.

14

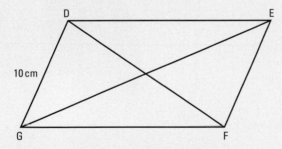

Diagram **NOT**
accurately drawn

DEFG is a parallelogram with DG = 10 cm.
The diagonals DF and EG are of length 16 cm and 24 cm respectively.

a Calculate the size of angle DGE.

b Calculate the length of DE.

15 The diagram shows a pyramid HABCD standing on
horizontal ground. The points A, B, C and D are the
corners of its square base. The length of a side of
the square is 12 m and its diagonals intersect at O.
Each sloping edge makes an angle of 28° with the
ground. Calculate:

a the height, OH, in metres to 3 significant figures

b the size, to the nearest degree, of the angle
which the plane HCB makes with the ground.

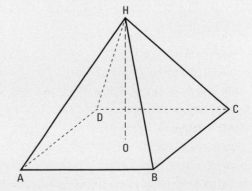

289

18 VECTORS

Netball players have to decide on the exact direction and power needed to pass a ball to a teammate without other players intercepting. The direction and strength of the throw can be described using vectors. Vectors are used to describe any quantity that requires both a direction and a size. Many physical problems can be explored and solved using vectors.

◎ Objectives

In this chapter you will:
- understand and use vector notation
- calculate, and represent graphically, the sum of two vectors, the difference of two vectors and a scalar multiple of a vector
- calculate the resultant of two vectors
- learn how to solve geometrical problems in two dimensions and apply vector methods for simple geometrical proofs.

◔ Before you start

You should be able to:
- understand bearings
- plot points on a graph
- use Pythagoras' Theorem.

18.1 Vectors and vector notation

◉ Objective

◉ You can understand and use vector notation.

⊘ Why do this?

You can describe journeys using vectors, for example, a trip from London to Brighton is a vector with magnitude 60 km and direction south.

◈ Get Ready

1. The points A, B, C and D are the vertices of a parallelogram where A is (1, 0), B is (4, 0) and C is (5, 2). Draw points A, B and C on squared paper to find the coordinates of point D.

◷ Key Points

◉ In mathematics, there are many quantities that need a **direction** as well as a size in order to describe them completely.

◉ For example, to describe a change in position or a **displacement**, it is necessary to give the direction of the movement as well as the distance moved.

◉ Similarly, when describing a force it is important to state the direction. This is because how an object moves when it is pushed or pulled will depend on the direction of the push or pull as well as the size or **magnitude** of the push or pull.

◉ Displacements and forces that need a magnitude and a direction to describe them are examples of vectors. In this chapter, only displacement vectors will be considered but the results apply to other vectors as well.

◉ As vectors need magnitude and direction to describe them, vectors are equal only when they have equal magnitudes and the same direction. For example:

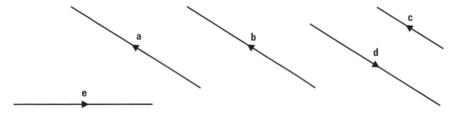

◉ The vectors **a** and **b** are **equal vectors**, that is **a** = **b**. They have the same magnitude and direction.

◉ The vectors **a** and **c** are not equal. Although they have the same direction, they do not have the same magnitude.

◉ The vectors **a** and **d** are not equal. Although they have the same magnitude and are parallel, they are in opposite directions and so do not have the same direction.

◉ The vectors **a** and **e** are not equal. They have the same magnitude but they do not have the same direction.

◉ The displacement from A to B is 4 cm on a bearing of 030°.

◉ This displacement is written $\overrightarrow{AB}$ to show that it is a vector and it has a direction from A to B.

◉ In the diagram the line from A to B is drawn 4 cm long in a direction of 030° and it is marked with an arrow to show that the direction is from A to B.

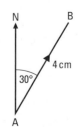

- Vectors can also be labelled with single bold letters such as **a**, **b** and **c**.
  For example, the vector **b** has been drawn on a grid.
- When hand writing the vector **a** you can use <u>a</u> to represent it.
- The displacement represented by **b** can be described as 4 to the right and 2 up.

  As with translations this can be written as the column vector $\begin{pmatrix} 4 \\ 2 \end{pmatrix}$.

  So we can write $\mathbf{b} = \begin{pmatrix} 4 \\ 2 \end{pmatrix}$.

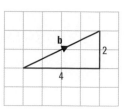

Example 1

a Point A has coordinates (1, 6) and point B has coordinates (4, 1).
Write $\overrightarrow{AB}$ as a column vector.

b The point C is such that $\overrightarrow{BC} = \begin{pmatrix} -2 \\ 3 \end{pmatrix}$. Find the coordinates of C.

a

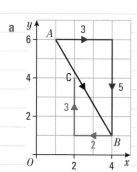

⟵ Mark the points A and B on a grid.

⟵ To move from A to B go 3 to the right and 5 down.

$\overrightarrow{AB} = \begin{pmatrix} 3 \\ -5 \end{pmatrix}$

b The coordinates of C are (2, 4). ⟵ For $\overrightarrow{BC} = \begin{pmatrix} -2 \\ 3 \end{pmatrix}$, from B go 2 to the left and 3 up to find C.

Exercise 18A

Questions in this chapter are targeted at the grades indicated.

A

1 On squared paper draw and label the following vectors.

a $\mathbf{a} = \begin{pmatrix} 1 \\ 2 \end{pmatrix}$
b $\mathbf{b} = \begin{pmatrix} 4 \\ -2 \end{pmatrix}$
c $\mathbf{c} = \begin{pmatrix} -5 \\ -3 \end{pmatrix}$
d $\overrightarrow{AB} = \begin{pmatrix} -4 \\ 3 \end{pmatrix}$
e $\overrightarrow{CD} = \begin{pmatrix} 0 \\ 5 \end{pmatrix}$

2 The point A is (1, 3), the point B is (6, 9) and the point C is (5, −3).
a Write as column vectors:

i $\overrightarrow{AB}$ **ii** $\overrightarrow{BC}$ **iii** $\overrightarrow{AC}$

b What do you notice about your answers in **a**?

A03

3 The points A, B, C and D are the vertices of a quadrilateral where A has coordinates (2, 1),
$\overrightarrow{AB} = \begin{pmatrix} 2 \\ 4 \end{pmatrix}$, $\overrightarrow{BC} = \begin{pmatrix} 3 \\ 1 \end{pmatrix}$ and $\overrightarrow{CD} = \begin{pmatrix} 4 \\ -2 \end{pmatrix}$.

a On squared paper draw quadrilateral ABCD.
b Write as a column vector $\overrightarrow{AD}$.
c What type of quadrilateral is ABCD?
d What do you notice about $\overrightarrow{BC}$ and $\overrightarrow{AD}$?

4 The points A, B, C and D are the vertices of a parallelogram.

A has coordinates (0, 1), $\overrightarrow{AB} = \begin{pmatrix} 4 \\ 0 \end{pmatrix}$ and $\overrightarrow{AD} = \begin{pmatrix} 2 \\ 3 \end{pmatrix}$.

  a On squared paper draw the parallelogram ABCD.

  b Write as a column vector **i** $\overrightarrow{DC}$ **ii** $\overrightarrow{CB}$

  c What do you notice about **i** $\overrightarrow{AB}$ and $\overrightarrow{DC}$ **ii** $\overrightarrow{AD}$ and $\overrightarrow{CB}$?

A03   **A**

5 Here are eight vectors.

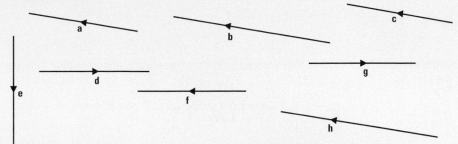

    There are three pairs of equal vectors. Name the equal vectors.

18.2 The magnitude of a vector

◉ Objective

◉ You can calculate the magnitude of a vector.

② Why do this?

When designing high-speed trains, aerodynamic modellers use vectors of different magnitudes to represent air resistance and friction.

◆ Get Ready

1. Work out   **a** $\sqrt{24^2 + 7^2}$ **b** $\sqrt{(-8)^2 + (-5)^2}$.

🕵 Key Points

◉ The magnitude of the vector **a** is written a or $|a|$.

◉ The magnitude of the vector $\overrightarrow{AB}$ is AB, that is, the length of the line segment AB.

◉ In general, the magnitude of the vector $\begin{pmatrix} x \\ y \end{pmatrix}$ is $\sqrt{x^2 + y^2}$.

🔍 Example 2   Find the magnitude of the vector $\mathbf{a} = \begin{pmatrix} 4 \\ -6 \end{pmatrix}$.

Give your answer: **i** as a surd and **ii** correct to 3 significant figures.

$\begin{pmatrix} 4 \\ -6 \end{pmatrix}$ means 4 to the right and 6 down.

Draw a right-angled triangle to show this.

$a^2 = 4^2 + 6^2 = 16 + 36$

$a^2 = 52$

i $a = \sqrt{52} = 2\sqrt{13}$

ii $a = 7.21$ (to 3 s.f.)

Use Pythagoras' Theorem to find the length, a, of the hypotenuse.

Example 3 Find the magnitude of the vector $\overrightarrow{AB} = \begin{pmatrix} -3 \\ -4 \end{pmatrix}$.

$AB = \sqrt{(-3)^2 + (-4)^2}$ ← Substitute $x = -3$ and $y = -4$ into $\sqrt{x^2 + y^2}$.

$\quad\ = \sqrt{9 + 16}$

$AB = 5$

Exercise 18B

A

1 Work out the magnitude of each of these vectors. (Where necessary, answers may be left as surds.)

a $\mathbf{a} = \begin{pmatrix} 5 \\ 12 \end{pmatrix}$ b $\mathbf{b} = \begin{pmatrix} 12 \\ -5 \end{pmatrix}$ c $\mathbf{c} = \begin{pmatrix} 1 \\ 3 \end{pmatrix}$

d $\mathbf{d} = \begin{pmatrix} -5 \\ -7 \end{pmatrix}$ e $\overrightarrow{AB} = \begin{pmatrix} 8 \\ -15 \end{pmatrix}$ f $\overrightarrow{PQ} = \begin{pmatrix} -8 \\ 4 \end{pmatrix}$

A*

2 In triangle ABC, $\overrightarrow{AB} = \begin{pmatrix} -20 \\ -15 \end{pmatrix}$ and $\overrightarrow{AC} = \begin{pmatrix} 24 \\ -7 \end{pmatrix}$.

a Work out the length of the side AB of the triangle.

b Show that the triangle is an isosceles triangle.

A03

3 In quadrilateral ABCD, $\overrightarrow{AB} = \begin{pmatrix} 3 \\ 4 \end{pmatrix}$, $\overrightarrow{BC} = \begin{pmatrix} 5 \\ 0 \end{pmatrix}$, $\overrightarrow{CD} = \begin{pmatrix} -3 \\ -4 \end{pmatrix}$, $\overrightarrow{DA} = \begin{pmatrix} -5 \\ 0 \end{pmatrix}$.

What type of quadrilateral is ABCD?

18.3 Addition of vectors

Objectives

● You can calculate, and represent graphically, the sum of two vectors.

● You can calculate the resultant of two vectors.

Why do this?

When kayaking against a current you would need to allow for the strength and direction of the current in order to reach your destination.

Get Ready

1. Translate shape A by the vector $\begin{pmatrix} 2 \\ 0 \end{pmatrix}$. Label this new shape B.

Translate shape B by the vector $\begin{pmatrix} -1 \\ 2 \end{pmatrix}$. Label this new shape C.

What single translation will map shape A onto shape C?

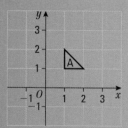

Key Points

● The two-stage journey from A to B and then from B to C has the same starting point and the same finishing point as the single journey from A to C.

That is, A to B followed by B to C is equivalent to A to C,

or $\overrightarrow{AB}$ followed by $\overrightarrow{BC}$ is equivalent to $\overrightarrow{AC}$.

This is written as $\overrightarrow{AB} + \overrightarrow{BC} = \overrightarrow{AC}$.

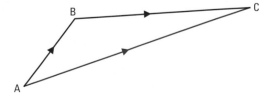

● Notice the pattern here AC + BC gives AC.
   This leads to the triangle law of vector addition.

Results Plus
Watch Out!

This does not mean that AB + BC = AC.
The sum of the lengths of AB and BC is
not equal to the length of AC.

Triangle law of vector addition

● Let $\overrightarrow{AB}$ represent the vector **a** and $\overrightarrow{BC}$ represent the vector **b**.
   Then if $\overrightarrow{AC}$ represents the vector **c**,
   a + **b** = **c**.

Parallelogram law of vector addition

● PQRS is a parallelogram.
   In a parallelogram, opposite sides are equal in length and are parallel.
● So since $\overrightarrow{PQ}$ and $\overrightarrow{SR}$ are also in the same direction $\overrightarrow{PQ} = \overrightarrow{SR}$ (= **a**).

   Similarly $\overrightarrow{PS} = \overrightarrow{QR}$ (= **b**).

   From the triangle law $\overrightarrow{PQ} + \overrightarrow{QR} = \overrightarrow{PR}$ so that $\overrightarrow{PR} = $ **a** + **b**.

   Hence $\overrightarrow{PR} = \overrightarrow{PQ} + \overrightarrow{PS}$ as $\overrightarrow{PQ} = $ **a** and $\overrightarrow{PS} = $ **b**.

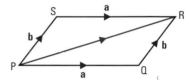

● So if in parallelogram PQRS, $\overrightarrow{PQ}$ represents the vector **a** and $\overrightarrow{PS}$ represents the vector **b**, the diagonal $\overrightarrow{PR}$ of
   the parallelogram represents the vector **a** + **b**.
● When **c** = **a** + **b** the vector **c** is said to be the **resultant vector** of the two vectors **a** and **b**.
● $\begin{pmatrix} a \\ b \end{pmatrix} + \begin{pmatrix} c \\ d \end{pmatrix} = \begin{pmatrix} a + c \\ b + d \end{pmatrix}$

Example 4   Find, by drawing, the sum of the vectors **a** and **b**.

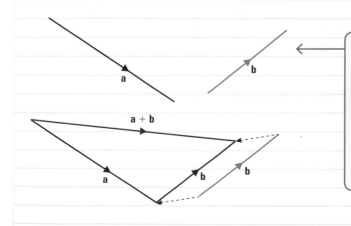

Use the triangle law of vector addition.

Move vector **b** to the end of vector **a** so
that the arrows follow on.

Draw and label the vector **a** + **b** to
complete the triangle.

a + **b** could also have been found by
moving the vector **a** to the beginning of
vector **b**. The answer is the same as the
two triangles are congruent.

Example 5

In the quadrilateral ABCD, $\overrightarrow{AB} = \mathbf{a}$, $\overrightarrow{BC} = \mathbf{b}$ and $\overrightarrow{CD} = \mathbf{c}$.
Find the vectors **i** $\overrightarrow{AC}$ **ii** $\overrightarrow{AD}$.

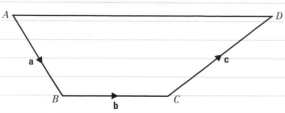

i $\overrightarrow{AC} = \overrightarrow{AB} + \overrightarrow{BC}$

so $\overrightarrow{AC} = \mathbf{a} + \mathbf{b}$

> Use the triangle law of vector addition.
> Make sure that the Bs follow each other.

ii $\overrightarrow{AD} = \overrightarrow{AC} + \overrightarrow{CD}$

so $\overrightarrow{AD} = (\mathbf{a} + \mathbf{b}) + \mathbf{c}$

$\overrightarrow{AD} = \mathbf{a} + \mathbf{b} + \mathbf{c}$

> Use $\overrightarrow{AC} = \mathbf{a} + \mathbf{b}$.
> Vector expressions like this can be treated as in ordinary algebra.
> The brackets can be removed.

Example 6

$\overrightarrow{AB} = \begin{pmatrix} 3 \\ 5 \end{pmatrix}$ and $\overrightarrow{BC} = \begin{pmatrix} 8 \\ -4 \end{pmatrix}$

Find $\overrightarrow{AC}$.

$\overrightarrow{AC} = \overrightarrow{AB} + \overrightarrow{BC}$

> Use the triangle law of vector addition.

$\overrightarrow{AC} = \begin{pmatrix} 3 \\ 5 \end{pmatrix} + \begin{pmatrix} 8 \\ -4 \end{pmatrix}$

> Draw a sketch.

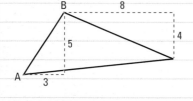

> From A to B is 3 to the right.
> From B to C is 8 to the right.
> So from A to C is $3 + 8 = 11$ to the right.
> From B to C is 4 down.
> So from A to C is $5 + -4 = 1$ up.

$\overrightarrow{AC} = \begin{pmatrix} 11 \\ 1 \end{pmatrix}$

Example 7

$\mathbf{a} = \begin{pmatrix} 5 \\ -6 \end{pmatrix}$ and $\mathbf{b} = \begin{pmatrix} -3 \\ 4 \end{pmatrix}$ Find $\mathbf{a} + \mathbf{b}$.

$\mathbf{a} + \mathbf{b} = \begin{pmatrix} 5 \\ -6 \end{pmatrix} + \begin{pmatrix} -3 \\ 4 \end{pmatrix} = \begin{pmatrix} 5 + -3 \\ -6 + 4 \end{pmatrix}$

> Add across.

$\mathbf{a} + \mathbf{b} = \begin{pmatrix} 2 \\ -2 \end{pmatrix}$

Exercise 18C

A

1   A vector **a** has magnitude 5 cm and direction 030°. A vector **b** has magnitude 7 cm and direction 140°.
Draw the vector   **a** **a**   **b** **b**   **c** **a** + **b**.

2   Work out.

a $\begin{pmatrix} 2 \\ 6 \end{pmatrix} + \begin{pmatrix} 4 \\ 2 \end{pmatrix}$ **b** $\begin{pmatrix} 6 \\ 3 \end{pmatrix} + \begin{pmatrix} -2 \\ 5 \end{pmatrix}$ **c** $\begin{pmatrix} -5 \\ 8 \end{pmatrix} + \begin{pmatrix} 3 \\ -4 \end{pmatrix}$ **d** $\begin{pmatrix} 6 \\ 0 \end{pmatrix} + \begin{pmatrix} 3 \\ -5 \end{pmatrix}$ **e** $\begin{pmatrix} -5 \\ 3 \end{pmatrix} + \begin{pmatrix} -3 \\ -6 \end{pmatrix}$

3 $\overrightarrow{PQ} = \begin{pmatrix} 3 \\ 1 \end{pmatrix}$ $\overrightarrow{QR} = \begin{pmatrix} 7 \\ -6 \end{pmatrix}$
Work out $\overrightarrow{PR}$.

4 $\mathbf{p} = \begin{pmatrix} 3 \\ 6 \end{pmatrix}$ $\mathbf{q} = \begin{pmatrix} 1 \\ -3 \end{pmatrix}$ $\mathbf{r} = \begin{pmatrix} 4 \\ 7 \end{pmatrix}$
  a Work out **i** $\mathbf{p} + \mathbf{q}$ **ii** $\mathbf{q} + \mathbf{p}$
  b What do you notice?
  c Work out **i** $(\mathbf{p} + \mathbf{q}) + \mathbf{r}$ **ii** $\mathbf{p} + (\mathbf{q} + \mathbf{r})$
  d What do you notice?

5 ABCDEF is a regular hexagon.
  $\overrightarrow{AB} = \mathbf{n}$
  a Explain why $\overrightarrow{ED} = \mathbf{n}$.
  $\overrightarrow{BC} = \mathbf{m}$ $\overrightarrow{CD} = \mathbf{p}$
  b Find **i** $\overrightarrow{AC}$ **ii** $\overrightarrow{AD}$.
  c What is $\overrightarrow{FD}$?

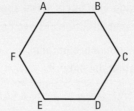

A02
A03

18.4 Parallel vectors

◎ **Objectives**

◉ You can calculate, and represent graphically, the difference of two vectors.
◉ You can use the scalar multiple of a vector.

Why do this?

Planes flying in formation would need to use parallel vectors.

Get Ready

1. Work out   **a** $\begin{pmatrix} 3 \\ 5 \end{pmatrix} + \begin{pmatrix} 3 \\ 5 \end{pmatrix}$ **b** $\begin{pmatrix} -2 \\ 0 \end{pmatrix} + \begin{pmatrix} -7 \\ -3 \end{pmatrix}$ **c** $\begin{pmatrix} -4 \\ 6 \end{pmatrix} + \begin{pmatrix} 4 \\ -6 \end{pmatrix}$.

Key Points

◉ The ordinary rules of algebra state that $a + a = 2a$. This can also be applied to vectors.
  For example, here is the vector **a**.

  Here are $\mathbf{a} + \mathbf{a}$ and $2\mathbf{a}$.

◉ $2\mathbf{a}$ is a vector in the same direction as **a** and with twice the magnitude.
  For $\mathbf{a} = \begin{pmatrix} 2 \\ 5 \end{pmatrix}$, $\mathbf{a} + \mathbf{a} = \begin{pmatrix} 2 \\ 5 \end{pmatrix} + \begin{pmatrix} 2 \\ 5 \end{pmatrix} = \begin{pmatrix} 2 + 5 \\ 5 + 5 \end{pmatrix} = \begin{pmatrix} 2 \times 2 \\ 2 \times 5 \end{pmatrix}$
  that is, $2\mathbf{a} = 2\begin{pmatrix} 2 \\ 5 \end{pmatrix} = \begin{pmatrix} 2 \times 2 \\ 2 \times 5 \end{pmatrix} = \begin{pmatrix} 4 \\ 10 \end{pmatrix}$.

◉ Similarly, $3\mathbf{a}$ is a vector in the same direction as **a** and with magnitude 3 times the magnitude of **a**.
  And $3\mathbf{a} = 3\begin{pmatrix} 2 \\ 5 \end{pmatrix} = \begin{pmatrix} 3 \times 2 \\ 3 \times 5 \end{pmatrix} = \begin{pmatrix} 6 \\ 15 \end{pmatrix}$.

◉ The vector $\overrightarrow{AB}$ is the displacement from A to B, and $\overrightarrow{BA}$ is the displacement from B to A.

◉ These displacements have the same magnitudes but are in opposite directions, so $\overrightarrow{AB}$ followed by $\overrightarrow{BA}$ is the zero displacement (0) as there is no overall change in position.
This is written $\overrightarrow{AB} + \overrightarrow{BA} = \mathbf{0}$.

◉ Using the usual rules of algebra, it follows that $\overrightarrow{BA} = -\overrightarrow{AB}$.

◉ A negative sign in front of a vector reverses the direction of the vector.
$$\overrightarrow{AB} = \begin{pmatrix} 3 \\ -5 \end{pmatrix} \text{ so } \overrightarrow{BA} = -\begin{pmatrix} 3 \\ -5 \end{pmatrix} = -1\begin{pmatrix} 3 \\ -5 \end{pmatrix} = \begin{pmatrix} -1 \times 3 \\ -1 \times -5 \end{pmatrix} = \begin{pmatrix} -3 \\ 5 \end{pmatrix}$$
showing that the reverse of 3 to the right and 5 down is 3 to the left and 5 up.

◉ The vector $-\mathbf{a}$ has the same magnitude as $\mathbf{a}$ but is in the opposite direction.
The vector $-3\mathbf{a}$ has the same magnitude as $3\mathbf{a}$ but is in the opposite direction.
So the vector $-3\mathbf{a}$ has 3 times the magnitude as $\mathbf{a}$ but is in the opposite direction.
Vectors that are parallel either have the same direction or have opposite directions.

◉ For any non-zero value of k, the vectors $\mathbf{a}$ and $k\mathbf{a}$ are parallel.
The number k is called a **scalar**; it has magnitude only.
If $\mathbf{a} = \begin{pmatrix} p \\ q \end{pmatrix}$ then $k\mathbf{a} = k\begin{pmatrix} p \\ q \end{pmatrix} = \begin{pmatrix} kp \\ kq \end{pmatrix}$

◉ With the origin O, the vectors $\overrightarrow{OA}$ and $\overrightarrow{OB}$ are called the **position vectors** of the points A and B. In general, the point (p, q) has position vector $\begin{pmatrix} p \\ q \end{pmatrix}$.

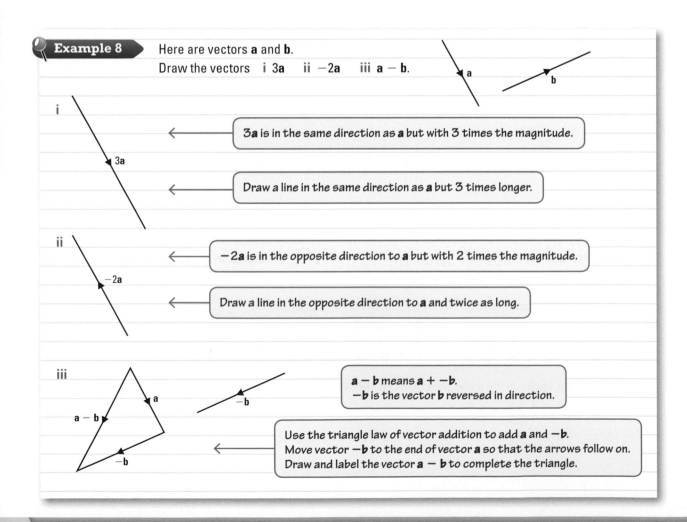

Example 8

Here are vectors **a** and **b**.
Draw the vectors   **i** 3**a**   **ii** −2**a**   **iii** **a** − **b**.

i

3**a**

> 3**a** is in the same direction as **a** but with 3 times the magnitude.

> Draw a line in the same direction as **a** but 3 times longer.

ii

−2**a**

> −2**a** is in the opposite direction to **a** but with 2 times the magnitude.

> Draw a line in the opposite direction to **a** and twice as long.

iii

a − **b**

a

−**b**

−**b**

> **a** − **b** means **a** + −**b**.
> −**b** is the vector **b** reversed in direction.

> Use the triangle law of vector addition to add **a** and −**b**.
> Move vector −**b** to the end of vector **a** so that the arrows follow on.
> Draw and label the vector **a** − **b** to complete the triangle.

Example 9

With origin O, the points A, B, C and D have coordinates (1, 3), (2, 7), (−6, −10) and (−1, 10) respectively.

a Write down as a column vector **i** $\overrightarrow{OA}$ **ii** $\overrightarrow{OB}$.

b Work out **i** $\overrightarrow{AB}$ as a column vector   **ii** $\overrightarrow{CD}$ as a column vector.

c What do these results show about AB and CD?

a i $\overrightarrow{OA} = \begin{pmatrix} 1 \\ 3 \end{pmatrix}$ ⟵ From O to A is 1 across and 3 up.

ii $\overrightarrow{OB} = \begin{pmatrix} 2 \\ 7 \end{pmatrix}$ ⟵ From O to B is 2 across and 7 up.

b i

Method 1

$\overrightarrow{AB} = \begin{pmatrix} 1 \\ 4 \end{pmatrix}$ ⟵ A to B, that is, (1, 3) to (2, 7), is 1 across and 4 up.

Method 2

$\overrightarrow{AB} = \overrightarrow{AO} + \overrightarrow{OB}$

$\overrightarrow{AB} = -\overrightarrow{OA} + \overrightarrow{OB}$ $\overrightarrow{AO} = -\overrightarrow{OA}$ ⟵ Another way to obtain $\overrightarrow{AB}$ is to use the triangle law of vector addition.

$\overrightarrow{AB} = -\begin{pmatrix} 1 \\ 3 \end{pmatrix} + \begin{pmatrix} 2 \\ 7 \end{pmatrix}$

$= \begin{pmatrix} -1 \\ -3 \end{pmatrix} + \begin{pmatrix} 2 \\ 7 \end{pmatrix}$

$= \begin{pmatrix} -1 + 2 \\ -3 + 7 \end{pmatrix}$

$\overrightarrow{AB} = \begin{pmatrix} 1 \\ 4 \end{pmatrix}$

ii $\overrightarrow{CD} = \begin{pmatrix} 5 \\ 20 \end{pmatrix}$ ⟵ Using Method 1, C to D, that is, (−6, −10) to (−1, 10), is 5 to the right and 20 up.

c $\overrightarrow{CD} = \begin{pmatrix} 5 \\ 20 \end{pmatrix} = 5\begin{pmatrix} 1 \\ 4 \end{pmatrix}$

$\overrightarrow{CD} = 5\overrightarrow{AB}$

The lines CD and AB are parallel and the length of ⟵ **a** and k**a** are parallel vectors.
the line CD is 5 times the length of the line AB.

Example 10

Simplify **i** $3\mathbf{a} + 5\mathbf{b} + 2\mathbf{a} - 3\mathbf{b}$ **ii** $2\mathbf{a} + \frac{1}{2}(4\mathbf{a} - 2\mathbf{b})$.

i $3\mathbf{a} + 5\mathbf{b} + 2\mathbf{a} - 3\mathbf{b}$ ⟵ $3\mathbf{a} + 2\mathbf{a} = 5\mathbf{a}$

$= 5\mathbf{a} + 2\mathbf{b}$ ⟵ $5\mathbf{b} - 3\mathbf{b} = 2\mathbf{b}$

The ordinary rules of algebra can be applied to vector expressions like this.

ii $2\mathbf{a} + \frac{1}{2}(4\mathbf{a} - 2\mathbf{b})$ ⟵ $\frac{1}{2}(4\mathbf{a} - 2\mathbf{b}) = 2\mathbf{a} - \mathbf{b}$

$= 2\mathbf{a} + 2\mathbf{a} - \mathbf{b}$

$= 4\mathbf{a} - \mathbf{b}$

Example 11 ABC is a straight line where BC = 3AB.

$\overrightarrow{OA} = \mathbf{a}$ $\overrightarrow{AB} = \mathbf{b}$

Express $\overrightarrow{OC}$ in terms of **a** and **b**.

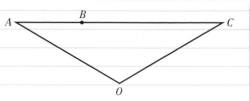

$\overrightarrow{OC} = \overrightarrow{OA} + \overrightarrow{AC}$
$\overrightarrow{OC} = \overrightarrow{OA} + 4\overrightarrow{AB}$
$\overrightarrow{OC} = \mathbf{a} + 4\mathbf{b}$

> Use the triangle law of vector addition.

> As BC = 3AB, $\overrightarrow{AC} = 4\overrightarrow{AB}$.

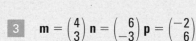

Exercise 18D

A

1. The vector **a** has magnitude 4 cm and direction 130°.
   The vector **b** has magnitude 5 cm and direction 220°.
   Draw the vector   **a a**   **b b**   **c** −**b**   **d a** − **b**.

2. Here is the vector **p**.
   Draw the vector   **a** 2**p**   **b** $-\frac{1}{2}$**p**.

A*

3. $\mathbf{m} = \begin{pmatrix} 4 \\ 3 \end{pmatrix}$ $\mathbf{n} = \begin{pmatrix} 6 \\ -3 \end{pmatrix}$ $\mathbf{p} = \begin{pmatrix} -2 \\ 6 \end{pmatrix}$
   a Find as a column vector.   i 5**m**   ii −2**n**   iii 4**m** + 3**p**   iv 2**m** − 4**n** + 5**p**
   b Find   i the magnitude of the vector **m**   ii the magnitude of the vector 2**m** − **p**.

A02
A03

4. The points P, Q, R and S have coordinates $(-2, 5)$, $(3, 1)$, $(-6, -9)$ and $(14, -25)$ respectively.
   a Write down the position vector, $\overrightarrow{OP}$, of the point P.
   b Write down as a column vector.   i $\overrightarrow{PQ}$ ii $\overrightarrow{RS}$.
   c What do these results show about the lines PQ and RS?

A02
A03

5. The point A has coordinates $(1, 3)$, the point B has coordinates $(4, 5)$, the point C has coordinates
   $(-2, -4)$. Find the coordinates of the point D where $\overrightarrow{CD} = 6\overrightarrow{AB}$.

6. $\overrightarrow{OA} = \mathbf{a}$ $\overrightarrow{OB} = \mathbf{b}$
   a Express $\overrightarrow{AB}$ in terms of **a** and **b**.
   b Where is the point C such that $\overrightarrow{OC} = \frac{1}{2}\mathbf{b}$?

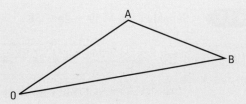

A02

7. Here are five vectors.
   $\overrightarrow{AB} = 2\mathbf{m} + 4\mathbf{n}$, $\overrightarrow{CD} = 6\mathbf{m} - 12\mathbf{n}$, $\overrightarrow{EF} = 4\mathbf{m} + 8\mathbf{n}$, $\overrightarrow{GH} = -\mathbf{m} - 2\mathbf{n}$, $\overrightarrow{IJ} = 6\mathbf{m} + 16\mathbf{n}$
   a Three of these vectors are parallel. Which are the parallel vectors?
   b Simplify.   i 8**p** + 5**q** − 3**p** − 8**q**   ii $2(2\mathbf{m} - 5\mathbf{n}) + \frac{2}{3}(3\mathbf{m} - 6\mathbf{n})$

8   Here is a regular hexagon ABCDEF. In the hexagon, FC is parallel to AB and twice as long.

$\overrightarrow{AB} = \mathbf{m}$

   a   Express $\overrightarrow{FC}$ in terms of **m**.

$\overrightarrow{CD} = \mathbf{n}$

   b   Express $\overrightarrow{FD}$ in terms of **m** and **n**.

$\overrightarrow{BC} = \mathbf{x}$

   c   Express $\overrightarrow{AC}$ in terms of **m** and **x**.

The lines AC and FD are parallel and equal in length.

   d   Find an expression for **x** in terms of **m** and **n**.

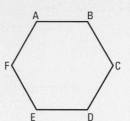

A02
A03   A*

18.5 Solving geometric problems in two dimensions

◉ Objectives

◉ You can solve geometric problems in two dimensions using vector methods.

◉ You can apply vector methods for simple geometric proofs.

◈ Why do this?

Videogame programmers solve 2D and 3D geometric problems using vectors, in order to create virtual worlds.

◈ Get Ready

1.   The point A has coordinates $(0, 4)$, the point B has coordinates $(7, -3)$, and the point C has coordinates $(-8, -1)$.

   Write down as a column vector   **a** $\overrightarrow{AB}$ **b** $\overrightarrow{AC}$ **c** $\overrightarrow{BC}$

◈ Key Points

To solve geometric problems the following results are useful:

◉ Triangle law of vector addition so that $\overrightarrow{PQ} + \overrightarrow{QR} = \overrightarrow{PR}$.

◉ When $\overrightarrow{PQ} = \mathbf{a}$, $\overrightarrow{QP} = -\mathbf{a}$.

◉ When $\overrightarrow{PQ} = k\overrightarrow{RS}$, k is a scalar (number), the lines PQ and RS are parallel and the length of PQ is k times the length of RS.

◉ When $\overrightarrow{PQ} = k\overrightarrow{PR}$ then the lines PQ and PR are parallel. But these lines have the point P in common so that PQ and PR are part of the same straight line. That is, the points P, Q and R lie on the same straight line.

Example 14   In triangle OAB the point M is the midpoint of OA and the point N is the midpoint of OB.

$\overrightarrow{OA} = 2\mathbf{a}$ $\overrightarrow{OB} = 2\mathbf{b}$

   i   Express $\overrightarrow{AB}$ in terms of **a** and **b**.

   ii   Express $\overrightarrow{MN}$ in terms of **a** and **b**.

   iii   Explain what the answers in **i** and **ii** show about AB and MN.

> Use the triangle law of vector addition.

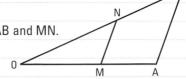

i $\overrightarrow{AB} = \overrightarrow{AO} + \overrightarrow{OB}$
$\overrightarrow{AB} = -2a + 2b$

⟵ $\overrightarrow{OA} = 2a$ so $\overrightarrow{AO} = -2a$.

ii $\overrightarrow{OM} = \frac{1}{2}2a = a$

⟵ M is the midpoint of OA so $\overrightarrow{OM} = \frac{1}{2}\overrightarrow{OA}$.

Similarly, $\overrightarrow{ON} = b$
$\overrightarrow{MN} = \overrightarrow{MO} + \overrightarrow{ON}$
$\overrightarrow{MN} = -a + b$

Use the triangle law of vector addition.

⟵ $\overrightarrow{OM} = a$ so $\overrightarrow{MO} = -a$

iii $\overrightarrow{AB} = 2\overrightarrow{MN}$

⟵ $\overrightarrow{AB} = -2a + 2b$ and $\overrightarrow{MN} = -a + b$

This means that AB and MN are parallel and that the length of AB is twice the length of MN.

Example 15 ▶ OABC is a quadrilateral in which
$\overrightarrow{OA} = a$, $\overrightarrow{OB} = a + 2b$ and $\overrightarrow{OC} = 4b$.

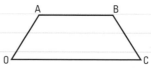

  i Find $\overrightarrow{AB}$ in terms of **a** and **b** and explain what this answer means.
  ii Find $\overrightarrow{CB}$ in terms of **a** and **b**.
  D is the point such that $\overrightarrow{BD} = \overrightarrow{OC}$, and X is the midpoint of BC.
  Find in terms of **a** and **b** iii $\overrightarrow{OD}$ iv $\overrightarrow{OX}$ and   v explain what these results mean.

i $\overrightarrow{AB}$ $= \overrightarrow{AO} + \overrightarrow{OB}$

⟵ Express $\overrightarrow{AB}$ in terms of known vectors using the triangle law of vector addition.

$\overrightarrow{AB}$ $= -a + a + 2b = 2b$

⟵ $\overrightarrow{OA} = a$ so $\overrightarrow{OA} = -a$.

$\overrightarrow{OC}$ $= 2\overrightarrow{AB}$

⟵ $\overrightarrow{OC} = 4b$

OC and AB are parallel and the length of OC is twice the length of AB.

ii $\overrightarrow{CB}$ $= \overrightarrow{CO} + \overrightarrow{OB}$

⟵ Express $\overrightarrow{CB}$ in terms of known vectors using the triangle law of vector addition.

$\overrightarrow{CB}$ $= -4b + a + 2b$
       $= a - 2b$

⟵ $\overrightarrow{OC} = 4b$ so $\overrightarrow{CO} = -4b$.

$\overrightarrow{CB} = \overrightarrow{CO} + \overrightarrow{OA} + \overrightarrow{AB}$
could also have been used.

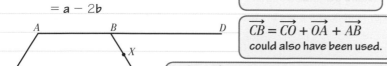

$\overrightarrow{BD} = \overrightarrow{OC}$ means that the point D is on AB extended so that BD and OC have the same length. Redraw the diagram with BD in and X the midpoint of BC.

iii $\overrightarrow{OD}$ $= \overrightarrow{OA} + \overrightarrow{AD}$

⟵ Use the triangle law of vector addition for $\overrightarrow{OD}$.

$\overrightarrow{OD}$ $= a + 6b$

⟵ $\overrightarrow{OA} = a$, $\overrightarrow{BD} = \overrightarrow{OC} = 4b$.

$\overrightarrow{AD} = \overrightarrow{AB} + \overrightarrow{BD} = 2b + 4b = 6b$.

iv As X is the midpoint of BC

$\overrightarrow{CX} = \frac{1}{2}\overrightarrow{OD}$ ← $\overrightarrow{CB} = \mathbf{a} - 2\mathbf{b}.$

$\phantom{\overrightarrow{CX}} = \frac{1}{2}(\mathbf{a} - 2\mathbf{b})$

$\overrightarrow{CX} = \frac{1}{2}\mathbf{a} - \mathbf{b}$

$\overrightarrow{OX} = \overrightarrow{OC} + \overrightarrow{CX} = 4\mathbf{b} + \frac{1}{2}\mathbf{a} - \mathbf{b}$ ← Use the triangle law of vector addition for $\overrightarrow{OX}$.

$\overrightarrow{OX} = \frac{1}{2}\mathbf{a} + 3\mathbf{b}$

v $\overrightarrow{OD} = 2\overrightarrow{OX}$ ← $\overrightarrow{OD} = \mathbf{a} + 6\mathbf{b}$

$\overrightarrow{OX} = \frac{1}{2}\mathbf{a} + 3\mathbf{b}$

So the lines OD and OX are parallel with the point O in common.

This means that OX and OD are part of the same straight line.

That is, OXD is a straight line such that the length of OD is 2 times the length of OX.

In other words, X is the midpoint of OD.

Exercise 18E

1 The points A, B and C have coordinates (2, 13), (5, 22) and (11, 40) respectively.

    a Find as column vectors   **i** $\overrightarrow{AB}$ **ii** $\overrightarrow{AC}$.

    b What do these results show about the points A, B and C?

2 In triangle OAB, $\overrightarrow{OA} = \mathbf{a}$ and $\overrightarrow{OB} = \mathbf{b}$.

    a Find in terms of **a** and **b** the vector $\overrightarrow{AB}$.

    P is the midpoint of AB.

    b Find in terms of **a** and **b** the vector $\overrightarrow{AP}$.

    c Find in terms of **a** and **b** the vector $\overrightarrow{OP}$.

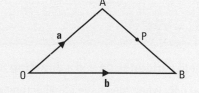

3 OACB is a parallelogram with $\overrightarrow{OA} = \mathbf{a}$ and $\overrightarrow{OB} = \mathbf{b}$.

    P is the midpoint of AB.

    a Use the result of question **3** to write down $\overrightarrow{OP}$ in terms of **a** and **b**.

    b Express $\overrightarrow{OC}$ in terms of **a** and **b**.

    Q is the midpoint of OC.

    c Express $\overrightarrow{QO}$ in terms of **a** and **b**.

    d What do your answers to **a** and **c** show about the points P and Q?

    e What property of a parallelogram has been proved in this question?

4 KLMN is a quadrilateral where $\overrightarrow{KL} = \mathbf{k}$, $\overrightarrow{LM} = \mathbf{m}$, $\overrightarrow{MN} = \mathbf{n}$ and $\overrightarrow{KN} = 3\mathbf{m}$.

    a What type of quadrilateral is KLMN?

    b Express **n** in terms of **k** and **m**.

A

A02
A03   A*

A02
A03

A02
A03

A*
AO2
AO3

5 OACB is a parallelogram with $\overrightarrow{OA} = $ **a** and $\overrightarrow{OB} = $ **b**.

E is the point on AC such that $AE = \frac{1}{4}AC$.

F is the point on BC such that $BF = \frac{1}{4}BC$.

a Find in terms of **a** and **b**.

i $\overrightarrow{AB}$ **ii** $\overrightarrow{AE}$ **iii** $\overrightarrow{OE}$ **iv** $\overrightarrow{OF}$ **v** $\overrightarrow{EF}$

b Write down two geometric properties connecting EF and AB.

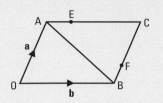

AO2
AO3

6 In triangle OMN, $\overrightarrow{OM} = $ **m** and $\overrightarrow{ON} = $ **n**.

The point P is the midpoint of MN and Q is the point such that $\overrightarrow{OQ} = \frac{3}{2}\overrightarrow{OP}$.

a Find in terms of **m** and **n**. **i** $\overrightarrow{OP}$ **ii** $\overrightarrow{OQ}$ **iii** $\overrightarrow{MQ}$

The point R is such that $\overrightarrow{OR} = 3\overrightarrow{ON}$.

b Find in terms of **m** and **n** the vector $\overrightarrow{MR}$.

c Explain why MQR is a straight line and give the value of $\dfrac{MR}{MQ}$.

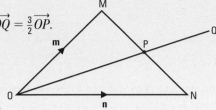

AO2
AO3

7 In the diagram $\overrightarrow{OR} = 6$**a**, $\overrightarrow{OP} = 2$**b** and $\overrightarrow{PQ} = 3$**a**.

The point M is on PQ such that $\overrightarrow{PM} = 2$**a**.

The point N is on OR such that $\overrightarrow{ON} = \frac{1}{3}\overrightarrow{OR}$.

The midpoint of MN is the point S.

a Find in terms of **a** and/or **b** the vector $\overrightarrow{NM}$.

b Find in terms of **a** and/or **b** the vector $\overrightarrow{OS}$.

T is the point such that $\overrightarrow{QT} = $ **a**.

c Find in terms of **a** and **b** the vector $\overrightarrow{OT}$.

d Give a geometric fact about the point S and the line OT.

e When **a** $= \begin{pmatrix} 8 \\ 2 \end{pmatrix}$ and **b** $= \begin{pmatrix} 3 \\ 15 \end{pmatrix}$ find the length of QR.

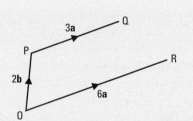

Chapter review

⦿ A vector needs both a **magnitude** and a **direction** to describe it completely.

⦿ Vectors are equal only when they have equal magnitudes and the same direction.

⦿ Vectors can be labelled with single bold letters such as **a**, **b** and **c**.

⦿ When hand-writing the vector **a** you can use <u>a</u> to represent it.

⦿ The **displacement** represented by **b** can be described as 4 to the right and 2 up.

This can be written as the column vector $\begin{pmatrix} 4 \\ 2 \end{pmatrix}$.

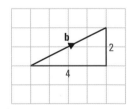

⦿ The magnitude of the vector **a** is written a or $|a|$.

⦿ The magnitude of the vector $\overrightarrow{AB}$ is AB, that is, the length of the line segment AB.

⦿ In general, the magnitude of the vector $\begin{pmatrix} x \\ y \end{pmatrix}$ is $\sqrt{x^2 + y^2}$.

⦿ $\overrightarrow{AB} + \overrightarrow{BC} = \overrightarrow{AC}$, or **a** + **b** = **c**.

This is the triangle law of vector addition.

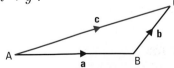

- PQRS is a parallelogram.

  $\overrightarrow{PQ} = \overrightarrow{SR} = \mathbf{a}, \overrightarrow{PS} = \overrightarrow{QR} = \mathbf{b}$.

  From the triangle law, $\overrightarrow{PQ} + \overrightarrow{QR} = \overrightarrow{PR}$ so that $\overrightarrow{PR} = \mathbf{a} + \mathbf{b}$.

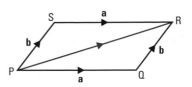

- So the diagonal $\overrightarrow{PR}$ of the parallelogram represents the vector $\mathbf{a} + \mathbf{b}$.

  This is the parallelogram law of vector addition.

- When $\mathbf{c} = \mathbf{a} + \mathbf{b}$ the vector $\mathbf{c}$ is said to be the **resultant** of the two vectors $\mathbf{a}$ and $\mathbf{b}$.

- $\begin{pmatrix} a \\ b \end{pmatrix} + \begin{pmatrix} c \\ d \end{pmatrix} = \begin{pmatrix} a + c \\ b + d \end{pmatrix}$

- $\overrightarrow{AB} + \overrightarrow{BA} = \mathbf{0}$ (the zero displacement).

- $\overrightarrow{BA} = -\overrightarrow{AB}$.

- A negative sign in front of a vector reverses the direction of the vector.

- For any non-zero value of k, the vectors $\mathbf{a}$ and $k\mathbf{a}$ are parallel.

  The number k is called a **scalar**; it has magnitude only.

  If $\mathbf{a} = \begin{pmatrix} p \\ q \end{pmatrix}$ then $k\mathbf{a} = k\begin{pmatrix} p \\ q \end{pmatrix} = \begin{pmatrix} kp \\ kq \end{pmatrix}$

- With the origin O, the vectors $\overrightarrow{OA}$ and $\overrightarrow{OB}$ are called the **position vectors** of the points A and B.

  In general, the point (p, q) has position vector $\begin{pmatrix} p \\ q \end{pmatrix}$.

- To solve geometric problems the following results are useful:
  - Triangle law of vector addition, so that $\overrightarrow{PQ} + \overrightarrow{QR} = \overrightarrow{PR}$.
  - When $\overrightarrow{PQ} = \mathbf{a}, \overrightarrow{QP} = -\mathbf{a}$.
  - When $\overrightarrow{PQ} = k\overrightarrow{RS}$, k is a scalar (number), the lines PQ and RS are parallel and the length of PQ is k times the length of RS.
  - When $\overrightarrow{PQ} = k\overrightarrow{PR}$ then the lines PQ and PR are parallel. But these lines have the point P in common so that PQ and PR are part of the same straight line. That is, the points P, Q and R lie on the same straight line.

Review exercise

1 The diagram shows two vectors $\mathbf{a}$ and $\mathbf{b}$.

$\overrightarrow{PQ} = \mathbf{a} + 2\mathbf{b}$

Copy the grid and draw the vector $\overrightarrow{PQ}$.

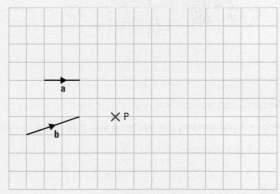

March 2005

2 **a** A is the point (1, 3). $\overrightarrow{AB} = \begin{pmatrix} 3 \\ 2 \end{pmatrix}$

Find the coordinates of B.

b C is the point (4, 3). BD is a diagonal of the parallelogram ABCD.

Express $\overrightarrow{BD}$ as a column vector.

c $\overrightarrow{CE} = \begin{pmatrix} 1 \\ -3 \end{pmatrix}$

Calculate the length of AE.

A

A02
A03

A★

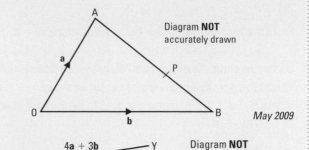

A★
AO2
AO3

3 OAB is a triangle.

$\overrightarrow{OA} = \mathbf{a}$ $\overrightarrow{OB} = \mathbf{b}$

a Find the vector $\overrightarrow{AB}$ in terms of **a** and **b**.

P is the point on AB such that AP : PB = 3 : 2.

b Show that $\overrightarrow{OP} = \frac{1}{5}(2\mathbf{a} + 3\mathbf{b})$.

Diagram **NOT** accurately drawn

May 2009

AO2
AO3

4 $\overrightarrow{OX} = 2\mathbf{a} + \mathbf{b}$ $\overrightarrow{OY} = 4\mathbf{a} + 3\mathbf{b}$

a Express the vector $\overrightarrow{XY}$ in terms of **a** and **b**.
Give your answer in its simplest form.

XYZ is a straight line.

XY : YZ = 2 : 3

b Express the vector $\overrightarrow{OZ}$ in terms of **a** and **b**.
Give your answer in its simplest form.

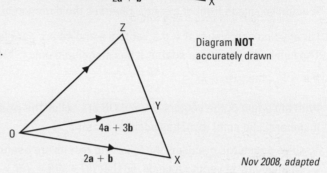

Diagram **NOT** accurately drawn

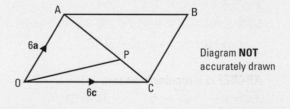

Diagram **NOT** accurately drawn

Nov 2008, adapted

AO2
AO3

5 **a** A is the point (1, 4) and B is the point (−3, 1).

i Write $\overrightarrow{AB}$ as a column vector.

ii Find the length of the vector $\overrightarrow{AB}$.

b D is the point such that $\overrightarrow{BD}$ is parallel to $\begin{pmatrix} 0 \\ 1 \end{pmatrix}$ and the length of $\overrightarrow{AD}$ = the length of $\overrightarrow{AB}$.

O is the point (0, 0).

Find $\overrightarrow{OD}$ as a column vector.

c C is the point such that ABCD is a rhombus.

AC is a diagonal of the rhombus.

Find the coordinates of C.

AO2
AO3

6 OABC is a parallelogram.

P is the point on AC such that AP = $\frac{2}{3}$AC.

$\overrightarrow{OA} = 6\mathbf{a}$ $\overrightarrow{OC} = 6\mathbf{c}$

a Find the vector $\overrightarrow{OP}$.

Give your answer in terms of **a** and **c**.

The midpoint of CB is M.

b Prove that OPM is a straight line.

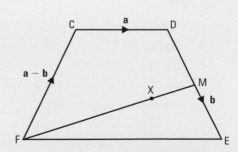

Diagram **NOT** accurately drawn

June 2004

AO2
AO3

7 CDEF is a quadrilateral with

$\overrightarrow{CD} = \mathbf{a}$, $\overrightarrow{DE} = \mathbf{b}$ and $\overrightarrow{FC} = \mathbf{a} - \mathbf{b}$.

a Express $\overrightarrow{CE}$ in terms of **a** and **b**.

b Prove that FE is parallel to CD.

M is the midpoint of DE.

c Express $\overrightarrow{FM}$ in terms of **a** and **b**.

X is the point on FM such that FX : XM = 4 : 1.

d Prove that C, X and E lie on the same straight line.

8 PQRS is a kite.

The diagonals PR and QS intersect at M.

$\overrightarrow{PM} = 4\mathbf{p}$ $\overrightarrow{QM} = \mathbf{q}$

$\overrightarrow{MR} = \mathbf{p}$ $\overrightarrow{QM} = \overrightarrow{MS}$

   a Find expressions in terms of **p** and/or **q** for

   i $\overrightarrow{PR}$

   ii $\overrightarrow{QS}$

   iii $\overrightarrow{PQ}$.

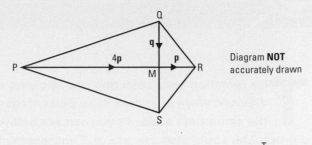

Diagram **NOT** accurately drawn

SR and PQ are extended to meet at point T.

Q is the midpoint of PT.

   b Find $\overrightarrow{RT}$ in terms of **p** and **q**.

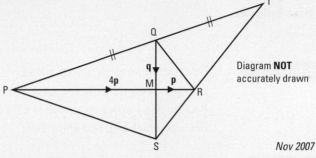

Diagram **NOT** accurately drawn

Nov 2007

9 OPQ is a triangle.

R is the midpoint of OP.

S is the midpoint of PQ.

$\overrightarrow{OP} = \mathbf{p}$ $\overrightarrow{OQ} = \mathbf{q}$

   a Find $\overrightarrow{OS}$ in terms of **p** and **q**.

   b Show that RS is parallel to OQ.

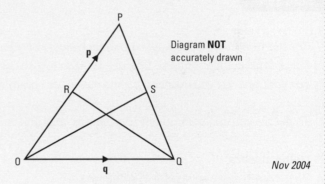

Diagram **NOT** accurately drawn

Nov 2004

10 OPQR is a trapezium with PQ parallel to OR.

$\overrightarrow{OP} = 2\mathbf{b}$ $\overrightarrow{PQ} = 2\mathbf{a}$ $\overrightarrow{OR} = 6\mathbf{a}$

M is the midpoint of PQ and N is the midpoint of OR.

   a Find the vector $\overrightarrow{MN}$ in terms of **a** and **b**.

   b X is the midpoint of MN and Y is the midpoint of QR.

     Prove that XY is parallel to OR.

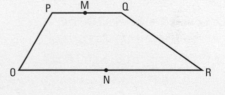

June 2005

11 ABCDEF is a regular hexagon.

$\overrightarrow{AB} = \mathbf{a}$ $\overrightarrow{BC} = \mathbf{b}$ $\overrightarrow{AD} = 2\mathbf{b}$

   a Find the vector $\overrightarrow{AC}$ in terms of **a** and **b**.

$\overrightarrow{AC} = \overrightarrow{CX}$

   b Prove that AB is parallel to DX.

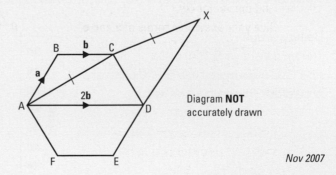

Diagram **NOT** accurately drawn

Nov 2007

PRICE COMPARISONS

This question tests selecting and applying a method (AO2) as there are a number of different ways of approaching percentages and ensuring both prices are compared over the same time scale. As you can see below, it can be solved using algebra or by drawing a graph. You could also use trial and improvement – try this method on one of the questions that follows.

Elsa is comparing electricity prices.

Energee has a standing charge of £65 per annum and a unit rate of 8.5p.

Powero has a standing charge of £116 per annum and charge 7.5p per unit.

Discuss which firm Elsa should use.

Solution

The best deal will depend upon how much electricity Elsa uses.

For a small number of units Energee is obviously cheaper, but for large amounts Powero will cost less.

For your answer you will need to find out after how many units Powero becomes the best option.

To decide you could use algebra, graphs or trial and improvement.

Using algebra

Write down expressions for the charges.

$$\text{Energee} = 65 + 0.085u$$
$$\text{Powero} = 116 + 0.075u$$
$$65 + 0.085u = 116 + 0.075u$$
$$u = 5100$$

> Find the value of u for which the prices are the same.

A customer using more than 5100 units is better off with Powero.

Using a graph

The lines meet at 5100 units, so if Elsa uses more than 5100 units she should use Powero.

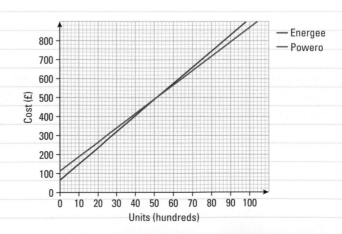

Now try these

* **1** Ahmed is trying to find the cheapest provider for gas in his area.

Cogas has a standing charge of £63 per annum with a charge of 3.5p per unit.

Ourgas has a standing charge of £120 per annum and charges 2p per unit.

Discuss which company Ahmed should use.

2 The cost of installing cable broadband is £30. The monthly cost of the contract is £5.

The monthly cost of broadband using a wireless router is £6.50 for a minimum period of 18 months.

The router is free and there is no installation charge.

    a Work out an expression in m for each company for the cost of broadband over m months.

    b Investigate which of the two broadband deals is cheaper.

3 The cost of hiring a car from ACARS and BMotors is shown in the advertisements below.

    a Work out an expression in x for the cost of hiring a car for a day and travelling x miles.

    b Investigate which of the two firms is cheaper.

ACARS	**BMotors**
£60 per day	£50 per day
32p per mile	40p per mile

* **4** Anna works in a small business and has to decide which courier her company should use when sending parcels.

Parcels Fly delivers parcels at a cost of £5.50 if they weigh less than 2 kg. For heavier parcels, it charges 85p per 250 g.

Quick Delivery charges £3 for the first kg then £1.90 per 500 g.

Investigate which company Anna should use.

* **5** Tom works for a builder and has to order the concrete for drives and paths.

His firm uses two different suppliers.

Pete's Mix sells concrete for £70 per cubic metre with a delivery charge of £80.

Concrete Sue sells concrete at £85 per cubic metre with a delivery charge of £30.

Investigate which firm he should use.

The following question tests your ability to analyse and interpret problems (AO3).

> **Example**

Here is a diagram of a perfume bottle.
The bottle is in the shape of a square-based pyramid.

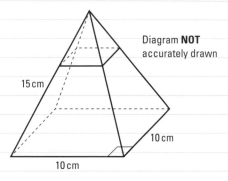

Diagram **NOT**
accurately drawn

15 cm

10 cm

10 cm

The lengths of the edges of the base are 10 cm.
The lengths of all the four slant edges of the pyramid are 15 cm.
The bottle is to be sold in a box in the shape of a cuboid.
Find the height of the smallest box that could be used.
Give your answer to 3 significant figures.

> **Solution**

The pyramid-shaped bottle must fit into a box. The minimum height of the box will be the height of the pyramid. You need to find the vertical height of a square-based pyramid given the base lengths and the slant heights.

Diagonal of base $= \sqrt{10^2 + 10^2} = 14.142$ cm

> The only sensible method of solution is to use Pythagoras' theorem.

> Firstly, find the diagonal of the square base.

Half the diagonal $= 7.071$ cm

> Then work out the height on a vertical right-angled triangle which has the slant edge of the pyramid as the hypotenuse.

Height $= \sqrt{15^2 - 7.071^2} = 13.2$ cm

Now try these

1   Jenny has a pencil tin in the shape of a cuboid.

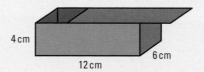

4 cm

12 cm

6 cm

The dimensions for the inside of the tin are 12 cm, 6 cm and 4 cm.
What is the length of the longest pencil that Jenny can fit into her tin?

2   Miriam has a stick that is 30 cm long.
She uses the stick to stir paint. She leaves the stick in the paint tin with
some of it sticking out at the top.
The tin has a diameter of 10 cm and height of 15 cm.
What is the shortest length of stick that could stick out of the tin?

3   Dave has a garden shed that is 6 ft long by 4 ft wide. Its walls are 6 ft high.
The tallest point of the roof is 7 ft from the ground.

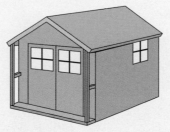

Dave wants to store some bean poles in the shed.
The poles are 9 ft long.
Explain, by showing your working, that it is possible to store the bean poles in the shed.

GOING ON HOLIDAY

FS

British people take more than 60 million holidays abroad each year. Of these, 75% are taken during the months of July and August, when many people travel to southern Europe and the Mediterranean. Some reward card companies allow people to collect points when they spend money in particular stores. You can use the points to pay for flights abroad.

QUESTION

1. Jared wants to book a return flight from London to Valetta in Malta. The number of points he needs is calculated using the formula:

$$p = \frac{d}{c^2}$$

p = points required
d = distance of flight (km)
c = class

Class

First class: $c=1$
Business class: $c=2$
Economy class: $c=3$

The distance of the journey is calculated in a straight line for the purpose of the reward points. Approximately how many points does he need to save up to travel in economy class?

2. The exchange rate is £1 = 1.12€. Jared has seen a camera in England priced at £475. The shop is offering a 10% discount on this camera. While in Malta, he sees the same camera priced at 420€, plus VAT at 15%. Will it be cheaper to buy the camera in Malta or back in England?

- For **Question 1** you need to be able to use your knowledge of scale drawings and maps from **Chapter 14**.
- You need to work out the percentage increase and decrease for **Question 2**, you learnt about them in **Chapter 3**.
- You learnt about bounds in **Chapter 2**, you will use this in **Question 3**.

3. Jared's plane uses 12 litres of fuel for every kilometre it flies. The pilot has chosen to carry an extra 5% of the fuel required, for safety reasons. The plane's fuel tank can be filled to the nearest 500 litres. What is the lowest bound for filling the tank?

FS

Ships navigate using bearings. They can also calculate their position according to their bearings from two known points. Trigonometry, loci and circle theorems can all be helpful tools when understanding and solving bearings problems.

1. A ship sails around an island from a port on the west coast to a harbour on the north shore. The harbour is 11 046 metres away on a bearing on 065°. There is a lighthouse on an outcrop of rocks on the north-west tip of the island, which is on a bearing of 042° from the port and at a distance of 6000 metres. To avoid the rocks, the ship must sail no closer than 720 metres from the lighthouse, passing to the north. The harbour is also 6000 metres from the lighthouse. Work out the total distance sailed.

2. Two coastguard stations are 8000 metres apart with one due east of the other. Simultaneously, they receive a call for help from a ship a sea. They are able to identify the direction from which the call is made but this is subject to a possible error of ±5° due to the fact that the ship is still moving and the accuracy of their equipment.

   One coastguard station estimates the bearing from which the call was made to be 065° whilst the other estimates the bearing to be 310°. Draw a scale diagram to identify the search area to which helicopters and lifeboats should be sent.

3. Distances at sea are normally given in nautical miles. A nautical mile is slightly longer than a mile. Ship's speeds are measured in knots: the number of nautical miles per hour.

At midday a ship's captain sees a radio mast that he knows is 8 nautical miles away on a bearing of 020°. The ship is sailing on a bearing of 045° at a speed of 15 knots. At what time will the ship be nearest to the radio mast?

LINKS

● For **Question 1** you need to be able to use bearings and Pythagoras' Theorem in your calculations. You learnt how to do this in **Chapter 14** and **Chapter 16**.

● You learnt how to draw scale diagrams in **Chapter 14**. You will need to use this for **Question 2**.

● For **Question 3** you need to be able to use bearings and Pythagoras' Theorem in your calculations. You learnt how to do this in **Chapter 14** and **Chapter 16**.

Answers

Chapter 1 Answers

1.1 Get Ready

a 20 000 **b** 100 **c** 49

Exercise 1A

1 a 31 **b** 6.4 **c** 17 **d** 1.5
  e 32
2 a 39.36 **b** 32.65 **c** 5.76 **d** 155.125
3 a 219.5 **b** 305.7 **c** 22.6 **d** 410.9
4 a 5.17 **b** 5.34 **c** 3.16 **d** 1.67
5 a 2.77 **b** 7.68 **c** 205 **d** 455 000
6 a 0.917 **b** 1.08 **c** 8.67 **d** 15.8

Exercise 1B

1 a 0.25 **b** 1.6 **c** 0.156 25 **d** $2^3 = 8$

1.2 Get Ready

a $\frac{53}{63}$ **b** £400 **c** 19.278

Exercise 1C

1 £280 **2** £47.98 **3** £4.20
4 6 **5** £147.80 **6** £65.05
7 £224.74

Exercise 1D

1 £3.24
2 a 4 days **b** 5 days
3 2 hours **4** 9 hours **5** 9 days
6 8 **7** 70 cm

1.3 Get Ready

1 a terminating **b** recurring **c** recurring

Exercise 1E

1 $\frac{7}{9}$ **2** $\frac{34}{99}$ **3** $\frac{305}{333}$
4 $\frac{2}{11}$ **5** $\frac{317}{999}$ **6** $\frac{1}{18}$
7 $\frac{323}{990}$ **8** $\frac{347}{495}$ **9** $\frac{7}{30}$
10 $6\frac{83}{99}$ **11** $2\frac{7}{66}$ **12** $7\frac{317}{900}$

1.4 Get Ready

1 a 4×10^4 **b** 7×10^{-4} **c** 5.67×10^4 **d** 5.03×10^{-1}
2 a 2000 **b** 90 000 **c** 840 000 **d** 0.003 8
3 a 10^{15} **b** 10^1

Exercise 1F

1 a 5×10^3 **b** 3×10^{-3} **c** 8×10^{-4} **d** 5×10
  d 6×10^{-1} **f** 5×10 **g** 4×10^2 **h** 1×10^{-4}
2 a 3.6×10^8 **b** 3.2×10^8 **c** 3.5×10^3 **d** 1×10^{14}
3 a 1000 **b** 9×10^{-8} **c** 500

4 9.3×10^7, 3×10^5
  Time $= (9.3 \times 10^7 \times 1.6) \div (3 \times 10^5) = 496$ seconds
5 160 m

Exercise 1G

1 a 2.1×10^8 **b** 2.4×10^{-5} **c** 5.2×10^0
  d 4.416×10^{-3} **e** 2.684×10^4 **f** 9.84×10^3
  g 9.6×10^{13} **h** 9.84×10^{-17} **i** 9.6×10^{-15}
2 a 5.38×10^4 **b** 6.20×10^{14} **c** 5.38×10^{16}
  d 6.20×10^{-6}
3 a 7.45×10^3 **b** 1.36×10^{19} **c** 4.24×10^5
  d 3.28×10^{12}
4 a 6×10^{-5} **b** 1.99×10^{-5} **c** 1.6×10^{-5}
5 1 : 30.1
6 6.28×10^{24}

Review exercise

1 a $0.08\dot{3}$ **b** $2.\dot{6}$ **c** 0.4 **d** $0.\dot{3}$
2 621
3 225
4 2.33 pm or 14:33
5 1.258 048 316
6 75 cm
7 1.4×10^{10}
8 a $\frac{4}{9}$ **b** $\frac{1}{6}$ **c** $\frac{3}{11}$ **d** $\frac{24}{77}$
9 a 2100 **b** 1225 **c** 6×10^{-4}
10 9.43×10^{12}
11 5.8×10^{-4}
12 a 1.44×10^6 **b** 1700
13 a 5.72×10^6 **b** 1.4×10^{-7}

Chapter 2 Answers

2.1 Get Ready

1 a 6.1 **b** 7.0 **c** 6.5 **d** 6.5
2 a 0.3 **b** 0.3 **c** 0.3 **d** 0.3

Exercise 2A

1 a 84.5, 83.5 **b** 84.05, 83.95 **c** 84.005, 83.995
2 a 0.95, 0.85 **b** 0.905, 0.895 **c** 0.095, 0.085
3 a 118.5 cm **b** 117.5 cm
4 a 6450 g **b** 6350 g
5 a 48.05 l **b** 47.95 l
6 a 1.005 m **b** 0.995 m

2.2 Get Ready

1 10.1
2 2.1
3 a e and f **b** c and f **c** e and f **d** c and f

Exercise 2B

1 a 535 **b** 26 612.25 **c** 229 920.25
2 a 6.88 **b** 10.823 575 **c** 24.35

3 a 11 275 568.625 **b** 151 474.75
4 a 11 **b** 1.3606…
5 a 0.8 **b** 1.3137…
6 a −31.3375 **b** −29.8275 **c** −30 (1 s.f.)
7 2.05×10^{17} (3 s.f.), 2.28×10^{17} (3 s.f.)
8 a 50 **b** 15.5%

Review exercise

1 a 17.1 **b** 31.95 **c** 60.1425 **d** 24.5025
2 a 200 **b** 645 **c** 1.8163… **d** 0.0638…
3 a 7.45 **b** 8.05 **c** 6.435 **d** 465
 e 3455
4 a 8.25 **b** 9.95 **c** 7.995 **d** 44.5
 e 3995
5 lower bound for length is 199.5 cm, so rod may fit into slot of length 199.8 cm
6 8.75 km/l
7 LB of cylinder's capacity = 325 ml, so the cylinder always contains more than stated on the label.
8 15 cm
9 14.2 m

Chapter 3 Answers

3.1 Get Ready

1 £120 **2** £30 **3** $\frac{2}{5}$

Exercise 3A

1 a 1.64 **b** 1.03 **c** 1.14 **d** 1.4
 e 1.134 **f** 1.125 **g** 1.15 **h** 1.0236
2 a 1.4 **b** £21.56
3 Helen £12 504, Tom £25 008, Sandeep £33 344
4 £621
5 a £144 **b** 70 kg **c** 2.784 m
 d £1370.20 **e** 128.52 cm

Exercise 3B

1 a 0.93 **b** 0.8 **c** 0.84 **d** 0.73
 e 0.944 **f** 0.975 **g** 0.9275 **h** 0.992
2 a £255 **b** £34 **c** £1020
3 77.9 kg
4 £748
5 a £5840 **b** £4672

3.2 Get Ready

1 £12.75 **2** £1300 **3** $\frac{9}{40}$

Exercise 3C

1 a 50% **b** 25% **c** 40% **d** 20%
 e 25% **f** 25% **g** 60% **h** 12.5%
2 90%
3 45%
4 a 20% **b** 60% **c** 11.25% **d** 8.75%

Exercise 3D

1 a +50% **b** +60% **c** −12% **d** −8%
2 15%

3 40%
4 Shop C – as percentage increases are A 5.19%, B 4.77%, C 5.50%
5 20% profit
6 2.7%

3.3 Get Ready

1 2^5 **2** 0.3 **3** 1.25

Exercise 3E

1 a 1.728 **b** 0.6561 **c** 1.0608 **d** 0.52
2 £1102.50
3 a 1.1136 **b** £66 816
4 No. It is the same as an increase of 68%.
5 8 years

3.4 Get Ready

1 1.15 **2** 0.85 **3** 1.04 **4** 0.96

Exercise 3F

1 £24 000 **2** £280 **3** £620
4 £180 **5** 421 000 **6** £270

Review exercise

1 18 years
2 CompuSystems (Able £23 000, Beta £23 400, CompuSystems £24,240, Digital £24 000)
3 72%
4 a 62.5% **b** $\frac{1}{4}$
5 19.9%
6 a It will be worth 32.8% of its original value.
 b 0.64
7 £275
8 £8400
9 £1600
10 £665
11 B (A 9.2% over two years, B 9.2025% over two years)

Chapter 4 Answers

4.1 Get Ready

1 a $4 + 2p + 5q$ **b** $-3 + 6z$ **c** $12m^2 + 36m$

Exercise 4A

1 $a = 4$ **2** $b = 7$
3 $c = 2.5$ **4** $d = 1.8$
5 $e = -1.5$ **6** $f = -4$
7 $g = 6$ **8** $h = 0.5$
9 $k = -2.6$ **10** $m = -3.5$

4.2 Get Ready

1 a $x = 6$ **b** $b = 4$ **c** $q = 8$

Exercise 4B

1 $x = 3$ **2** $y = 3$
3 $x = 2.25$ **4** $y = 2$

5 $x = 0.25$

7 $z = -0.25$

9 $x = 1$

6 $w = -4$

8 $x = -\frac{1}{9}$

10 $y = \frac{13}{3}$

4.3 Get Ready

1 a $x = 8$ **b** $a = 2$ **c** $b = 2$

Exercise 4C

1 $a = 2.5$

3 $c = 3$

5 $e = 1$

7 $x = 15$

9 $x = -3$

2 $b = -2.5$

4 $d = -2$

6 $f = -\frac{9}{10} = 0.9$

8 $x = -2$

10 $x = \frac{3}{5} = 0.6$

4.4 Get Ready

1 a $x = 2$ **b** $x = -\frac{6}{7}$ **c** $x = 0.9$

Exercise 4D

1 $p = 20$

3 $m = 30$

5 $y = \frac{53}{8}$

7 $n = \frac{5}{42}$

9 $x = 6.5$

2 $q = 10$

4 $x = 24$

6 $x = 33$

8 $t = -44$

10 $y = \frac{76}{43}$

4.5 Get Ready

1 a $a = -2$ **b** $b = -\frac{1}{2}$ **c** $x = 11\frac{2}{3}$

Exercise 4E

1 $x = 5$

2 31

3 Jessica 80%, Mason 60%, Zach 70%

4 $x = 8$

5 $23\frac{1}{4}$ hours

6 a $(x - 4) = \dfrac{3(x + 6)}{5}$ **b** 19 cm

7 149 units

Review exercise

1 $t = 4.5$

2 $x = 4.5$

3 a $x = 2.5$ **b** $y = -2.5$

4 58°

5 57 cm

6 Uzma £18, Hajra £38, Mabintou £76

7 A £8, B £12, C £4

8 $x = \frac{44}{3} = 14\frac{2}{3}$

9 $x = 5.5$

10 $y = 10$

Chapter 5 Answers

5.1 Get Ready

1 a 2.8   **b** 0.4   **c** −0.8   **d** −2.6

Exercise 5A

1 a

b

c

d

e

2 a $x \leqslant 4$ **b** $x > -1$ **c** $x \leqslant 5$ and $x > -2$

   d $x < 0$ and $x > -3$ **e** $x \leqslant 3$ and $x \geqslant -5$

   f $x < 5$ and $x \geqslant 1$

5.2 Get Ready

1 $x = 10$

2 $x = \frac{1}{2}$

3

4

Exercise 5B

1 a $x > 4$

    b $x \leqslant 1$

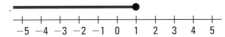

    c $x \leqslant -2$

    d $x > 1.6$

2 a $x < 4.5$ **b** $x > 4$ **c** $x \leqslant 2.5$ **d** $x > 6.5$

3 a $x \leqslant 3.25$ **b** $x > 23$ **c** $x \geqslant -\frac{6}{11}$ **d** $x \geqslant -\frac{1}{19}$

5.3 Get Ready

1 $x \geqslant 4$ **2** $x > 4$ **3** $x \leqslant 2$ **4** $x > 3$

Exercise 5C

1 a $-1, 0, 1, 2, 3, 4, 5$ **b** $-4, -3, -2, -1, 0, 1$
   c $1, 2, 3$ **d** $-5, -4, -3, -2, -1, 0, 1, 2, 3, 4$
2 a $-3, -2, -1, 0, 1, 2, 3$
   b $-4, -3, -2, -1, 0, 1, 2, 3, 4, 5, 6, 7$
   c $0, 1, 2, 3, 4$
   d $-3, -2, -1, 0, 1, 2, 3, 4, 5, 6, 7, 8, 9$
3 a $-2, -1, 0, 1, 2, 3$
   b $-1, 0, 1, 2, 3, 4$
   c $-1, 0, 1, 2, 3, 4$
   d $-7, -6, -5, -4, -3, -2, -1, 0, 1, 2, 3, 4$
4 a $-2, -1, 0, 1, 2, 3, 4, 5, 6$
   b $-4, -3, -2, -1, 0, 1, 2, 3, 4, 5, 6, 7$
   c $0, 1, 2, 3$
   d $-4, -3, -2, -1, 0, 1, 2, 3, 4, 5, 6, 7, 8, 9, 10$

3

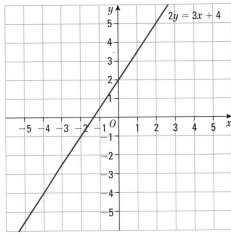

5.4 Get Ready

1

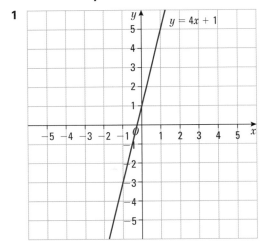

2

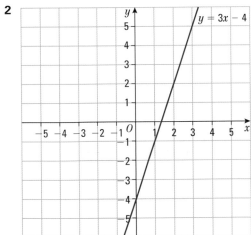

Exercise 5D

1 $(1, 1), (1, 0), (1, -1), (0, 0), (0, -1), (-1, -1)$

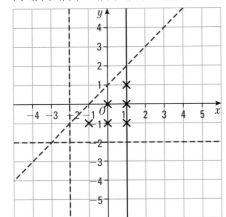

2 a

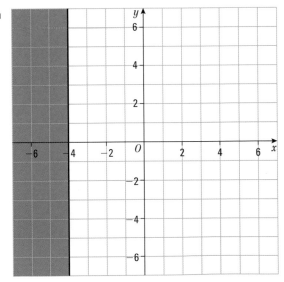

Answers

b

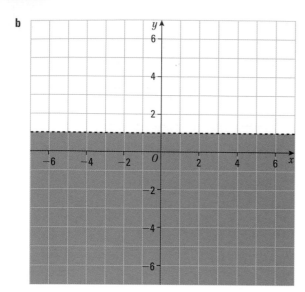

3 a

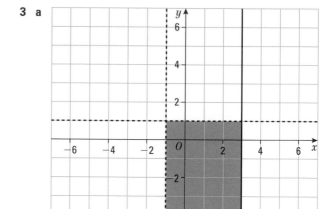

c

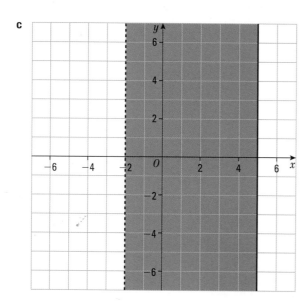

b

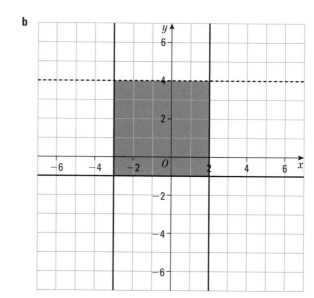

d

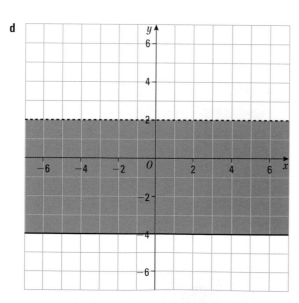

c

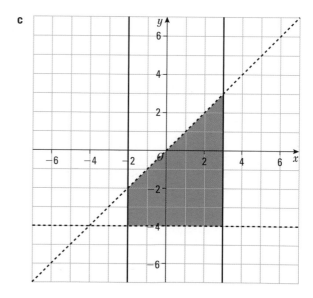

d

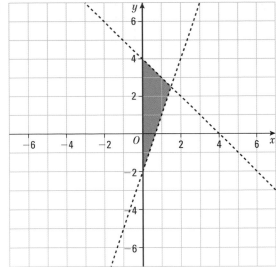

4 **a** $y = 3, y = 2x + 3, x + y = -1$
 b $y < 3, y \geqslant 2x + 3, x + y > -1$
 c $(-2, 2), (-1, 2)$ and $(-1, 1)$

5 **a** $y = 2, x = 3, y = \frac{1}{2}x - 3, y = -x$
 b $y \leqslant 2, x < 3, y > \frac{1}{2}x - 3, y \geqslant -x$
 c -1

5.5 Get Ready

1 60 miles   **2** 1.8×10^{17}

Exercise 5E

1 $x = \dfrac{y - 3}{5}$ **2** $d = \dfrac{c + 2}{5}$

3 $a = \dfrac{v - u}{t}$ **4** $x = \dfrac{P - y}{5y}$

5 $m = \dfrac{4 - E}{3}$ **6** $g = \dfrac{f + 30}{3}$

7 $x = 7T - 2$ **8** $n = \dfrac{2Y + 3m}{3}$

9 $p = \dfrac{3W - 2y}{2y}$ **10** $w = 4 - 3A$

5.6 Get Ready

1 $u = v - at$ **2** $R = \dfrac{V}{I}$ **3** $m = \dfrac{E}{c^2}$

Exercise 5F

1 $R = \sqrt{\dfrac{A}{\pi}}$ **2** $x = P^2 + y$

3 $u = \sqrt{v^2 - 2as}$ **4** $g = \dfrac{2s}{T^2}$

5 $x = \sqrt{\dfrac{75 - y^2}{50}}$ **6** $g = \dfrac{f + 4}{4}$

7 $m = \dfrac{1}{T - 3}$ **8** $p = \dfrac{q - 12}{q + 4}$

9 $c = \dfrac{a - 3}{b + 7}$ **10** $T = \dfrac{7}{2W^2 - 3}$

Review exercise

1 **a** 8 **b** $p = \dfrac{S - 3q}{4}$
2 **a** $1 < x < 4$ **b** $-1 < x \leqslant 5$ **c** $x < 2$
3 **a**

 b
 c
 d

4 $-3, -2, -1, 0, 1$
5 **a** $t < 5.5$ **b** 5
6 $-3, -2, -1, 0, 1, 2$
7 $p < 5$
8 $0 < x < 4$
9 **a** $L = x + (x + 4) + 2(x + 4) = x + x + 4 + 2x + 8$
     $= 4x + 12$
 b $4x + 12 < 50$ **c** $0 < x < 9.5$

10

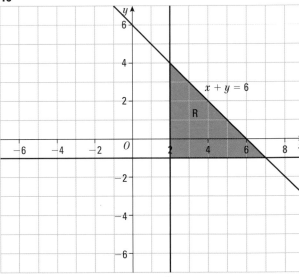

11 **a** $\dfrac{10}{7}$ **b** $u = \dfrac{vf}{v - f}$
12 $x = \dfrac{4y + 15}{5 + 3y}$
13 **a** 49.6 **b** $r = \dfrac{P - 2a}{\pi + 2}$
14 **a** Any two of: $(1, 1), (1, 2), (2, 1)$
 b

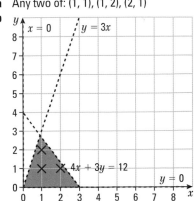

15 a $x \geqslant 2, y \geqslant 3, 2x + 3y < 20$

b

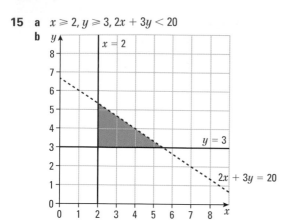

Chapter 6 Answers

6.1 Get Ready

1

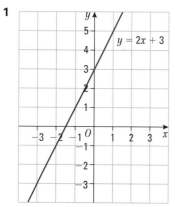

2 a -4 **b** 4

3 a 2   **b** 18

Exercise 6A

1 a

x	-3	-2	-1	0	1	2	3
y	6	1	-2	-3	-2	1	6

b

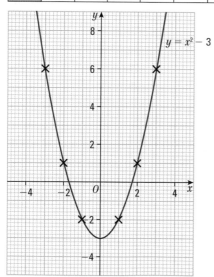

c $x = 0$ **d** $(0, -3)$

2 a

x	-3	-2	-1	0	1	2	3
y	-5	0	3	4	3	0	-5

b

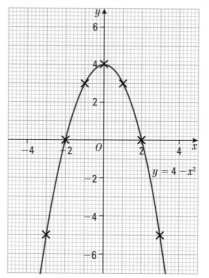

c $(0, 4)$ **d** $x = 2$ and $x = -2$

3 a

x	-3	-2	-1	0	1	2	3
y	20	10	4	2	4	10	20

b

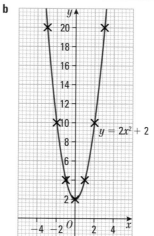

c i $y = 6.5$ **ii** $x = 2.1$ and $x = -2.1$

4 a

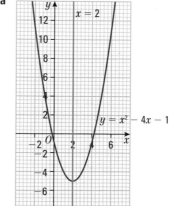

i $x = 4.2$ and $x = -0.2$ **ii** $x = 2$

b

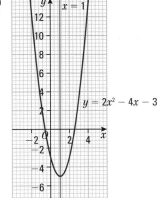

$y = 2x^2 - 4x - 3$

i $x = 2.6$ and $x = -0.6$ **ii** $x = 1$

c

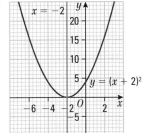

$y = (x + 2)^2$

i touches the x-axis at -2 **ii** $x = -2$

d

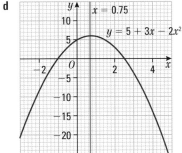

$x = 0.75$

$y = 5 + 3x - 2x^2$

i $x = 2.5$ and $x = -1$ **ii** $x = 0.75$

Exercise 6B

1 a $x = 0$ and $x = 3$ **b** $x = -1$ and $x = 4$
    c $x = 2.8$ and $x = -1.3$ **d** $x = 3.6$ and $x = -1.6$

2 a $x = 0.4$ and $x = 2.6$ **b** $x = 0.4$ and $x = 2.6$
    c $x = 3.1$ and $x = -1.6$ **d** $x = 3.2$ and $x = -1.2$

3 a

x	-2	-1	0	1	2	3	4
y	-7	-2	1	2	1	-2	-7

b

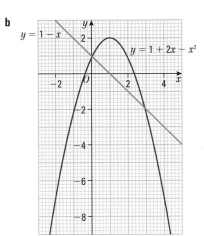

$y = 1 - x$

$y = 1 + 2x - x^2$

c $x = 0$ and $x = 3$

4 a

x	-3	-2	-1	0	1	2	3
y	32	16	6	2	4	12	26

b

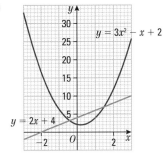

$y = 3x^2 - x + 2$

$y = 2x + 4$

c $x = 1.5$ and $x = -0.5$

6.2 Get Ready

1 1, 8, 27, 64, 125
2 a 1 000 000   **b** -1000

Exercise 6C

1 a

x	-3	-2	-1	0	1	2	3
y	-25	-6	1	2	3	10	29

b

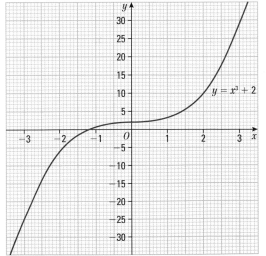

$y = x^3 + 2$

c $y = 17.6$

Answers

2 a

x	−4	−3	−2	−1	0	1	2	3	4
y	−28	0	10	8	0	−8	−10	0	28

b

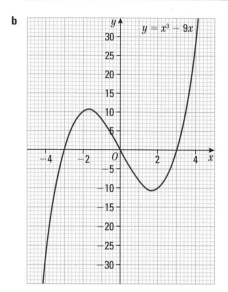

c $x = -3$, $x = 0$ and $x = 3$

3 a

x	−3	−2	−1	0	1	2	3	4
y	+45	+4	−7	0	+13	+20	+9	−32

b

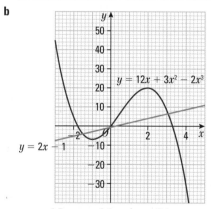

c $x = -1.5$, $x = -0.1$ and $x = 3.1$

4 i C   **ii** A   **iii** D   **iv** B

6.3 Get Ready

1 a $\frac{1}{4}$ **b** 4   **c** 0.4   **d** 2.5

2 As the value of x gets bigger, the value of $\frac{1}{x}$ gets smaller.

Exercise 6D

1 a

x	0.2	0.4	0.5	1	2	4	5	10	20
y	25	12.5	10	5	2.5	1.25	1	0.5	0.25

b

x	−20	−10	−5	−4	−2	−1	−0.5	−0.4	−0.2
y	−0.25	−0.5	−1	−1.25	−2.5	−5	−10	−12.5	−25

c

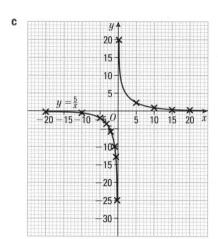

2

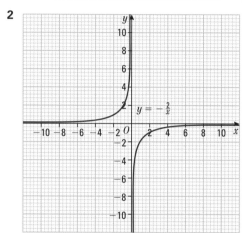

3 a

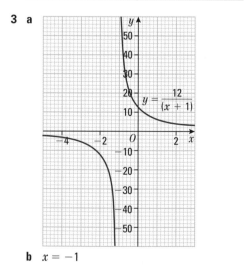

b $x = -1$

6.4 Get Ready

1 a 81   **b** 1   **c** $\frac{1}{9}$

2 a $x = 4$ **b** $x = 2$ **c** $x = 3$

Exercise 6E

1 a

x	−3	−2	−1	0	1	2	3
y	0.04	0.111	0.33	1	3	9	27

b

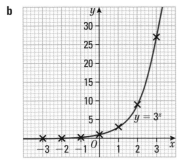

c i 5.2　　**ii** 2.5

2　A is $y = \left(\frac{1}{2}\right)^x$, B is $y = 3^{-x}$, C is $y = 5^x$, D is $y = 2^x$

3　**a** 10　　　**b** 320　　　**c** 17 minutes

4　$p = 0.625$, $q = 16$

6.5 Get Ready

1　**a** 4.6　　　**b** 2.2　　　**c** 1.5

2　When $x = 1$, $x + \dfrac{1}{x} = 1 + \dfrac{1}{1} = 2$

3　**a** -2　　　**b** 2

Exercise 6F

1　**a** 1 and 2　　**b** 0 and 1　　**c** 0 and 1, -1 and 0, 4 and 5
　　d 2 and 3

2　**a** 1.7　　　**b** -1.3　　　**c** 4.8 and 0.2

3　**a** 3.04　　　**b** 2.17　　　**c** 4.72

4　**a** $x(x + 2)(x - 2) = x^3 - 4x$　　**b** $x = 8.6$
　　c 10.6 cm, 6.6 cm, 8.6 cm

Review exercise

1　**a**

t	0	2	4	6	8	10
s	0	49	116	201	304	425

b

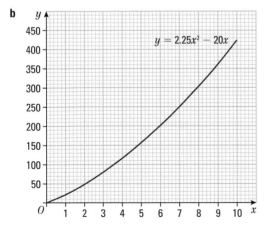

c i 80.25 m　　**ii** 9.6 seconds

2　**a**

x	-4	-3	-2	-1	0	1	2
y	8	3	0	-1	0	3	8

b

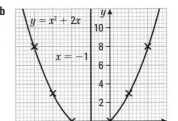

c $x = -1$

d i $y = 1.25$　　**ii** $y = 1.6$ and $y = -3.6$

3　**a**

x	-3	-2	-1	0	1	2	3
y	-10	-4	0	2	2	0	-4

b

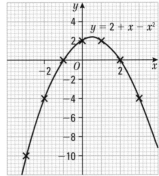

c i $x = -1$ and $x = 2$　　**ii** 2.8 and -1.8

d (0.5, 2.25)

4　**a** $x^2 - 3x - 2 = x - 2$
　　$x^2 - 3x - 2 - x + 2 = 0$
　　$x^2 - 4x = 0$

　　b $x = 0$ and $x = 4$　　　**c** $y = 2 - x$

5　**a** When $x = -0.5$, $x^2 + \dfrac{2}{x} + 3 = -0.75$

　　When $x = -1$, $x^2 + \dfrac{2}{x} + 3 = 2$

　　So there must be a solution to $x^2 + \dfrac{2}{x} + 3 = 0$
　　between $x = -0.5$ and $x = -1$.

　　b 0.60

6　**a**

x	-2	-1	0	1	2	3	4
y	-8	1	0	-5	-8	-3	16

b

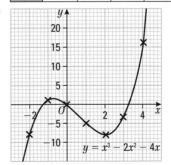

c i $x = 0$, $x = 3.2$ and $x = -1.2$
　　ii $x = -1.8$, $x = 1$ and $x = 2.8$

7 a

x	−3	−2	−1	−0.5	−0.1	0.1	1	2	3
y	3.7	4	5	7	23	−17	1	2	2.3

b

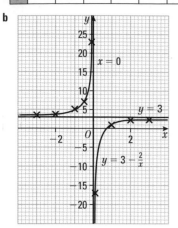

c $x = 0, y = 3$

8 a

x	0.5	1	2	3	4
y	8	4	2	1.3	1

b, c

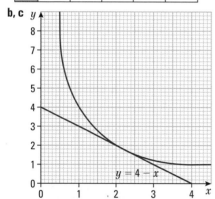

d $x = 2$

9 a

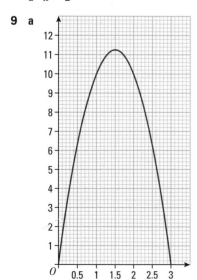

b 1.5 seconds
c 2.85 seconds

10 a

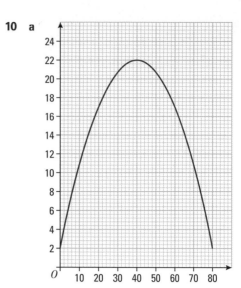

b i 22 metres
ii 9 metres and 71 metres
iii 82 metres

11 a

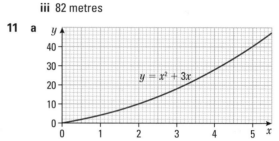

b 3.2 cm
12 7.63 cm
13 a C   **b** D   **c** A   **d** E   **e** B
14 $k = 0.25$

Chapter 7 Answers

7.1 Get Ready

1 a $x = -1.5$ **b** $y = 4$
2 a -4 **b** 21

Exercise 7A

1 $x = 4, y = 1$ **2** $x = 5, y = 3$
3 $x = 1, y = -2$ **4** $x = 2, y = 1$
5 $x = 1, y = 4$ **6** $x = 1, y = 2$
7 $x = -1, y = -1$ **8** $x = -2, y = -1$
9 $x = 0.5, y = 2$ **10** $x = 0.5, y = -4$
11 $x = 3, y = 1$ **12** $x = 2, y = -0.5$
13 $x = -3, y = 5$ **14** $x = 1, y = -3$
15 $x = 0.5, y = -1$

7.2 Get Ready

1 $x = 2, y = 3$
2 $x = 1, y = 2$
3 $x = 3, y = 1$

Exercise 7B

1 7, 12
2 £18.50
3 £8.80
4 Adult = £4.50, child = £3
5 14 g
6 **a** $4a = 3b + 1$ or $4a - 3b = 1$
     $2a - 3 = 5 - b$ or $2a + b = 8$
   b $a = 2.5\,\text{cm}, b = 3\,\text{cm}$
7 Ann = £9/day, Mary = £12/day
8 $x = 3, y = 2$

7.3 Get Ready

1 $4a = 3b + 4$
   $3a = 24 - 3b$
2 $a = 4, b = 4$
3 56

Exercise 7C

1 **a** B $y = 2x$, A $x + y = 3$, C $x - 2y = 3$
   b **i** $x = 1, y = 2$ **ii** $x = 3, y = 0$ **iii** $x = -1, y = -2$
2 **a** $x = 2, y = 0$ **b** $x = 3, y = 3$ **c** $x = 2, y = 8$

7.4 Get Ready

1 **a** $x(x - 4)$ **b** $y(2y + 5)$ **c** $(2x - 1)(x + 3)$
   d $(3y + 4)(2y - 5)$

Exercise 7D

1 **a** 0, 4   **b** 3, −5   **c** 0.5, 2.25
   d 0, −2   **e** 0, 1   **f** 0, 1.75
2 **a** 2, 4   **b** −1, −6   **c** 3, −4
   d 3   **e** 9, −4   **f** 4, −4
   g −5   **h** 10, −10
3 **a** $-5, -\frac{1}{5}$ **b** $3, \frac{2}{3}$ **c** $\frac{1}{2}, -4$
   d $-3, \frac{1}{5}$
4 **a** 3, −2   **b** 5, −2   **c** 7, −3
   d 7, −5   **e** 3, −2.5   **f** $4\frac{1}{2}, -\frac{2}{3}$
   g 8, −1   **h** $1, -\frac{9}{16}$

7.5 Get Ready

1 $x^2 + 6x + 9$
2 $x^2 - 10x + 25$
3 $x^2 + 2ax + a^2$

Exercise 7E

1 **a** $(x + 2)^2 - 4$ **b** $(x + 5)^2 - 25$
   c $(x + 6)^2 - 36$ **d** $(x - 1)^2 - 1$
   e $(x - 7)^2 - 49$ **f** $(x - 12)^2 - 144$
   g $(x + 0.5)^2 - 0.25$ **h** $(x - 1.5)^2 - 2.25$
   i $(x + 2)^2 + 3$ **j** $(x + 4)^2 + 1$
   k $(x + 5)^2 - 45$ **l** $(x - 3)^2 + 2$
   m $(x - 10)^2 - 20$ **n** $(x - 13)^2 - 170$
   o $(x - 0.5)^2 + 1.25$ **p** $(x + 2.5)^2 - 11.25$
2 **a** $2(x + 3)^2 - 18$ **b** $2(x - 1)^2 + 3$
   c $3(x - 2)^2 - 2$ **d** $5(x + 5)^2 - 25$
3 **a** $p = 4, q = 8$ **b** 8

4 **a** (0, 10)   **b** (−3, 1)
5 **a** $r = 5, s = 2$ **b** 5   **c** $x = 2$

7.6 Get Ready

1 $x = 2, x = -2$
2 $x = 3, x = -2$
3 $x = 7, x = -4$
4 **a** $2\sqrt{3}$ **b** $4\sqrt{2}$

Exercise 7F

1 **a** $x = 3 \pm \sqrt{11}$ **b** $x = -2 \pm \sqrt{3}$
   c $x = -5 \pm \sqrt{37}$ **d** $x = 1 \pm 2\sqrt{2}$
   e $x = \dfrac{3 \pm \sqrt{15}}{2}$ **f** $x = \dfrac{-6 \pm \sqrt{21}}{5}$
2 **a** $x = -0.68, x = -7.32$ **b** $x = 8.27, x = 0.73$
   c $x = 2.37, x = -3.37$ **d** $x = 0.87, x = -2.87$
   e $x = 0.88, x = -0.38$ **f** $x = 0.93, x = -0.43$

7.7 Get Ready

1 $x = \pm\sqrt{14} - 2$
   $x = 1.74, x = -5.74$
2 $x = \pm\sqrt{14} - 3$
   $x = 0.74, x = -6.74$
3 $x = \pm\sqrt{8} + 1$
   $x = 3.83, x = -1.83$
4 **a** 8   **b** 7   **c** 149

Exercise 7G

1 −0.586, −3.41   **2** −0.807, −6.19
3 0.606, −6.61   **4** 2.70, −3.70
5 5.32, −1.32   **6** 4.30, 0.697
7 −0.293, −1.71   **8** 1.54, 0.260
9 1.64, −0.811   **10** 0.322, −0.622
11 1.39, 0.360   **12** 3.45, −1.45
13 0.372, −5.37   **14** 5.11, −4.11
15 1.72, −1.52

7.8 Get Ready

1 $2x^2 + 9x - 18$
2 $6x^2 - 18x + 12$
3 $2x + 12$

Exercise 7H

1 $x = 0, x = 5$ **2** $x = 2, x = -1.8$
3 $x = 1.5, x = -2$ **4** $x = 1, x = -\frac{5}{9}$
5 $x = 0.5, x = -0.6$ **6** $x = 2, x = -0.75$
7 $x = 2.30, x = -1.30$ **8** $x = 4.48, x = -3.73$
9 $x = 5.32, x = -1.32$ **10** $x = 3.27, x = -4.27$
11 $x = -0.576, x = -10.4$ **12** $x = 3.41, x = 0.586$

7.9 Get Ready

1 $4x - 2x^2$
2 $x^2 + 2x + 1$
3 $8 + 10x - 3x^2$

Answers

Exercise 7I

1. 4, −6
2. 4
3. 12 and 48
4. 7.22 m
5. **a** $x^2 + (x + 1)^2 = 41$
   $2x^2 + 2x − 40 = 0$
   $x^2 + x − 20 = 0$
   b 4, 5 or −4, −5
6. 12.35 cm, 15.72 cm and 20 cm
7. 10 cm by 10 cm
8. 14 m or 16 m
9. $x = 5$
10. 54 km/h

7.10 Get Ready

1. **a**, **c**, **d** and **e**
2, 3

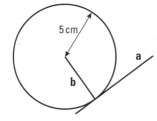

Exercise 7J

1.

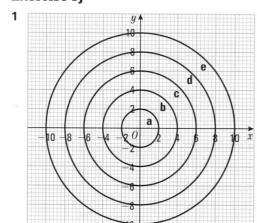

2.

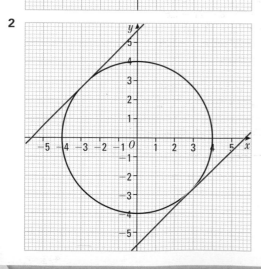

(2.83, −2.83) and (−2.83, 2.83)

3. **a** $y = 1, y = 11$ **b** $y = −9, y = 1$
   c $x = −2, x = 8$ **d** $x = −10, x = 0$
4. $y = x + 12.58, y = x − 4.48$

7.11 Get Ready

1.

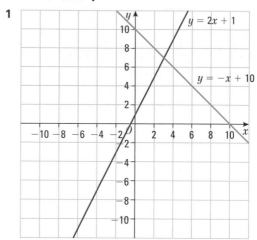

$x = 3, y = 7$

2.

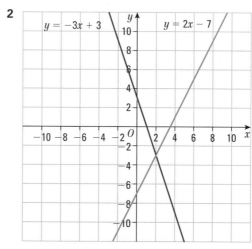

$x = 2, y = −3$

Exercise 7K

a graphs of the equations
b :
1. $x = 2, y = 2; x = −3, y = −3$
2. $x = 4, y = 7; x = −2, y = 1$
3. $x = 3, y = −0.5; x = −1, y = 1.5$
4. $x = 1, y = 1; x = −1.67, y = 6.33$
5. $x = 1, y = 2; x = 2, y = −1$
6. $x = 3.5, y = 21; x = −0.5, y = −3$

7.12 Get Ready

1

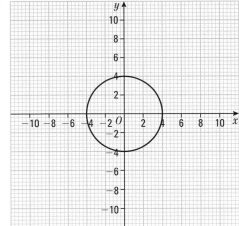

2

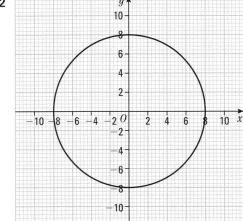

3

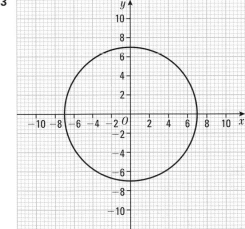

Exercise 7L

1 $x = 2.7, y = 5.4; x = -2.7, y = -5.4$
2 $x = 1.2, y = 4.2; x = -4.2, y = -1.2$
3 $x = 6.8, y = -1.8; x = -1.8, y = 6.8$
4 **a** $x = 2, y = 3; x = -3, y = -2$
  b $x = 4, y = -2; x = -2, y = 4$
  c $x = 2.2, y = 5.4; x = -3, y = -5$
5 **a** $x = 0.449, y = 4.45; x = -4.45, y = -0.449$
  b $x = 1.49, y = 5.46; x = -2.09, y = -5.26$
  c $x = 5.63, y = 8.26; x = -3.23, y = -9.46$

Review exercise

1 **a** $x = \pm3$ **b** $x = \pm6$ **c** $x = \pm7.35$
2 **a** $y = \pm2$ **b** $t = \pm2$ **c** $p = \pm3$
3 $3\,cm$
4 **a** $y = 1$ and $y = 6$ **b** $t = 2$ and $t = -2$
  c $p = 1$ and $p = -4$
5 **a** $x = \frac{4}{3}, y = \frac{25}{3}$ **b** $x = -1, y = 1$
  c $x = -2, y = -5$
6 **a** $x = 2, y = 2$ **b** $x = 0.8, y = 2.4$
7 $29.12\,m$
8 **a** $h = 0$ when $t = 0$ or 6, so the ball was in the air for
    6 seconds.
  b By symmetry of graph, maximum height is when $t = 3$,
    giving height of $45\,m$
  c $h = 25$ when $t = 1$ or $t = 5$, so the ball was $25\,m$ above
    the ground after 1 second and 5 seconds.
9 **a** $k = 0.5$ and $k = 5$ **b** $m = -0.5$ and $m = 1.5$
  c $n = 2$ and $n = -2.2$
10 **a** $p = 4, q = -11$ **b** $x = -0.683$ and $x = -7.32$
11 $x = 6.27$ and $x = -1.27$
12 **a, b** $x = 2.41$ and $x = -0.414$
13 **a** $(x - 1)^2 + 2$ **b** 2 **c** $x = 1$
  d

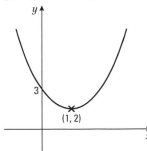

14 $4(x + 3)^2 - 36.\ a = 4, p = 3, q = -36$
15 **a** $x = 2, y = 2$ **b** $x = 0, y = 2$
  c $x = -4, y = 3$
16 $F = 12.5, g = 5$
17 $y = 2x - 3$ and $2y + x = -6$
18 $x = -5, y = 8$
19 **a** $p = 4, q = 7$ **b** $(4, 7)$
20 $50p$
21 $x = 1, y = 0$ and $x = -1.33, y = 9.33$
22 **a** $5(x - 1) = (4 - 3x)(x + 2)$
    $5x - 5 = 8 - 2x - 3x^2$
    $3x^2 + 7x - 13 = 0$
  b $x = 1.22$ and $x = -3.55$
23 **a** $5x + x(2x + 1) = 95$
    $2x^2 + 6x - 95 = 0$
  b $x = 5.55$ and $x = -8.55$
24 **a** Time for first $10\,km = \dfrac{10}{x}$
    Time for second $10\,km = \dfrac{10}{x - 1}$
    Total time = 4, so $\dfrac{10}{x} + \dfrac{10}{x - 1} = 4$
  b $10(x - 1) + 10x = 4x(x - 1)$
    $4x^2 - 24x + 10 = 0$
  c $x = 5.55$ and $x = 0.450$
  d $4.55\,km/h$

25 **a** $x = -4.30$, $y - 0.697$ and $x = -0.697$, $y - 4.30$
   b $x = 0.350$, $y = 6.70$ and $x = -5.15$, $y = -4.30$
26 $x = 3.41$ and $x = 0.586$
27 $x^2 + y^2 = 16$ is a circle with radius 4, centre the origin, so it intersects the x- and y-axes at ± 4. The point $(1, 2)$ must lie inside this circle. Therefore any straight line that passes through $(1, 2)$ must intersect the circle twice.

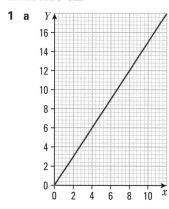

28 $x = 1.83$, $y = 4.65$ and $x = -2.63$, $y = -4.25$

Chapter 8 Answers

8.1 Get Ready

1 **a** 50p   **b** 75p   **c** £1   **d** £1.25
2 **a** 20 seconds   **b** 40 seconds
   c 80 seconds   **d** 160 seconds

Exercise 8A

1 **a**
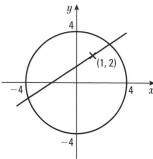

   b Yes, the graph is a straight line through the origin
2 $k = 0.4$
3 **a** $\frac{9}{4} = \frac{22.5}{10} = \frac{36}{16} = \frac{45}{20}$ **b** $k = 2.25$
   c 72
4 **a** Y

   b $Y = \frac{1}{2}t$
   c $Y = 50$
5 **a** **i** £3.60   **ii** £5.40   **iii** £7.20
   b $C = 0.45v$ **c** 13 cm³

8.2 Get ready

1 **a** $x = 18$ **b** $x = 72$ **c** $x = 75$

Exercise 8B

1 $k = 1.5$
2 $k = 0.7$
3 **a** $P \propto h$, so $P = kh$
      When $P = 40.5$, $h = 18$, giving $k = \frac{9}{4}$
      So $P = \frac{9}{4} h$
   b 72   **c** 12
4 **a** $V \propto c$, so $V = kc$
      When $V = 10$, $c = 0.04$, giving $k = 250$
      So $V = 250c$
   b 0.036 amps
5 **a** $E = \frac{7}{150} m$ **b** 21 mm   **c** 1125 g
6 $66\frac{2}{3}$ cm³

8.3 Get Ready

1 **a** $V = kT$ **b** $p = kT$

Exercise 8C

1 **a** **i** $M \propto n$ **ii** $M = kn$
   b **i** $L \propto h^2$ **ii** $L = kh^2$
   c **i** $P \propto t^3$ **ii** $P = kt^3$
   d **i** $Q \propto \sqrt{y}$ **ii** $Q = k\sqrt{y}$
   e **i** $W \propto \sqrt[3]{x}$ **ii** $W = k\sqrt[3]{x}$
   f **i** $A \propto \frac{1}{b}$ **ii** $A = \frac{k}{b}$
   g $H \propto \frac{1}{g^2}$ **ii** $H = \frac{k}{g^2}$
   h $U \propto \frac{1}{f^2}$ **ii** $U = \frac{k}{f^2}$
   i $E \propto \frac{1}{w^3}$ **ii** $E = \frac{k}{w^3}$
   j $V \propto \frac{1}{\sqrt{r}}$ **ii** $V = \frac{k}{\sqrt{r}}$
2 **a** $N \propto d^2$ **b** $N = kd^2$
3 $H = kc^2$
4 $Q = kt^3$
5 F is inversely proportional to the square of r
6 T is proportional to the square root of l

8.4 Get Ready

1 **a** 225   **b** 512   **c** 121   **d** 9   **e** 8   **f** 64

Exercise 8D

1 0.8
2 16
3 $Z = 1.28b^2$
4 $L = 1.875v^3$
5 4.5
6 **a** 11.25 m   **b** 14.1 m/s
7 2.51 mm
8

Mass (grams)	Length (cm)
20	5
34.56	6
43.94	6.5
50	6.79

8.5 Get ready

1 $P = \dfrac{1}{T}$

Exercise 8E

1 45
2 a 2.5   **b** 6
3 a 3.6   **b** 2.67
4 2.67 m³
5 a 384 hertz   **b** 0.96 m
6 a 3.27   **b** 6.32
7 a $S = \dfrac{8000}{f^2}$ **b** 500

Review exercise

1 $0.75 \div 0.5 = 2.25 \div 1.5 = 5.25 \div 3.5 = 6.75$
$\div 4.5 = 9.0 \div 6.0 = 1.5$
2 a 1600   **b** 560
3 a $T \propto b$ **b** R is directly proportional to a
   c $P \propto m^2$ **d** $Z \propto g^3$
   e H is inversely proportional to y
4 a $S = 1.5p$
   b

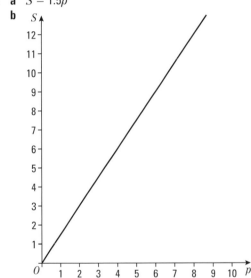

5 £70 125
6 Yes, the pressure will be 6 bars.
7 10.2 units
8 a $\dfrac{D}{t^2} = \dfrac{5}{1^2} = \dfrac{20}{2^2} = 5$

t	0	1	2	3	4	5
D	0	5	20	45	80	125

   b 1.73 seconds

9

Radius (r mm)	Resistance (R ohms)
10	500
15	222.22
17.5	163.27
14.14	250

10 a $X \propto \sqrt{h}$, so $X = k\sqrt{h}$
   When $X = 3$, $h = 16$, giving $k = 0.75$
   So $X = 0.75\sqrt{h}$
   b 3.75   **c** 64
11 a $T \propto \dfrac{1}{\sqrt{m}}$, so $T = \dfrac{k}{\sqrt{m}}$
   When $T = 20$, $m = 50$, giving $k = 100\sqrt{2}$
   So $T = 100\sqrt{\dfrac{2}{m}}$
   b 88.9 g
12 a A   **b** C
13 a 540   **b** 1.82
14 a $q = \dfrac{k}{l^2}$ **b** 5.44
15 51.6
16 a $y = kz$
   b $u = kv^2$ $v = \dfrac{l}{\sqrt{w}}$
               $v^2 = \dfrac{l^2}{w}$
   $u = k \times \dfrac{l^2}{w}$
   $uw = kl^2$
   kl^2 is a constant so uw is a constant.

Chapter 9 Answers

9.1 Get Ready

1 18   **2** $x = \frac{11}{4}$ **3** $y = (t + 1)^2$

Exercise 9A

1 a 12   **b** 0   **c** 48
   d 1   **e** −4   **f** 8
2 a 96   **b** 6.4   **c** 4
   d 28
3 a 8   **b** 2
4 a i −4   **ii** −3   **iii** 0   **b** 4, −4
5 a i 18   **ii** −12   **iii** −30   **b** 3, −4
6 a i −3   **ii** −4   **iii** 10   **b** 0, 4   **c** −1, 5
7 a 16   **b** $x^2 − 4$ **c** $(x − 4)^2$
8 a −8   **b** $4(2x + 1)$ **c** $12(x + 1)$

9.2 Get Ready

1 $(2, 5)$ **2** $(4, 3)$ **3** $(−1, 3)$

Exercise 9B

1 a

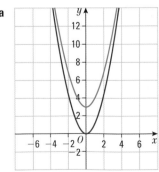

Answers

b

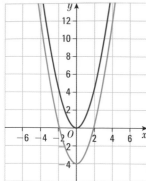

c

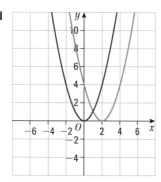

d

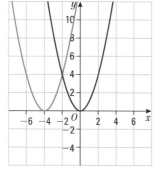

2 a i

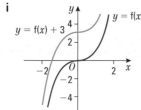

ii

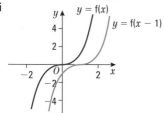

b i (0, 3)   **ii** (1, 0)

3 a

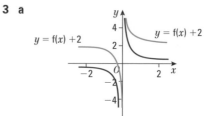

b (2, 2.5)   **c i** $y = f(x) + 2$ **ii** $y = \frac{1}{x} + 2$

d

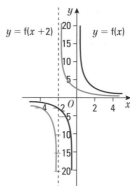

e $p = -1, q = 4$ **f i** $y = f(x + 2)$ **ii** $y = \frac{4}{x + 2}$

4 a (3, 2)   **b** (1, 5)

5 a Translation by +3 units parallel to the y-axis.
 b $y = f(x) + 3$ **c** $y = 4x - x^2 + 3$

6 a Translation by +2 units parallel to the x-axis.
 b $y = f(x - 2)$ **c** $y = (x - 2)^2$

7 a $a = 2, b = 5$ **b** Translation of $\begin{pmatrix} -2 \\ 5 \end{pmatrix}$

c, d

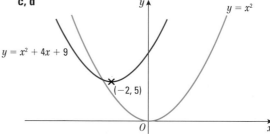

8 Translation by $\begin{pmatrix} 0 \\ -6 \end{pmatrix}$

9.3 Get Ready

1

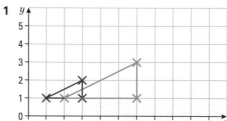

The first triangle has been stretched by a factor of 2 parallel to the x-axis.

332

2

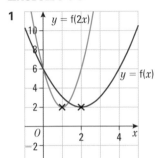

The first triangle has been stretched by a factor of 3 parallel to the y-axis.

Exercise 9C

1

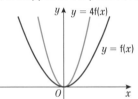

$(2, 2)$ is mapped to the point $(1, 2)$

2 a

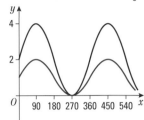

b $y = 4x^2$

c Stretch of magnitude 4 parallel to the y-axis

Stretch of magnitude $\frac{1}{2}$ parallel to the x-axis

3 a

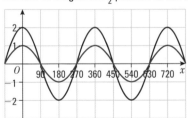

$y = 2 \cos x°$

b

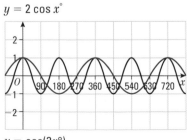

$y = \cos(2x°)$

4 a

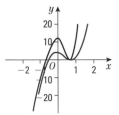

b $(0, 4)$ is mapped to $(0, 12)$. $-\frac{2}{3}, 0$ is mapped to $-\frac{2}{3}, 0$

5 a

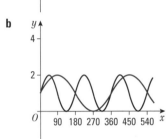

b

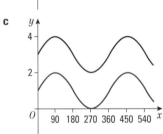

c

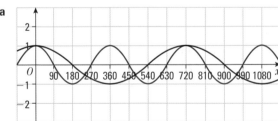

6 a

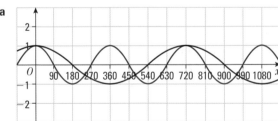

b 1 solution

7 a Stretch of magnitude 2 parallel to the x-axis.

b

c i $y = f\left(\frac{x}{2}\right)$ **ii** $y = \frac{x}{2}\left(\frac{x}{2} - 4\right)$

333

Answers

8 a $p = 4$, $q = -11$ **b** $(0, 5)$

c Stretch parallel to the y-axis of magnitude 0.6.
$0.6(x^2 - 8x + 5)$

Exercise 9D

1 a

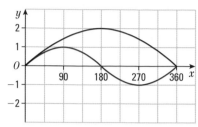

b $y = 2\sin\left(\dfrac{x}{2}\right)$

2 a

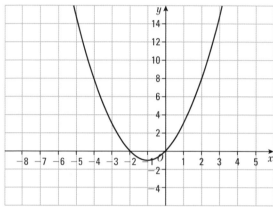

b $y = 2f\left(\dfrac{x}{2}\right)$

c

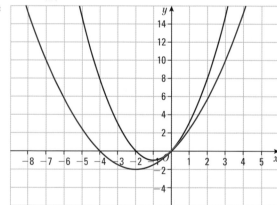

3 a, b

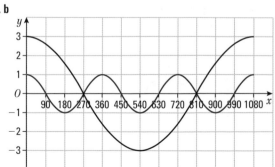

4 a $4\left(\dfrac{x^2}{16} + 3\right)$ **b** $4\left(\dfrac{4}{x} + 1\right)$

c $4 \times 2^{\wedge\frac{x}{4}} = 2^{\wedge\frac{x}{4} + 2}$ **d** $8\sin\left(\dfrac{x}{2}\right)$

9.4 Get Ready

1

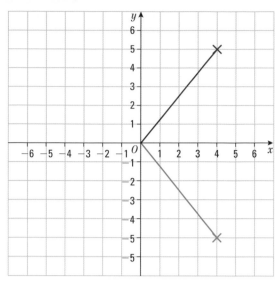

2

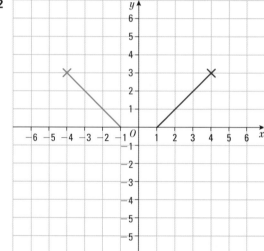

3

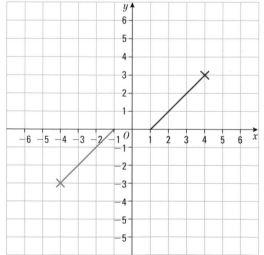

Exercise 9E

1 a

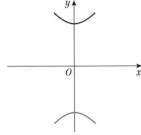

b $y = -(x^2 + 3)$

2 a

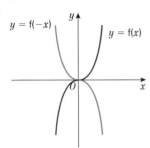

b $p = -2, q = 8$ **c** $y = -x^3$

3 a

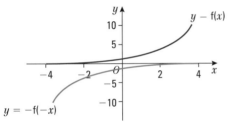

b $(r = -1, t = -2)$ **c** $y = -2^{-x}$

Review exercise

1 a 6   **b** 11   **c** $a = 0$

2 $f(x - 1) = (x - 1)^2 + 3(x - 1) = x^2 + x - 2$
$= (x + 2)(x - 1)$

3

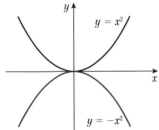

4 a

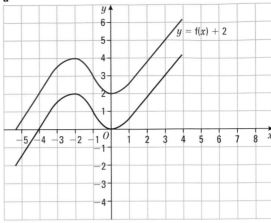

b

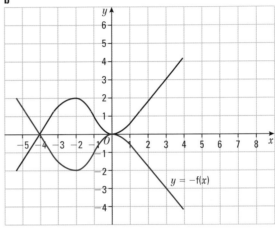

5 $f(x - 4)$

6 a, b

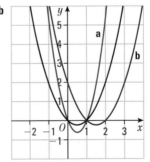

c $x = 1$ and $x = 2$

7 a $(180, 0)$

b

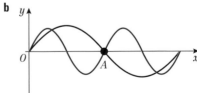

8

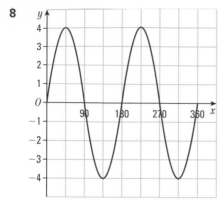

9 Translation by $\begin{pmatrix} 4 \\ -5 \end{pmatrix}$

10 **a, b** Stretch parallel to the y-axis of magnitude 4
Stretch parallel to the x-axis of magnitude 0.5

11 **a** $p = 2, q = 12$
b $(2, 12)$
c Stretch parallel to the y-axis of magnitude 2
d $y = 2\mathrm{f}(x)$

Chapter 10 Answers

10.1 Get Ready

1

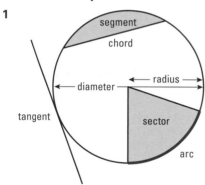

Exercise 10A

1 **a** 22.0 cm   **b** 40.5 mm   **c** 17.6 cm
   d 126 cm   **e** 68.8 m
2 **a** 145 cm   **b** 25.1 cm
3 6.00 cm
4 **a** 210 cm   **b** 2.1 km   **c** 2859
5 **a** 7.54 cm   **b i** 45.2 cm   **ii** 2.51 cm
6 **a** 408 cm   **b** 14.6 cm
7 71.7 cm

Exercise 10B

1 **a** 201 cm²   **b** 507 cm²   **c** 2550 mm²
   d 297 cm²   **e** 499 m²
2 **a** 452 cm²   **b** 54.1 cm²   **c** 0.709 m²
   d 2680 mm²   **e** 262 cm²
3 **a** 1260 cm²   **b** 2590 cm²   **c** £5050.50
4 **a** 10 800 m²   **b** £2484
5 70.7 cm²
6 13 cm
7 193 cm²

10.2 Get Ready

1 **a** 6   **b** $\frac{1}{6}$ **c** $\frac{1}{9}$

Exercise 10C

1 **a** 1.75 cm   **b** 17.6 cm   **c** 64.9 m
2 **a** 55.4 cm²   **b** 9.25 cm²   **c** 230 cm²
3 **a** 19.3 cm   **b** 41.6 m
4 70°
5 5 cm
6 **a** 1710 m²   **b** 48.9 m
7 **a** 57.9 cm   **b** 235 cm²

10.3 Get Ready

1 **a** $10a$ **b** $3b$
2 $x = 4.5$

Exercise 10D

1 **a** 9π cm   **b** π m   **c** 16π m
2 **a** 4π cm²   **b** 100π m²   **c** 30.25π cm²
3 **a** Area $= 17^2\pi - 15^2\pi = 64\pi$ cm²
   b 16π cm
4 **a** 8 cm   **b** 64π cm²
5 **a** 36π cm²   **b** 4π cm   **c** $36 + 4\pi$ cm
6 Perimeter $= \frac{1}{2}\pi Y + \frac{1}{2}\pi X + \frac{1}{2}\pi(X + Y) = \pi(X + Y)$

10.4 Get Ready

1 **a** 100   **b** 10 000   **c** 1 000 000
2 **a** 500   **b** 130 000   **c** 7 600 00
3 **a** 70   **b** 0.35   **c** 0.049

Exercise 10E

1 **a** 45 239 cm²   **b** 4.5239 m²
2 **a** 14 cm²   **b** 1400 mm²
3 **a** 40 000 cm²   **b** 69 000 cm²   **c** 6 cm²
   d 0.47 cm²
4 **a** 5 000 000 m²   **b** 300 000 m²   **c** 4 m²
   d 0.056 m²
5 **a** 1000   **b** 1 000 000   **c** 8 300 000 mm²
6 **a** 44 800 cm²   **b** 3.6 cm²

Review exercise

1 **a** 4.5 cm²   **b** 60 000 cm²
2 6.54 km²
3 $2 \times \pi \times 2.55 = 16$ m $= 32$ roses, so cost $=$ £134.40
4 Area $= 4 \times 4 - (4 \times 1 + 0.5 \times 1 \times 1.8 + \pi \times 1^2)$
     $= 7.958$ m²
     Cost $= 7.958 \times 4.60 =$ £36.61
5 **a** 308.5 m
   b Add 45.8 m to the straights making each one 105.8 m or
     add 29.1 m to the diameter of the bends, making each
     diameter 89.1 m
6 $200 - 50 \times \pi \times 1^2 = 42.9$ cm²
7 10.9 cm²
8 $12 + 4\pi$ cm
9 **a** $\frac{7\pi}{2}$ cm²   **b** $4 + \frac{7\pi}{2}$ cm
10 18.27 cm² to 2 d.p.

Chapter 11 Answers

11.1 Get Ready

1 a 28.3 cm² (3s.f.)   **b** 19.6 cm² (3s.f.)
   c 78.5 cm² (3s.f.)

Exercise 11A

1 a 251 cm³   **b** 13 600 000 mm³
   c 236 cm³   **d** 8930 cm³
2 a 360 π cm³   **b** 650 π cm³
   c 0.101 25 π m³
3 1600 π m³
4 114.4 cm³
5 31.92 mm
6 3.56 m³ = 3560 litres

11.2 Get Ready

1 a 2.211 cm²   **b** 14.56 mm²   **c** 65.0 mm² (to 3 s.f.)

Exercise 11B

1 a 60 m³   **b** 480 cm³   **c** 320 cm³
2 16 400 000 m³
3 a 100 cm³   **b** 125 cm³   **c** 0.0804 cm³
4 $53\frac{1}{3}\pi$ cm³
5 1590 cm³

11.3 Get Ready

1 a 10.24 π cm²   **b** 32.2 cm² (to 3 s.f.)
2 a 64   **b** 216   **c** 1331

Exercise 11C

1 268 cm³   **2** 4090 m³   **3** 8.31 cm
4 17.9 cm   **5** 1940 cm³   **6** 5.04 m

11.4 Get Ready

1 a 6.28 cm³ (to 3 s.f.)   **b** 576 cm³

Exercise 11D

1 a 223 cm³   **b** 177 m³   **c** 52.5 m³   **d** 812 cm³
2 120 π cm³   **3** 54 000 π cm³   **4** 1610 cm³
5 731.25 π cm³   **6** $\frac{8}{3}\pi r^3$ cm³

11.5 Get Ready

1 a 1 000 000 000   **b** 4620   **c** 91.875 m³

Exercise 11E

1 a 2 000 000 cm³   **b** 6 750 000 cm³   **c** 0.45 cm³
   d 0.0068 cm³
2 a 7000 mm³   **b** 3750 mm³   **c** 25 mm³
3 a 0.075 m³   **b** 0.0008 m³   **c** 0.000 125 m³
4 a 0.83 l **b** 5.6 l **c** 1000 l **d** 0.003 54 l
5 720 000 litres
6 1277 cubes
7 Volume A = 12.5 m³ and volume C = 0.375 m³ so shape A has the largest volume.

11.6 Get Ready

1 3.2 cm   **2** 3.6 cm   **3** 64.2 mm

Exercise 11F

1 a 240.3 cm²   **b** 3.3 m²
2 a 503 cm²   **b** 5890 mm²   **c** 302 cm²
3 a 100 π cm²   **b** 80 π cm²   **c** 356 π cm²
4 a 192 cm²   **b** 924 cm²

11.7 Get Ready

1 a 8.3   **b** 1.3   **c** 26.7   **d** 12.9

Exercise 11G

1 250 g/cm³   **2** 0.777 cm³   **3** 1.96 g/cm³
4 The aluminium block has the greater mass by 155 kg.

Review exercise

1 a Volume of cup = $\frac{\pi \times 10 \times 4.5^2}{4}$ = 159 ml, so the cup can hold 150 ml.
   b Volume of squash required = 30 $\times$ 150 $\times$ 3 = 13 500 ml
     = 13.5 litres
     1 bottle makes 0.8 $\times$ 7 = 5.6 litres
     $\frac{13.5}{5.6}$ = 3.2, so 3 bottles are needed.
   c £3.75
2 8 000 000 cm³
3 1.14 g/cm³
4 193 g
5 Volume of oil = $\pi \times 60^2 \times 180$ = 2 036 752 cm³,
   so mass of oil = 8754 kg
   Surface area of tank = $2\pi r^2 + 2\pi rh$ = 90 478 cm²,
   so mass of tank = 253 kg
   Total mass = 9007 kg = 9 tonnes
6 a 905 m³   **b** 4.92 m
7 9.3 cm
8 $h = \frac{10}{9}x^3 - 3x^2$
9 vol = 2.7 π litres
10 $h = 9x$
11 1700 cm³
12 Yes. Volume of a pyramid = $\frac{1}{3}$ $\times$ base area $\times$ height and volume of a cone = $\frac{1}{3}$ $\times$ base area $\times$ height, so if the base areas and volumes are the same, then the heights must also be equal.
13 a 315 cm³   **b** 0.6 g/cm³

Chapter 12 Answers

12.1 Get Ready

1 Not necessarily, one could be an enlargement of the other.
2 No
3 No

Exercise 12A

1 QS is the hypotenuse of both triangles
   angle QPS = angle SRQ = 90° (given)
   PS = QR (given)
   So the triangles are congruent (RHS)

Answers

2 angle YZX = angle WVX (alternate angles)
angle ZYX = angle VWX (alternate angles)
YZ = WV (given)
So the triangles are congruent (AAS)
And YX = XW, so X is the midpoint of WY

3 PQ = PR (given)
QS = RT (given)
angle PQS = angle PRT (isosceles triangle)
So triangle PQS is congruent to triangle PRT (SAS)
Therefore PS = PT and so triangle PST is isosceles

4 Let the point where the line from L cuts the base at right angles be X.
Now LM = LN (given)
LX is common to both triangles
angle LXM = angle LXN = 90° (given)
So triangles LMX and LNX are congruent (RHS)
As the triangles are congruent then MX = NX, so the line from L bisects the base.

5 AD = DB (as D is the midpoint of AB)
angle ADE = angle DBF (corresponding angles)
angle DAE = angle BDF (corresponding angles)
So the triangles ADE and DBF are congruent (AAS)

12.2 Get Ready

1 A and C

Exercise 12B

1 a Similar **b** Not similar
2 All corresponding angles are equal so the pentagons are similar
(angle CDE = angle HIJ = 150°)

Exercise 12C

1 15 cm
2 a 5.14 cm **b** 0.448 m
3 15 cm

12.3 Get Ready

By inspection

Exercise 12D

1 a i AC and FE, BC and DF, AB and DE
  ii angle ABC = angle EDF, angle BCA = angle DFE, angle BAC = angle DEF
 b i JK and GH, KL and HI, JL and GI
  ii angle KJL = angle HGI, angle JKL = angle GHI, angle KLJ = angle HIG
 c i PN and MN, QN and ON, PQ and MO
  ii angle NPQ = angle NMO, angle PQN = angle NOM, angle PNQ = angle MNO
 d i SR and WU, ST and WT, RT and UT
  ii angle STR = angle UTW, angle RST = angle TWU, angle SRT = angle TUW
2 a 5.525 cm **b** 7.225 cm
3 a angle DEB = angle ACB (given)
  angle DBE = angle ABC (same angle)
  Therefore angle BDE = angle BAC (angles in a triangle)
  So triangle ABC is similar to triangle DBE
 b 6.46 cm **c** 2.26 cm **d** 1.99 cm

4 a angle BDE = angle BAC (given)
  angle DBE = angle ABC (same angle)
  Therefore angle DEB = angle ACB (angles in a triangle)
  So triangle ABC is similar to triangle DBE
 b 6.36 cm **c** 10 cm
5 a angle ABM = angle CDM (alternate angles)
  angle BAM = angle DCM (alternate angles)
  angle AMB = angle CMD (vertically opposite angles)
  So triangle ABM is similar to triangle CDM
 b i 8 cm **ii** 12 cm

12.4 Get Ready

1 a 9 **b** 16 **c** 90
2 a 196 **b** 625 **c** 384.16

Exercise 12E

1 720 cm² **2** 270 cm² **3** 12 cm²
4 1125 cm²

Exercise 12F

1 a 18 **b** 15
2 8 cm

12.5 Get Ready

1 a 7 **b** 10
2 a 15.625 **b** 1331

Exercise 12G

1 270 cm³ **2** 135 cm³ **3** 1327 cm³ **4** 2.25 cm³

Exercise 12H

1 3 cm
2 a 18 **b** 7
3 a 28 **b** 1.2
4 1.95 cm
5 10 cm

12.6 Get Ready

1 2:9 **2** 4:5 **3** 1:6 **4** 1:1

Exercise 12I

1 1000 cm² **2** 3.91 cm³ **3** 2950 cm²
4 444 cm² **5 a** 292 cm² **b** 208 cm²
6 0.101 m²

Review exercise

1 A and C
2 Y and Z
3 2.25 cm
4 1.456 cm
5 a angle ADB = angle ADC = 90° (given)
  AB = AC (sides of equilateral triangle)
  AD is common
  So triangle ADC is congruent to triangle ADB (RHS)
 b BD = CD (corresponding sides)
  BD + CD = BC = AB
  So BD = $\frac{1}{2}$ BC

6 AB = BC (given)
AD = CD (given)
BD is common
So triangle ADB is congruent to triangle CDB (SSS)

7 angle ACB = angle CED (vertically opposite angles)
angle CAB = angle CED (alternate angles)
angle ABC = angle EDC (alternate angles)
So triangle ABC is similar to triangle EDC
a 12 cm **b** 9 cm

8 12.5 cm

9 a angle HJI = angle GJF (same angle)
angle IHJ = angle FGJ (corresponding angles)
angle HIJ = angle GFJ (corresponding angles)
So triangle HIJ is similar to triangle GFJ
b 3.43 cm

10 12.96 cm^2

11 a 20 cm **b** 6 cm

12 angle CAD = angle BAE (same angle)
angle ACD = angle ABE (corresponding angles)
angle ADC = angle AEB (corresponding angles)
So triangle ACD is similar to triangle ABE
a 2 cm **b** 5.25 cm

13 a 320 cm^2 **b** 7.5 cm^3

14 a 8 cm **b** 96 cm^3

15 3800 cm^2

16 a 1 : 4 **b** 1 : 8

17 a All the edges of a cube are the same length, so the
ratio between the edges of two cubes is a constant.
b 1 : 6.25
c The edges of a cuboid can be in any ratio, so two
cuboids may not have the same ratio between every
pair of edges.

18 a 11.9 cm **b** 25.6 cm^2

Chapter 13 Answers

13.1 Get Ready

1 a 28° **b** 49° **c** 57°

Exercise 13A

1 $a = 83°$

2 $b = 90°, c = 52°$

3 $d = 64°, e = 38°$

4 $f = 110°, g = 55°$

5 $h = 30°$

6 $i = 72°$

7 $j = 49°$
The line drawn from the centre of a circle to the midpoint
of a chord is perpendicular to that chord.

8 $k = 68°$
The angle at the centre of a circle is twice the angle at the
circumference, both subtended by the same arc.

9 $l = 90°$
The angle in a semicircle is a right angle.

10 a angle DAB = 65°
The angle at the centre of a circle is twice the angle at the
circumference, both subtended by the same arc.
b reflex angle DOB = 230°
Angles about a point add up to 360°

c angle BCD = 115°
The angle at the centre of a circle is twice the angle at the
circumference, both subtended by the same arc so reflex
angle BCD is twice angle BCD.

13.2 Get Ready

1 a 93° **b** 94° **c** 46°

Exercise 13B

1 $a = 40°$ **2** $b = 82°$ **3** $c = 72°$

4 $d = 50°, e = 100°, f = 40°$

5 $g = 58°$ **6** $h = 74°$ **7** $i = 71°$
Angles in the same segment are equal so angle SQR = 31°,
and angles in a triangle add up to 180°.

8 $j = 35°$
Opposite angles of a cyclic quadrilateral add up to 180° and
the base angles of an isosceles triangle are equal.

9 $k = 57°$
The angle between a tangent and a chord is equal to the
angle in the alternate segment and the angles on a straight
line (or in a triangle) add up to 180°.

Review exercise

1 angle OAD = 90° (angle between tangent and radius)
angle AOD = 180 − 90 − 36 = 54° (angle sum in a triangle)
angle ABC = $\frac{1}{2}$ × 54 = 27° (angle at centre is twice angle
at circumference)

2 a i 140°
ii The angle at the centre of a circle is twice the angle
at the circumference, both subtended by the same
arc.
b i 110°
ii The opposite angles of a cyclic quadrilateral add up
to 180°.

3 $a = 134°, b = 42°$

4 $d = 90°, e = 67°, f = 67°$

5 $g = 33°, h = 33°$

6 $o = 31°$ (angles in the same segment are equal)
$p = 46°$ (angles on a straight line add up to 180°; opposite
angles of a cyclic quadrilateral add up to 180°; angles in
the same segment are equal)

7 $s = 94°$ (angles on a straight line add up to 180°; opposite
angles of a cyclic quadrilateral add up to 180°)

8 $t = 52°$ (angle between a tangent and a radius is a right
angle; angles in a triangle add up to 180°)
$u = 26°$ (angle at the centre is twice the angle
at the circumference, both subtended by the same arc)

9 angle CBT = angle CBA (BC bisects angle ABT)
angle CBT = angle CAB (angle between tangent and
chord is equal to angle in alternate segment)
angle CBA = CAB
So triangle CAB is isosceles and CA = CB

10 angle PMA = $\frac{1}{2}$ × 132 = 66° (angle at the centre is twice
angle at circumference)
angle MAT = 66° (alternate angles, PM parallel to AT)
angle OAT = 90° (angle between tangent and radius)
angle OAM = 90 − 66 = 24°
OA = OM (radii), so triangle AOM is isosceles
angle OMA = 24° (base angles of isosceles triangle)

Chapter 14 Answers

14.1 Get Ready

Students' accurate drawings

Exercise 14A

1–5 Students' drawings

6 The sum of the two shorter sides is less than the longest side.

14.2 Get Ready

1
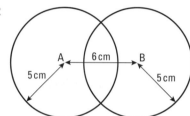

2

3

Exercise 14B

1

2 a
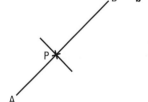

b
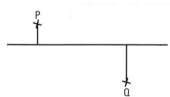

3

14.3 Get Ready

1
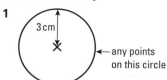

2,3
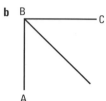

Exercise 14C

1 a
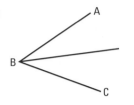

b

2 a
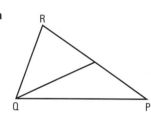

b
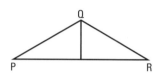

3 a   **b**   **c**

d   **e**

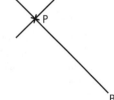

4

5
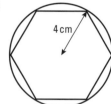

14.4 Get Ready

1
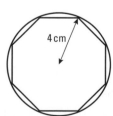

2

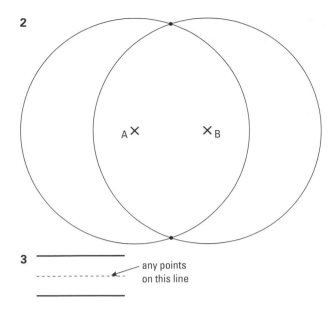

3

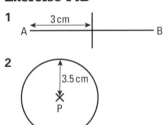

any points
on this line

Exercise 14D

1

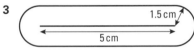

3 cm
A ———— B

2

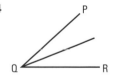

3.5 cm
⊗
P

3

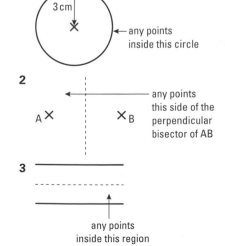

1.5 cm
5 cm

4

P
Q ———— R

14.5 Get Ready

1

3 cm
⊗

← any points
inside this circle

2

any points
this side of the
perpendicular
bisector of AB

A ✕   ✕ B

3

any points
inside this region

Exercise 14E

1

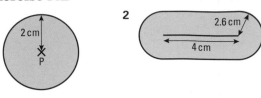

2 cm
✕
P

2

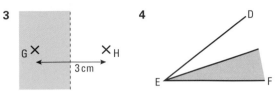

2.6 cm
4 cm

3

G ✕   ✕ H
3 cm

4

D
E   F

5

1.4 m
0.8 m
25 cm

14.6 Get Ready

1 a East   **b** South   **c** North-West

Exercise 14F

1 a 073°   **b** 225°   **c** 243°
2 a 070°   **b** 218°   **c** 102°
3 249°
4 312°
5 a 111°   **b** 239°

14.7 Get Ready

1 a 50 km   **b** 2.5 km
2 a 400 000 cm   **b** 30 000 cm

Exercise 14G

1 a 1 cm represents 2 km
   b i 6 km   **ii** 9.5 km   **iii** 4.2 km
2 a 13.5 cm   **b** 312 km
3 b 14.4 km
4

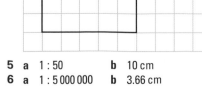

5 a 1 : 50   **b** 10 cm
6 a 1 : 5 000 000   **b** 3.66 cm

Answers

Review exercise

1–4 Students' accurate drawings

5 a 250 mm   **b** 324 m

6

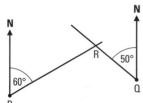

7 a 030°

b

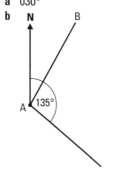

8

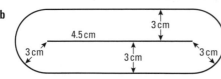

9

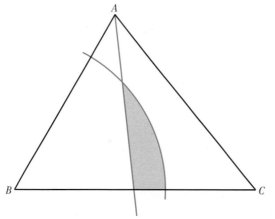

10 Construction of angle of 30° at P

11 a

b

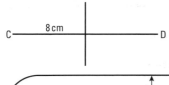

12 a Students' drawings

   b 034°   **c** 258°

13 Construction of bisector of angle ABC

14

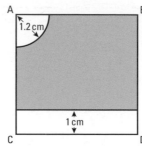

15 Construction of perpendicular bisector of a line 7 cm long

16 Construction of perpendicular to the line ST from a point above the line M

17 a

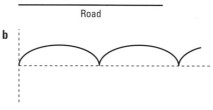

Locus of centre of wheel

Road

b

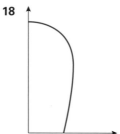

18

(diagram)

Chapter 15 Answers

15.1 Get Ready

1 arc, mean

   a Start at square K, go 1 to the right, then 1 down and stop. Then go 2 up, and 1 to the left and stop. Then go 4 to the right, and 3 down and stop.

   b Start at square C, go 1 up, then 1 left and stop. Then go 2 up and 1 to the right and stop. Then go 3 down and stop. Then go 1 to the left and 2 up and stop. Then go 3 to the right and 2 down and stop.

   c Start at square A, go 3 to the right and stop. Then go 1 up and then 1 down and stop.

   d Students' examples

Exercise 15A

1 a $\begin{pmatrix} 5 \\ 5 \end{pmatrix}$ **b** $\begin{pmatrix} 2 \\ -4 \end{pmatrix}$ **c** $\begin{pmatrix} 4 \\ 0 \end{pmatrix}$

   d $\begin{pmatrix} 0 \\ 4 \end{pmatrix}$ **e** $\begin{pmatrix} -2 \\ 3 \end{pmatrix}$ **f** $\begin{pmatrix} -2 \\ -5 \end{pmatrix}$

2

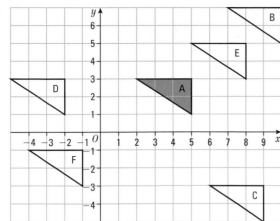

3 a

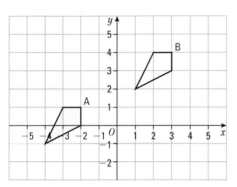

b $\begin{pmatrix} 5 \\ 3 \end{pmatrix}$

4 a, b

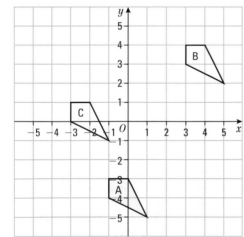

c $\begin{pmatrix} -2 \\ 4 \end{pmatrix}$ **d** $\begin{pmatrix} 2 \\ -4 \end{pmatrix}$

15.2 Get Ready

A: $x = 4$,
B: $y = -3$,
C: $y = x$,
D: $y = -x$

Exercise 15B

1 a, b

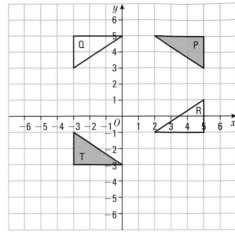

c Reflection in the line $y = 1$

2 a, b

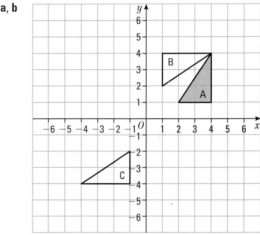

c Reflection in the line $y = x$

3 a i $x = 1$
   ii $y = -1$
 b Reflection in the line $x = 1$

15.3 Get Ready

1 a 9 **b i** 90° clockwise **ii** 180° **iii** 150° clockwise
2 a 90° clockwise
 b 90° anticlockwise
 c 180°
 d 150° anticlockwise

Answers

Exercise 15C

1

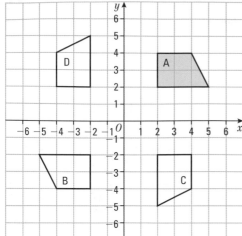

2

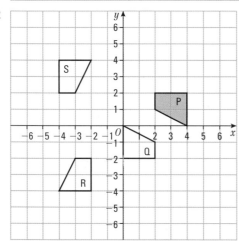

3 a i 90° anticlockwise about the origin
  ii 180° about the origin
  iii 90° clockwise about the origin
 b 90° clockwise about the origin
 c 180° about the origin
4 a i 180° about (3, 5)
  ii 90° clockwise about (2, 1)
  iii 180° about (0, 0)
  iv 90° anticlockwise about (0, 4)
  v 90° anticlockwise about (−2, 2)
 b 90° clockwise rotation about (2, 8)
 c i Translation $\begin{pmatrix} 6 \\ 10 \end{pmatrix}$ **ii** Translation $\begin{pmatrix} 4 \\ 0 \end{pmatrix}$

15.4 Get Ready

1, 2

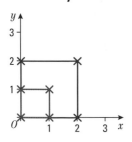

3 They are both squares.
 They have a common vertex at (0, 0).
 The second square is twice as large as the first square.

Exercise 15D

1 a 48 cm, 52 cm, 20 cm
 b The perimeter is also 4 times as long.
2 a, b

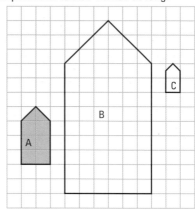

 c 6
3 a P is 4 cm by 2 cm, Q is 8 cm by 4 cm, R is 12 cm by 6 cm.
 b i 12 cm   **ii** 24 cm   **iii** 36 cm
 c i 8 cm^2 **ii** 32 cm^2 **iii** 72 cm^2
 d i 2, same as scale factor
  ii 3, same as scale factor
 e i 4, same as scale factor squared
  ii 9, same as scale factor squared
 f 96 cm

Exercise 15E

1 a

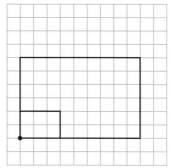

b

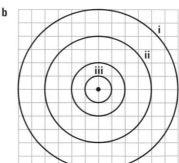

2 a, b

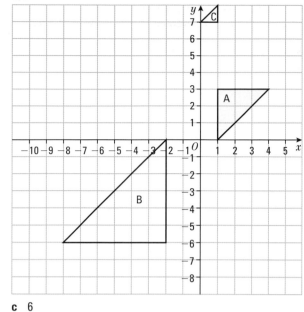

c Translation $\begin{pmatrix} 9 \\ 4 \end{pmatrix}$

d Translation $\begin{pmatrix} -9 \\ -4 \end{pmatrix}$

c 6

2 a, b

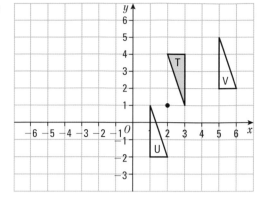

c Rotation 180° about (4, 3)

3 a

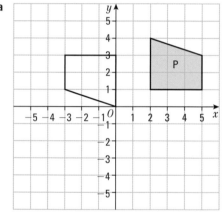

b Rotation 180° about (1, 2)

3 a, b

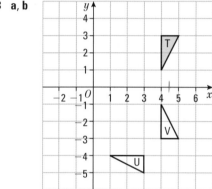

c Reflection in the x-axis

4 Reflection in the line $x = 5$

5 Reflection in the x-axis

Exercise 15F

1 a, b

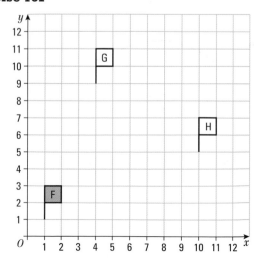

Review exercise

1

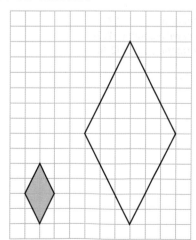

Answers

2 a

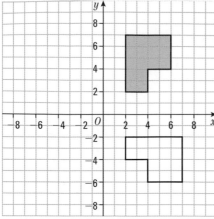

b Translation $\begin{pmatrix} 3 \\ -1 \end{pmatrix}$

3 a

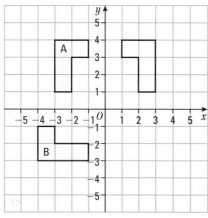

b 90° anticlockwise rotation about the origin

4

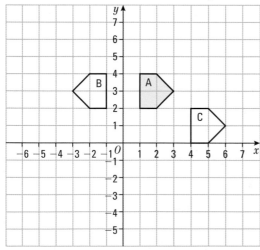

5 Students' tile designs

6 90° clockwise rotation about (−2, 3)

7

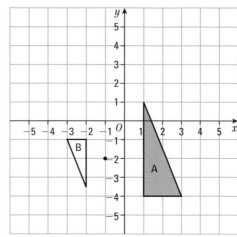

8 a Reflection in the line $y = x$

b

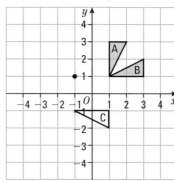

9 a

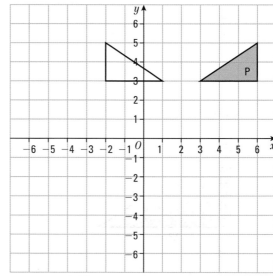

b Translation $\begin{pmatrix} 5 \\ -4 \end{pmatrix}$

10

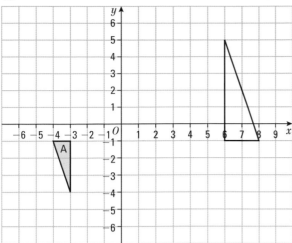

11 180° rotation about (1, 0)

12 180° rotation about the origin

Chapter 16 Answers

16.1 Get Ready

1 a 144   **b** 54.8

2 a 1680   **b** 277

3 a 225   **b** 103

4 a 25   **b** 12

5 a 6.44   **b** 8.91

Exercise 16A

1 a 13 cm   **b** 10 cm   **c** 10.4 cm   **d** 11.6 cm

2 a 12.6 cm   **b** 11.7 cm   **c** 14.4 cm   **d** 20.8 cm

Exercise 16B

1 a 7 cm   **b** 35 cm   **c** 12.3 cm   **d** 8.03 cm

2 a 15.2 cm   **b** 5.77 cm

   c i D

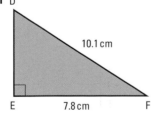

     ii 6.42 cm

16.2 Get Ready

1 The rounded lengths are 21 cm and 30 cm. The diagonal is approximately 36 cm.

Exercise 16C

1 a 15.0 m   **b** 6.92 cm   **c** 18.7 cm

2 3.47 m

3 52.4 cm²

4 a 3.20 m   **b i** 4.33 m   **ii** 2.33 m

   c 8.66 m

5 Triangles 1, 2, 4 and 6 are right-angled triangles.

6 2.85 cm

7 a 17 inches

   b Height = 18 inches, width = 32 inches

16.3 Get Ready

a 26   **b** 19.3   **c** 10.3

Exercise 16D

1 a 10   **b** 26   **c** $\sqrt{212} = 14.6$

   d $\sqrt{424} = 20.6$ **e** 29   **f** $\sqrt{117} = 10.8$

2 a i 5   **ii** $\sqrt{50} = 7.07$ **iii** 5

   b $5^2 + 5^2 = 50$, so triangle is right-angled at A.

3 a 13   **b** B, C, E lie on the circle.

16.4 Get Ready

1 a p **b** r **c** q

2 a e **b** f **c** d

3 a AC   **b** BC   **c** AB

Exercise 16E

1 a 0.3420   **b** 0.9542   **c** 0.5   **d** 0.9461

   e 1   **f** 15.8945   **g** −0.7408   **h** 0.0699

   i 0.7965   **j** 0.2538   **k** −0.3739   **l** 0.0471

2 a 53.1   **b** 25.5   **c** 60   **d** 43.8

   e 58.4   **f** 63.8   **g** 2.7   **h** 60

   i 45

Exercise 16F

1 a sine   **b** cosine   **c** tangent   **d** cosine

Exercise 16G

1 a 37.7°   **b** 46.2°   **c** 19.7°   **d** 40.1°

   e 47.1°   **f** 43.6°

2 a 36.3°   **b** 38.9°   **c** 32.3°

3 a 41.0°   **b** 57.2°   **c** 98°

16.5 Get Ready

1 a

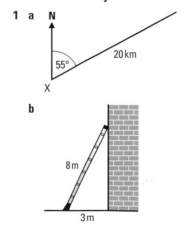

   b

2 cosine

Exercise 16H

1 a 5.47 cm   **b** 17.1 cm   **c** 11.6 cm

   d 26.1 cm   **e** 10.4 cm   **f** 11.3 cm

2 a 18.1 cm   **b** 13.4 m   **c** 10.8 cm
3 a i 9.96 cm   **ii** 8.36 cm
  b 5.30 cm   **c** 13.7 cm   **d** 68 cm²

Exercise 16I

1 a i 1.45 m   **ii** 5.44 m   **iii** 2.54 m
  b 65.0°
2 a 16.8 m   **b** 64.5°   **c** 24.5°
3 a 46.9 m   **b** 88.3 m
4 a 2.87 cm   **b** 16.5°
5 a i 13.0 km   **ii** 10.9 km
  b i 7.73 km   **ii** 23.8 km
  c i 20.7 km   **ii** 34.7 km
  d 40.4 km, 059.2°
6 a 3.84 cm   **b** 16.8 cm   **c** 43 cm²

Review exercise

1 6.36 cm
2 6.71 cm
3 230 km
4 19.4 cm
5 20.6°
6 a 51.3°   **b** 10.5 cm
7 21.7 m
8 033°
9 116 cm
10 a Alan
  b $5\frac{1}{3}$ seconds. Alan's time $= \dfrac{60 + 80}{5} = 28$ seconds,

    Bhavana's time $= \dfrac{\sqrt{60^2 + 80^2}}{3} = 33\frac{1}{3}$ seconds

11 a

  $x^2 + x^2 = 10^2$, so $x = 7.07$ cm
  b

  $x = 10$ cm
12

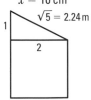

  a Area $= 4 \times 2.24 = 8.96$ m², so he needs to buy two
    rolls of felt.
  b Two rolls of felt cost £24.00, two tins of adhesive cost
    £13.98, so total cost = £37.98
13 No. It depends on the diagonal of the desk:
  $d^2 = 2.3^2 + 1.6^2$, $d = 2.80$ m, but the room is only 2.75 m
  wide.

Chapter 17 Answers

17.1 Get Ready

1 ABF, ABE, AEF, BEF, BCF, BCG, BFG, CFG, CDG, CDH, CHG,
  DHG, ADE, ADH, AEH, DEH, ABC, ABD, ACD, BCD, EFG, EFH,
  EGH, FGH

Exercise 17A

1 a i 8.94 cm   **ii** 13.6 cm   **iii** 15.3 cm   **iv** 15.8 cm
  b i 58.4°   **ii** 72.9°
2 $AG^2 = AC^2 + CG^2 = AB^2 + BC^2 + CG^2$
3 No. The longest needle that could fit in the box is 14 cm.
4 5 cm
5 a i 14.1 cm   **ii** 7.07 cm
  b 13.2 cm   **c** 61.9°   **d** 56.2°   **e** 14.1 cm
  f 70.5°

17.2 Get Ready

1 The diagrams could show the same pole viewed from
  different sides.

Exercise 17B

1 69.3°
2 a i 51.3°   **ii** 49.1°   **iii** 38.7°   **b** 90°
3 a 469 m   **b** 171 m   **c** 985 m   **d** 848 m
4 a 15.6 m
  b i 26.6°   **ii** 36.9°   **iii** 22.6°
5 a 86.5 cm   **b** 63.9°
6 a

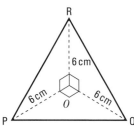

  b i 45°   **ii** 45°
  c 1692 cm³

17.3 Get Ready

1 a $(-u, v)$ **b** $(-u, -v)$ **c** $(u, -v)$
2 a 0.5 and 0.5   **b** 0.64 and -0.64
  c sin 30° and sin 150° give the same answer.
    cos 50° and cos 130° give the same value, but opposite
    signs.

Exercise 17C

1 a

b

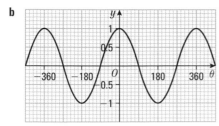

2 a 30, 150   **b** 84.3, 275.7
3 a $\sin \theta° = \frac{1}{3}$ so $\theta = 19.5$
  b $\theta = 19.5, 160.5, 379.5, 520.5$
4 a $\cos \theta° = -0.3$ so $\theta = 107.5$
  b $\theta = -252.5, -107.5, 107.5, 252.5$

17.4 Get Ready

1 a 30 cm²   **b** 54 cm²   **c** 18.4 cm²

Exercise 17D

1 a 21.9 cm²   **b** 29.2 cm²   **c** 15.4 cm²
  d 30.0 m²
2 16.8 m
3 33.3°
4 a 11.4 cm²   **b** 11.4 cm²
  c The answers are the same because sin 25° = sin 155°.
5 a 45°   **b** 12.7 cm²   **c** 102 cm²
6 3.41 cm²

17.5 Get Ready

1 a 0.866   **b** 13.6

Exercise 17E

1 a 8.06 cm   **b** 7.19 cm   **c** 6.35 cm
  d 9.01 cm   **e** 15.0 cm
  f $f = 6.06$ cm, $g = 11.4$ cm

17.6 Get Ready

1 52.0°

Exercise 17F

1 a 45.0°   **b** 63.6°   **c** 23.6°
  d $d = 43.4°$, $e = 63.7°$
2 a 13.3 cm   **b** 39.3°   **c** 154.7°
3 a 9.72 cm   **b** 9.89 cm   **c** 37.9 cm²
4 a 66.0°   **b** 7.63 cm
5 a i 42°   **ii** 55°
  b 9.79 km   **c** 11.9 km   **d** 6.55 km

17.7 Get Ready

1 a 11   **b** 3   **c** 22.4
2 3.82

Exercise 17G

1 a 8.80 cm   **b** 12.6 cm   **c** 5.01 cm
  d 8.42 cm   **e** 15.3 cm   **f** 16.1 cm
2 12.6 cm

17.8 Get Ready

1 a 50.8°   **b** 72.3°   **c** 54.8°

Exercise 17H

1 a 54.7°   **b** 81.2°   **c** 46.0°   **d** 131.2°
2 103.6°
3 a 68°   **b** 91.0 m
4 a 22.4 cm   **b** 127.7°   **c** 161 cm²
5 20.9 km
6 7.71 cm, 11.9 cm

17.9 Get Ready

1 a cosine rule   **b** sine rule
  c sine rule or cosine rule

Exercise 17I

1 a 70.5°, 59.0°, 50.5°   **b** 42.4 cm²
2 a 9.68 cm   **b** 43.1°   **c** 17.7 cm
3 a 7.32 cm   **b** 6.65 cm   **c** 78.7°
4 a 6.76 cm   **b** 75°   **c** 13.1 cm   **d** 19.9 cm
5 a 039° (or 321°)   **b** 148° (or 212°)
6 50°
7 a 70°   **b** 4.43 m
8 a 73.2°   **b** 26.2 m   **c** 67.2°   **d** 22.6°
9 20.5 km

Review exercise

1 13 cm
2 85.5°
3 5.89 cm
4 a 9.11 cm   **b** 19.2°
5 18.3 cm
6 76.3°
7 a $\dfrac{\sqrt{3}}{2}$ **b** $-\dfrac{\sqrt{3}}{2}$
8 42° and 318°
9 10.9 cm²
10 a 105.7°   **b** 10.7 cm²
11 22.2°
12 a 21.8 cm   **b** 66.4 cm²
13 11.1 cm
14 a 41.4°   **b** 17.8 cm
15 a 4.51 cm   **b** 37°

Chapter 18 Answers

18.1 Get Ready

1 (2, 2)

Exercise 18A

1

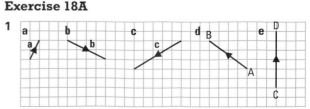

Answers

2 a i $\begin{pmatrix} 5 \\ 6 \end{pmatrix}$ **ii** $\begin{pmatrix} -1 \\ -12 \end{pmatrix}$ **iii** $\begin{pmatrix} 4 \\ 6 \end{pmatrix}$

  b They add up to $\begin{pmatrix} 0 \\ 0 \end{pmatrix}$.

3 a

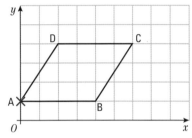

  b $\begin{pmatrix} 9 \\ 3 \end{pmatrix}$

  c trapezium

  d They are parallel and the length of AD is 3 times the length of BC.

4 a

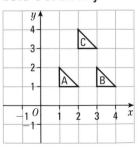

  b i $\begin{pmatrix} 4 \\ 0 \end{pmatrix}$ **ii** $\begin{pmatrix} -2 \\ -3 \end{pmatrix}$

  c i They are the same.

   ii They are parallel with the same length, but in opposite directions.

5 a and **c**, **b** and **h**, **d** and **g**

18.2 Get Ready

1 a 25   **b** 9.43

Exercise 18B

1 a 13   **b** 13   **c** $\sqrt{10}$ **d** $\sqrt{74}$
   e 17   **f** $4\sqrt{5}$

2 a 25
  b length AC $= \sqrt{(7^2 + 24^2)} = 25$, so AB = AC and the triangle is isosceles.

3 rhombus

18.3 Get Ready

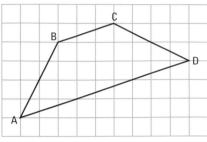

Translation by $\begin{pmatrix} 1 \\ 2 \end{pmatrix}$

Exercise 18C

1

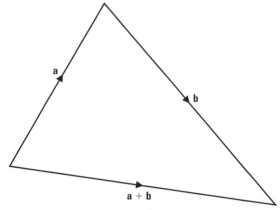

2 a $\begin{pmatrix} 6 \\ 8 \end{pmatrix}$ **b** $\begin{pmatrix} 4 \\ 8 \end{pmatrix}$ **c** $\begin{pmatrix} -2 \\ 4 \end{pmatrix}$ **d** $\begin{pmatrix} 9 \\ -5 \end{pmatrix}$ **e** $\begin{pmatrix} -8 \\ -3 \end{pmatrix}$

3 $\begin{pmatrix} 10 \\ -5 \end{pmatrix}$

4 a i $\begin{pmatrix} 4 \\ 3 \end{pmatrix}$ **ii** $\begin{pmatrix} 4 \\ 3 \end{pmatrix}$ **b** They are the same.

  c i $\begin{pmatrix} 8 \\ 10 \end{pmatrix}$ **ii** $\begin{pmatrix} 8 \\ 10 \end{pmatrix}$ **d** They are the same.

5 a $\overrightarrow{ED}$ is parallel to $\overrightarrow{AB}$, in the same direction and has the same length.

  b i n + m   **ii** n + m + p   **c** n + m

18.4 Get Ready

1 a $\begin{pmatrix} 6 \\ 10 \end{pmatrix}$ **b** $\begin{pmatrix} -9 \\ -3 \end{pmatrix}$ **c** $\begin{pmatrix} 0 \\ 0 \end{pmatrix}$

Exercise 18D

1 a   **b**

   c

   d

2 **a**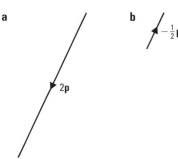

b $-\frac{1}{2}\mathbf{p}$

$2\mathbf{p}$

3 **a** **i** $\begin{pmatrix} 20 \\ 15 \end{pmatrix}$ **ii** $\begin{pmatrix} -12 \\ 6 \end{pmatrix}$ **iii** $\begin{pmatrix} 10 \\ 30 \end{pmatrix}$ **iv** $\begin{pmatrix} -26 \\ 48 \end{pmatrix}$

   b **i** 5 **ii** 10

4 **a** $\begin{pmatrix} -2 \\ 5 \end{pmatrix}$ **b** **i** $\begin{pmatrix} 5 \\ -4 \end{pmatrix}$ **ii** $\begin{pmatrix} 20 \\ -16 \end{pmatrix}$

   c They are parallel and the length of RS is 4 times the length of PQ.

5 (16, 8)

6 **a** $\mathbf{b} - \mathbf{a}$ **b** midpoint of OB

7 **a** $\overrightarrow{AB}, \overrightarrow{EF}$ and $\overrightarrow{GH}$

   b **i** $5\mathbf{p} - 3\mathbf{q}$ **ii** $6\mathbf{m} - 14\mathbf{n}$

8 **a** $2\mathbf{m}$ **b** $2\mathbf{m} + \mathbf{n}$ **c** $\mathbf{m} + \mathbf{x}$ **d** $\mathbf{x} = \mathbf{m} + \mathbf{n}$

18.5 Get Ready

1 **a** $\begin{pmatrix} 7 \\ -7 \end{pmatrix}$ **b** $\begin{pmatrix} -8 \\ -5 \end{pmatrix}$ **c** $\begin{pmatrix} -15 \\ 2 \end{pmatrix}$

Exercise 18E

1 **a** **i** $\begin{pmatrix} 3 \\ 9 \end{pmatrix}$ **ii** $\begin{pmatrix} 9 \\ 27 \end{pmatrix}$

   b ABC is a straight line such that the length of AC is 3 times the length of AB.

2 **a** $\mathbf{b} - \mathbf{a}$ **b** $\frac{1}{2}(\mathbf{b} - \mathbf{a})$ **c** $\frac{1}{2}(\mathbf{a} + \mathbf{b})$

3 **a** $\frac{1}{2}(\mathbf{a} + \mathbf{b})$ **b** $\mathbf{a} + \mathbf{b}$ **c** $\frac{1}{2}(\mathbf{a} + \mathbf{b})$

   d They are the same point.

   e The diagonals of a parallelogram bisect each other.

4 **a** trapezium **b** $\mathbf{n} = 2\mathbf{m} - \mathbf{k}$

5 **a** **i** $\mathbf{b} - \mathbf{a}$ **ii** $\frac{1}{4}\mathbf{b}$ **iii** $\mathbf{a} + \frac{1}{4}\mathbf{b}$ **iv** $\frac{1}{4}\mathbf{a} + \mathbf{b}$

   v $\frac{3}{4}(\mathbf{b} - \mathbf{a})$

   b EF and AB are parallel. The length of EF is $\frac{3}{4}$ times the length of AB.

6 **a** **i** $\frac{1}{2}(\mathbf{m} + \mathbf{n})$ **ii** $\frac{3}{4}(\mathbf{m} + \mathbf{n})$ **iii** $\frac{1}{4}(3\mathbf{n} - \mathbf{m})$

   b $3\mathbf{n} - \mathbf{m}$

   c MQ and MR are parallel with the point M in common, so MQ and MR are part of the same straight line.

   $\dfrac{MR}{MQ} = 4$

7 **a** $2\mathbf{b}$ **b** $2\mathbf{a} + \mathbf{b}$ **c** $4\mathbf{a} + 2\mathbf{b} = 2(2\mathbf{a} + \mathbf{b})$

   d S is the midpoint of OT.

   e 30

Review exercise

1

---

2 **a** (4, 5) **b** $\begin{pmatrix} -3 \\ -4 \end{pmatrix}$ **c** 5

3 **a** $\mathbf{b} - \mathbf{a}$

   b $\overrightarrow{AP} = \frac{3}{5}(\mathbf{b} - \mathbf{a})$

   $\overrightarrow{OP} = \overrightarrow{OA} + \overrightarrow{AP} = \mathbf{a} + \frac{3}{5}(\mathbf{b} - \mathbf{a}) = \frac{1}{5}(2\mathbf{a} + 3\mathbf{b})$

4 **a** $2(\mathbf{a} + \mathbf{b})$

   b $7\mathbf{a} + 6\mathbf{b}$

5 **a** **i** $\begin{pmatrix} -4 \\ -3 \end{pmatrix}$ **ii** 5

   b $\begin{pmatrix} -3 \\ 7 \end{pmatrix}$ **c** $(-7, 4)$

6 **a** $2(\mathbf{a} + 2\mathbf{c})$

   b $\overrightarrow{OM} = 6\mathbf{c} + 3\mathbf{a} = 3(\mathbf{a} + 2\mathbf{c}) = \frac{3}{2}\overrightarrow{OP}$

   OP and OM are parallel with the point O in common, so OP and OM are part of the same straight line.

7 **a** $\mathbf{a} + \mathbf{b}$

   b $\overrightarrow{FE} = \overrightarrow{FC} + \overrightarrow{CD} + \overrightarrow{DE} = \mathbf{a} - \mathbf{b} + \mathbf{a} + \mathbf{b} = 2\mathbf{a}$

   So FE is parallel to CD.

   c $2\mathbf{a} - \frac{1}{2}\mathbf{b}$

   d $\overrightarrow{FX} = \frac{4}{5}(2\mathbf{a} - \frac{1}{2}\mathbf{b}) = \frac{8}{5}\mathbf{a} - \frac{2}{5}\mathbf{b}$

   $\overrightarrow{CX} = (\frac{8}{5}\mathbf{a} - \frac{2}{5}\mathbf{b}) - (\mathbf{a} - \mathbf{b}) = \frac{3}{5}(\mathbf{a} + \mathbf{b})$

   $\overrightarrow{CE} = \mathbf{a} + \mathbf{b}$

   CX and CE are parallel with the point C in common, so CX and CE are part of the same straight line.

8 **a** **i** $5\mathbf{p}$ **ii** $2\mathbf{q}$ **iii** $4\mathbf{p} - \mathbf{q}$

   b $3\mathbf{p} - 8\mathbf{q}$

9 **a** $\frac{1}{2}(\mathbf{p} + \mathbf{q})$

   b $\overrightarrow{RS} = -\overrightarrow{OR} + \overrightarrow{OS} = \mathbf{p} + \frac{1}{2}(\mathbf{p} + \mathbf{q}) = \frac{1}{2}\mathbf{q}$

   So RS is parallel to OQ.

10 **a** $2\mathbf{a} - 2\mathbf{b}$

   b $\overrightarrow{QR} = -2\mathbf{a} - 2\mathbf{b} + 6\mathbf{a} = 4\mathbf{a} - 2\mathbf{b}$

   $\overrightarrow{XY} = -\frac{1}{2}\overrightarrow{MN} + \overrightarrow{MQ} + \frac{1}{2}\overrightarrow{QR} = -\frac{1}{2}(2\mathbf{a} - 2\mathbf{b}) + \mathbf{a}$

   $+ \frac{1}{2}(4\mathbf{a} - 2\mathbf{b}) = 2\mathbf{a}$

   So XY is parallel to OR.

11 **a** $\mathbf{a} + \mathbf{b}$

   b $\overrightarrow{AX} = 2(\mathbf{a} + \mathbf{b})$

   $\overrightarrow{DX} = -\overrightarrow{AD} + \overrightarrow{AX} = -2\mathbf{b} + 2(\mathbf{a} + \mathbf{b}) = 2\mathbf{a}$

   So AB is parallel to DX.

Price comparisons

1 Ahmed should use Cogas if he uses fewer than 3800 units per annum, otherwise he should use Ourgas.

2 **a** Cable: $C = 30 + 5m$; Broadband: $C = 6.5m$

   b Broadband is cheaper for up to 20 months.

3 **a** ACars: $C = 60 + 0.32x$; BMotors: $C = 50 + 0.4x$

   b ACars is cheaper for more than 125 miles a day.

4 Quick Delivery for parcels up to 1.5 kg and Parcels Fly for parcels over 1.5 kg

5 Pete's Mix if more than 3.33 m³, otherwise Concrete Sue

Trigonometry

1 14 cm

2 12.0 cm

3 From corner of shed to top of roof at opposite end
$= \sqrt{6^2 + 2^2 + 7^2} = 9.43$ ft, so the bean poles will fit in the shed.

Answers

Going on holiday

1. 155 points
2. England (England €478.80, Malta €483)
3. 18 500

All at sea

1. 12 664 km
2.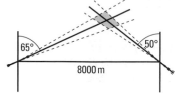
3. 12:29

Index

Index